The 1st International Conference on Maritime Education and Development

Sanja Bauk • Stojče Dimov Ilčev
Editors

The 1st International Conference on Maritime Education and Development

ICMED

23–24 November 2020, Durban, South Africa

Springer

Editors
Sanja Bauk
Maritime Studies Department
Durban University of Technology
Durban, South Africa

Stojče Dimov Ilčev
Space Science Centre (SSC)
Durban University of Technology (DUT)
Durban, South Africa

ISBN 978-3-030-64087-3 ISBN 978-3-030-64088-0 (eBook)
https://doi.org/10.1007/978-3-030-64088-0

This Springer imprint is published by the registered company Springer Nature Switzerland AG
The registered company address is: Gewerbestrasse 11, 6330 Cham, Switzerland

International Scientific Committee

Organizational Committee

1. Prof. Sanja Bauk
2. Yashoda Balkumar
3. Mthokozisi Buthelezi
4. Lindani Ntshangase
5. Litha Dalindyebo
6. Darren Fraser
7. Valentine Hlambisa
8. Ntombezinhle Kweyama
9. Balungile Ntshangase
10. Zenzile Gwamanda
11. Nkosinathi Michael Manqele
12. Margaret Balungile Masuku
13. Nhlanhla Rodgers Mtshali
14. Trywell Thabani Mtshali
15. Arshna Ramouthar
16. Aluwani Elijah Ramulifho
17. Nomalanga Sokhela
18. Asanda Zangwa
19. Bheka Zulu
20. Vincent Zulu

So who exactly is steering the ship?
(Egleman, D.)

Preface

It is our great pleasure to present the 1st volume of the proceedings related to the 1st International Conference in Maritime Education and Development (ICMED 2020), published by Springer.

The 1st International Conference on Maritime Education Development (ICMED 2020) was held virtually from 23 to 24 November 2020, organized jointly by Maritime Studies Department, Faculty of Applied Sciences (FAS) and Communication, Navigation and Surveillance (CNS) Centre of Durban University of Technology; and, Moses Kotane Institute. Co-organizers were Ethekwini Municipality and Ethekwini Maritime Cluster (EMC). The organizers and co-organizers are located in Durban in South Africa.

The Conference Proceedings are addressed to both scientists and professionals with focus on sharing their expert knowledge, experiences and research results concerning virtual smartness in maritime, related legal issues, maritime educational and training challenges, STCW requirements, voices of seafarers, maritime business and sustainability, safety and security, including the digital realm concerns and recent pandemic threats, and so on. In addition, gathering of researchers, scientists, academics, professionals, entrepreneurs, and other people involved in maritime education, business, industry and administration should bring research and education closer to the actual needs of maritime labor market.

The content of the ICMED 2020 Conference Proceedings is partitioned into eight parts entitled: *Virtual smartness* (covering nine chapters), *Legal concerns* (covering two chapters), *Educational and training challenges* (covering eleven chapters), *STCW concerns* (covering two chapters), *Voice of seafarers* (covering one comprehensive chapter), *Business vs sustainability* (covering three chapters), *Safety and security issues* (covering seven chapters) and *Miscellaneous* (covering five chapters). The papers are heterogeneous, but each contributes in its own way towards enriching the knowledge in maritime and beyond.

The authors and co-authors came from thirteen different countries in Africa, Asia, Europe and South America, i.e., from: Bosnia and Herzegovina, Brazil, Croatia, Estonia, Germany, Montenegro, Russia, Slovenia, South Africa, Spain, Sweden, Ukraine and United Kingdom.

We would like to express our gratitude to all distinguished authors and co-authors for their great contribution for success of the Conference and its Proceedings publication.

Durban, South Africa

Sanja Bauk
Stojče Dimov Ilčev

Contents

Contributors

John Abercrombie Maritime Studies Department, Faculty of Applied Sciences, Durban University of Technology, Durban, South Africa

Igor Afonin Sevastopol State University, Sevastopol, Russia

Anatoli Alop Centre of Maritime Studies, Estonian Maritime Academy of Tallinn University of Technology, Tallinn, Estonia

Zoran Ž. Avramović Pan-European University APEIRON, Banja Luka, Bosnia and Herzegovina

Gordan Bajić Pan-European University APEIRON, Banja Luka, Bosnia and Herzegovina

Michael Baldauf Department of Maritime Studies Warnemuende/Institute ISSIMS, Hochschule Wismar, University of Applied Sciences – Technology, Business and Design, Wismar, Germany

Department of Maritime Studies, Hochschule Wismar, University of Applied Sciences, Rostock, Germany

Mate Barić Maritime Department, University of Zadar, Zadar, Croatia

Sanja Bauk Maritime Studies, Durban University of Technology, Durban, South Africa

Administration for Maritime Safety and Port Management of Montenegro, Podgorica, Montenegro

Maritime Studies, Faculty of Applied Sciences, Durban University of Technology, Durban, South Africa

Anja Belojević Ministry of Transport and Maritime Affairs, Podgorica, Montenegro

Knud Benedict Department of Maritime Studies Warnemuende/Institute ISSIMS, Hochschule Wismar, University of Applied Sciences – Technology, Business and Design, Wismar, Germany

Department of Maritime Studies Warnemuende/Institute ISSIMS, Hochschule Wismar, University of Applied Sciences – Technology, Business and Design, Wismar, Germany

Bojan Beškovnik Faculty of Maritime Studies and Transport, University of Ljubljana, Portorož, Slovenia

Jean-Philippe Boisgard ECA Robotics SAS, Lannion, France

Sandra Tominac Coslovich Faculty of Maritime Studies, University of Rijeka, Rijeka, Croatia

Quentin N. Cox Teekay Shipping, Glasgow, UK

Jonas Don-yelee Dakora Department of Electrical Power Engineering, Durban University of Technology, Durban, South Africa

Innocent Davidson Springer-Verlag, Computer Science Editorial, Heidelberg, Germany

Innocent E. Davidson Department of Electrical Power Engineering, Durban University of Technology, Durban, South Africa

F. X. Martínez de Osés Department of Nautical Science and Engineering, Barcelona School of Nautical Studies, Technical University of Catalonia – BarcelonaTech, Barcelona, Catalonia, Spain

Vladimir Domazet Pan-European University Apeiron, Banja Luka, Bosnia and Herzegovina

Dalibor Drljača Pan-European University APEIRON, Banja Luka, Bosnia and Herzegovina

Igor Dugonjić Pan-European University APEIRON, Banja Luka, Bosnia and Herzegovina

Maksim Durmanov Sevastopol State University, Sevastopol, Russia

Maxim Durmanov Sevastopol State University, Sevastopol, Russia

Edilene Gasparini Fernandes Paula Souza Center – CEETESP, São Paulo State Technological College – FATEC, Sao Paulo, Brazil

Yuriy Gimpilevich Sevastopol State University, Sevastopol, Russia

Michael Gluch Department of Maritime Studies Warnemuende/Institute ISSIMS, Hochschule Wismar, University of Applied Sciences – Technology, Business and Design, Wismar, Germany

Luka Grbić Maritime Department, University of Zadar, Zadar, Croatia

Igor Grujić Nova Banka, Banja Luka, Bosnia and Herzegovina

Ana Gundić Maritime Department, University of Zadar, Zadar, Croatia

Maritime Department, University of Zadar, Zadar, Republic of Croatia

Ardian Haçkaj Cooperation & Development Institute, Tirana, Albania

Stojče Dimov Ilčev Durban University of Technology (DUT), Durban, South Africa

Vladimir Iskiv Sevastopol State University, Sevastopol, Russia

Dalibor Ivanišević Maritime Department, University of Zadar, Zadar, Croatia

Tomke Julia Janßen Technische Universität Berlin, Berlin, Germany

Elisenda Ventura Jariod Barcelona School of Nautical Studies, Technical University of Catalonia – BarcelonaTech, Barcelona, Catalonia, Spain

Nexhat Kapidani Administration for Maritime Safety and Port Management of Montenegro, Podgorica, Montenegro

Administration for Maritime Safety and port Management, Bar, Montenegro

Ernesa Kardović Ministry of Transport and Maritime Affairs, Podgorica, Montenegro

Matthias Kirchhoff ISSIMS GmbH, Rostock, Germany

Momoko Kitada World Maritime University, Malmö, Sweden

Slavojka Lazić Pan-European University APEIRON, Banja Luka, Bosnia and Herzegovina

Žarko Lukšić Administration for Maritime Safety and Port Management of Montenegro, Podgorica, Montenegro

Bheki Madonsela Springer-Verlag, Computer Science Editorial, Heidelberg, Germany

Ilcev Marijan Durban University of Technology (DUT), Durban, South Africa

Dražen Marinković Pan-European University APEIRON, Banja Luka, Bosnia and Herzegovina

Roger B. Mason Durban University of Technology, Durban, South Africa

Margaret B. Masuku Durban University of Technology, Maritime Studies, Durban, South Africa

Edesa Metaj Ministry of Infrastructure and Energy, Tirana, Albania

Yuri Mickhayluck Sevastopol State University, Sevastopol, Russia

Matlou Lesley Mokgobu Durban University of Technology, Durban, South Africa

Nhlanhla Rodgers Mtshali Durban University of Technology, Durban, South Africa

Chrispin Mulangu Springer-Verlag, Computer Science Editorial, Heidelberg, Germany

Gisela Müller-Plath Technische Universität Berlin, Berlin, Germany

Denis Nacharov Sevastopol State University, Sevastopol, Russia

Vladimir Nikishin Sevastopol State University, Russia

Oludayo Olugbara Durban University of Technology (DUT), Durban, South Africa

Đorđije Otašević Ministry of Transport and Maritime Affairs, Podgorica, Montenegro

Boris Pauković Pan-European University APEIRON, Banja Luka, Bosnia and Herzegovina

Nomzamo C. Phewa World Maritime University, Malmo, Sweden

Marija Pijaca Maritime Department, University of Zadar, Zadar, Republic of Croatia

Elena Redkina Sevastopol State University, Sevastopol, Russia

Boris Ribarić Pan-European University APEIRON, Banja Luka, Bosnia and Herzegovina

Saša Ristić Deloitte d.o.o. Banja Luka, Banja Luka, Bosnia and Herzegovina

Michèle Schaub Department of Maritime Studies Warnemuende/Institute ISSIMS, Hochschule Wismar, University of Applied Sciences – Technology, Business and Design, Wismar, Germany

Department of Maritime Studies Warnemuende/Institute ISSIMS, Hochschule Wismar, University of Applied Sciences – Technology, Business and Design, Wismar, Germany

Andrey Schekaturin Sevastopol State University, Sevastopol, Russia

Vladlen Shapo National University "Odessa Maritime Academy", Odessa, Ukraine

Gulshan Sharma Department of Electrical Power Engineering, Durban University of Technology, Durban, South Africa

Igor Shirokov Sevastopol State University, Sevastopol, Russia

Nedeljko Šikanjić Pan-European University APEIRON, Banja Luka, Bosnia and Herzegovina

Ivan Skorik Sevastopol State University, Sevastopol, Russia

Nedim I. Smailović Pan-European University APEIRON, Banja Luka, Bosnia and Herzegovina

Ivana Surian University of KwaZulu-Natal, Durban, South Africa

MACS Maritime Carrier Shipping (Pte) Ltd, Singapore, Singapore

Tijana Talić Pan-European University APEIRON, Banja Luka, Bosnia and Herzegovina

Sinisa Tomić Pan-European University APEIRON, Banja Luka, Bosnia and Herzegovina

Mihajlo Travar Regulatory Commission for Energy of Republic of Srpska, Trebinje, Bosnia and Herzegovina

Gerrit Tuschling Department of Maritime Studies Warnemuende/Institute ISSIMS, Hochschule Wismar, University of Applied Sciences – Technology, Business and Design, Wismar, Germany

Elen Twrdy Faculty of Maritime Studies and Transport, University of Ljubljana, Portorož, Slovenia

Dragan Vasiljević Pan-European University "Apeiron", Banja Luka, Bosnia and Herzegovina

Julijana Vasiljević Pan-European University "Apeiron", Banja Luka, Bosnia and Herzegovina

Semen Yankovskiy Sevastopol State University, Sevastopol, Russia

Bheka Clive Zulu Maritime Studies, Durban University of Technology, Durban, South Africa

Part I
Virtual Smartness

Chapter 1
Innovative SAMMON Teaching Method for Ship Handling: Element of the Project EURO-ZA Between South Africa and Europe

Knud Benedict, Michèle Schaub, Gerrit Tuschling, Michael Baldauf, Michael Gluch, and Matthias Kirchhoff

1.1 Introduction: Concept for Using Fast-Time Simulation and Intention for Use in the EU Project EURO-ZA

Need for Simulation Support and Fast-Time Simulation Software

Manoeuvring of ships is and will be a human-centred process despite expected further technological developments. Most important elements of this process are the human itself and the technical equipment to support its task. However, most of the work is to be done manually because even today nearly no automation support is available for complex manoeuvres. Up to now there is no electronic tool to demonstrate manoeuvring characteristics efficiently or, moreover, to design a manoeuvring plan effectively – even in briefing procedures for ship handling training, the potential manoeuvres will be "guessed" and drafted on paper or described by sketches and short explanations. The information on manoeuvring characteristics from shipyard trials goes into papers like Wheelhouse Posters instead of using it as math model for simulation on the bridge – this is waste of resources in the Digital Age!

K. Benedict (✉) · M. Schaub · G. Tuschling · M. Baldauf · M. Gluch
Department of Maritime Studies Warnemuende/Institute ISSIMS, Hochschule Wismar, University of Applied Sciences – Technology, Business and Design, Wismar, Germany
e-mail: knud.benedict@hs-wismar.de; michele.schaub@hs-wismar.de; gerrit.tuschling@hs-wismar.de; michael.baldauf@hs-wismar.de; michael.gluch@hs-wismar.de

M. Kirchhoff
ISSIMS GmbH, Rostock, Germany
e-mail: mkirchhoff@issim-gmbh.de

S. Bauk, S. D. Ilčev (eds.), *The 1st International Conference on Maritime Education and Development*, https://doi.org/10.1007/978-3-030-64088-0_1

However, due to the new demands, there is a need to prepare harbour approaches with complete berthing plans specifically in companies with high safety standards like cruise liners. These plans are necessary to agree on a concept within the bridge team and also for the discussion and briefing with the pilot. The plan for the potential manoeuvres must be developed – but still in a contemplative way by thinking ahead. By now there is no tool available to provide support for manoeuvring planning yet.

Ship handling simulation for simulator training has a proven high effect for the qualification in training. However, it is based on real-time simulation, and, i.e. 1 s calculation time by the computers represents 1 s manoeuvring time as in real world. This means despite all other advantages of full mission ship handling simulation, collecting/gathering of manoeuvring experiences remains an utmost time-consuming process. For instance, a training session for a berthing manoeuvre might take 1 h – if the first attempt fails or an alternative strategy should be tried, then the next session needs another hour: this is not very effective.

Therefore, the method of fast-time simulation (FTS) will be used in future for increasing the effectiveness of training and also the safety and efficiency for manoeuvring real ships. The basic principle of that FTS system is to represent the full information from Pilot Card, Wheelhouse Poster and Manoeuvring Booklet (and additional trial results), condensed in a ship dynamic simulation model, which is even capable of simulating wind, current and restricted water effects by using the innovative "Rapid Advanced Prediction & Interface Technology" (RAPIT). Even with standard computers, it can be achieved to simulate in 1 s computing time a manoeuvre lasting about 20 min. These RAPIT-based FTS tools were initiated in research activities of the Institute for Innovative Ship Simulation and Maritime System (ISSIMS) at the Maritime Simulation Centre Warnemuende, which is a part of the Department of Maritime Studies of Hochschule Wismar, University of Applied Sciences: Technology, Business and Design in Germany. They have been further developed by the start-up company Innovative Ship Simulation and Maritime Systems (ISSIMS GmbH [1, 2]).

A brief overview is given for the modules of the FTS tools and its potential application:

- SAMMON is the brand name of the innovative system for "Simulation Augmented Manoeuvring Design, Monitoring & Conning", consisting of the following software modules:
  - Manoeuvring Design & Planning Module: Design of ships manoeuvring concepts as "manoeuvring plan" for harbour approach and berthing manoeuvres (steered by virtual handles on screen by the mariner)
  - Manoeuvring Monitoring & Conning Module with Multiple Dynamic Manoeuvring Prediction: Monitoring of ships manoeuvres during simulator exercises or manoeuvres on a real ship using bridges handles, display of manoeuvring plan and predicted manoeuvres in parallel; calculation of various prediction tracks for full ships dynamic simulation and simplified path prediction as look ahead for the future ships motion

  - Manoeuvring Simulation Trial & Training Module: Ship handling simulation on laptop display to check and train the manoeuvring concept (providing the same functions as monitoring tool; steered by virtual handles on screen)

- These modules are made for both:

  - Application in maritime education and training to support lecturing for ship handling to demonstrate and explain more easily manoeuvring technology details and to prepare more specifically manoeuvring training in Ship Handling Simulator (SHS) environment, i.e. for developing manoeuvring plans in briefing sessions, to support manoeuvring during the exercise run and to help in debriefing sessions the analysis of replays and discussions of quick demonstration of alternative manoeuvres
  - Application on board to assist manoeuvring of real ships, e.g. to prepare manoeuvring plans for challenging harbour approaches with complex manoeuvres up to the final berthing/cast off of ships, to assist the steering by multiple prediction during the manoeuvring process and even to give support for analysing the result and for onboard training with the Simulation & Trial module
  - SIMOPT is a Simulation Optimiser software module based on FTS for optimising Standard Manoeuvres and modifying ship math model parameters both for simulator ships in SHS and for the SAMMON System both for lecturing / training and for on board application of the system [7]. SIMDAT is a software module for analysing simulation results both from simulations in SHS or SIMOPT / SAMMON and from real ship trials: the data for manoeuvring characteristics can be automatically retrieved, and comfortable graphic tools are available for displaying, comparing and assessing the results.

In this paper, the focus will be laid on the potential of the SAMMON software for supporting the lecturing and briefing/debriefing process with elements specifically for simulator training for advanced ship handling.

Partners for EU Project EURO-ZA and Objectives

At the conferences of IMLA in Durban in 2015 [3] and at the INSLC in Cape Town in 2017 [4], the "Simulation Augmented Manoeuvring Design, Monitoring & Conning" (SAMMON) software toolbox was introduced, and first ideas were described how this new technology could be used to improve the simulator training by FTS. The SAMMON system has matured over the years, and experiences how to use and the effectiveness [5] were made at the Maritime Simulation Centre Warnemuende (MSCW) and other centres. And now, specifically the use of the SAMMON manoeuvring planning tool will be an element of the transfer of knowledge within the current ERASMUS+ project EURO-ZA which sets out for

EURO-ZA Capacity building in the field of maritime education. It is a project between several partners from

- Europe (comprising universities HS Wismar/Germany, Solent, Southampton, UK; and SAMK, Rauma, Finland)
- South Africa (DUT, Durban; CPUT, Cape Town; and NMU, Port Elizabeth, as project coordinator)

All the EURO-ZA partners have a long history in maritime education with some European partners over 100 years. Therefore this is an opportunity to conduct a detailed analysis of the curriculums and facilities to truly evaluate any similarities, differences and opportunities for improvement for both the South African and European partners. Comparison of the curricula will lead to an understanding of similarities and differences. Out of this a comprehensive understanding can be established as to what the strengths and weaknesses are in each institution's curriculum. The results will allow for a possible alignment of qualifications allowing knowledge sharing and student, staff and research exchange ultimately recognising a global standard and an improved product to industry. But also it sets out for modernisation and development of online and distance learning pedagogical processes, e.g. improvement of training and assessment in e-learning, blended learning distance and simulator training. It is also focusing on the assessment of the facilities and resources at each institution by identifying the teaching and learning equipment to meet the aims and objectives for the twenty-first-century curriculum, resulting in a proposal and budgets for standardisation between the six institutions. One element of the potential improvement of equipment addresses the element of FTS for manoeuvring motion used for ship handling training. During the demonstrations of this new and innovative technology at the first project meeting in 2019 at the MSCW, it was seen that it could be one of the focus areas for improvement of equipment at the partners' institutions, both for new research projects and for improving the simulation basis for training.

In this paper the opportunities will be described on how to use the FTS in the SAMMON planning tool for teaching of ship's dynamic and training of ship handling elements. Additionally, some of the samples can be seen on YouTube [6].

1.2 Use of Fast-Time Simulation for Lecturing and Familiarisation Using the Example of Wind Impact on Ship's Motion

Planning Tool Interface

An important issue is the behaviour of the vessel under wind impact that can be easily explained and investigated by means of the SAMMON Manoeuvring Design & Planning tool. Some basic functions and interface displays for the FTS are shown in the next figures.

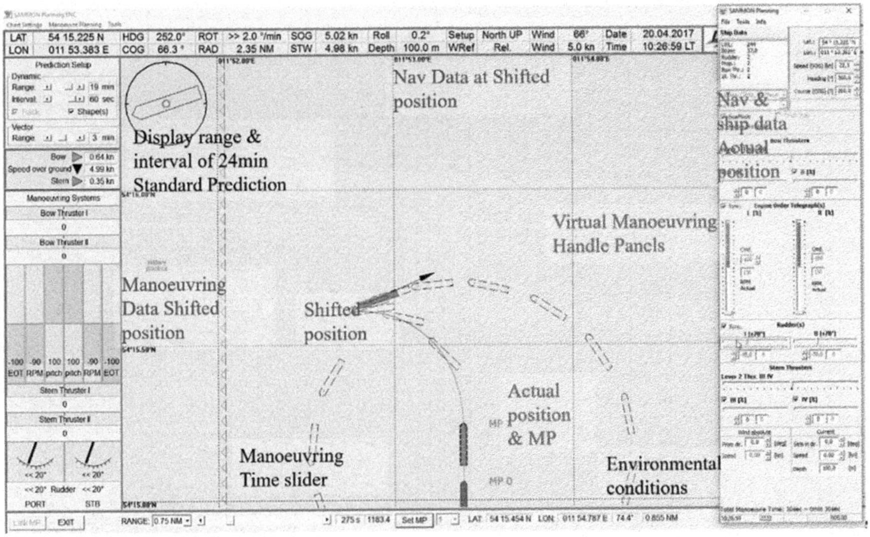

Fig. 1.1 SAMMON Manoeuvring Design & Planning tool – interface elements with explanations

Figure 1.1 explains the interface, which combines:

- The interface window for the steering panel of the ship (right) for adjusting the controls for the selected manoeuvring point (MP) (actual position in red) and to select the desired environment conditions down below, e.g. wind, current and water depths.
- The electronic navigational chart (ENC) window (centre) to visualise the simulated ship's motion: black contours indicating the positions in time intervals for the display range; by means of the time slider at the bottom, the reference position can be shifted to any position of the already predicted track where a new MP can be set and controls may be changed there.
- Interface elements to display the status of the current navigation data and actual ship manoeuvring controls (left and top) at the ship position on the track up to the positions of the next MP which is indicated as ship shape in blue colour in the ENC.

Samples for Drifting Under Wind with No Propulsion

In the following figures, the drift behaviour of a cruise ship with no propulsion in wind speed 30kn from North 0° will be shown, and the options will be discussed how the drifting speed and direction can be changed with bow and/or stern thrusters. The sample ship is the cruise liner AIDAblu of 181 m in length. In Fig. 1.2 the reference case is shown where the ship starts drifting from stop with zero speed. Also the blue shapes are highlighted after 1 and 2 min, where the full drifting speed of SOG=2.25kn

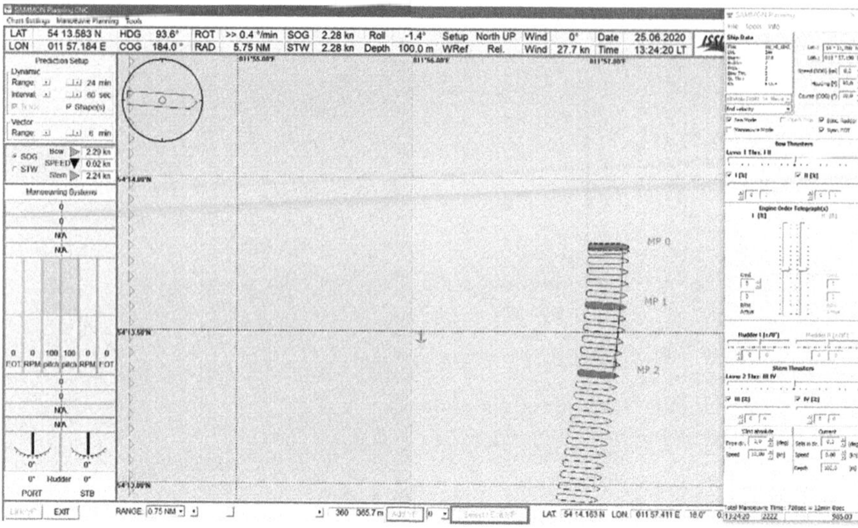

Fig. 1.2 Cruise ship AIDAblu transverse drifting speed SOG=2.25kn, HDG 94°, COG 184°

is being reached. The bow falls slightly down with the wind because the ship has its centre of the wind area ahead of midships position. If bow and stern thrusters are used fully against the wind, then the ship would drift only with SOG=1.51kn.

In Fig. 1.3 three different options are compared for using the thrusters to control the drift direction. Considering the procedure in the right part of the figure, it can be concluded that the drift direction can be changed +/- 48° from downwind direction: Maximum side drift can be achieved with bow thruster -100% to port (PT) and stern thruster +40% to starboard (SB) – i.e. the downwind speed component in 180° is less than the 1.51kn with both thrusters full against the wind and additionally with another course to potentially avoid an obstacle on lee side!

Another option is shown in Fig. 1.4: In the initial manoeuvring point MP0, the bow is brought into the wind combined with bow thruster to PT and stern thruster to SB. The thruster power was reduced in the following MP1 to MP3. And at the active MP4 (red contour), the ship will be steadied going astern at COG 180° keeping the bow into the wind with only small thruster control actions. The downwind drift speed is SOG=1.84kn, i.e. the ship is drifting slower than with direct beam wind and no thrusters against the wind, but a little more than the 1.51 with both thrusters full against the wind.

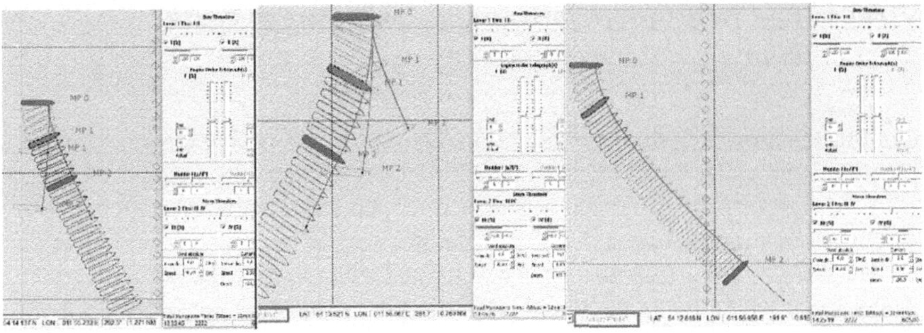

Fig. 1.3 Comparison of drifting behaviour with three different options:
• Left: With only bow thruster against the wind transverse speed SOG=1.87kn in COG 155°, i.e. downwind speed component is 1.87*cos(25°)=1,7kn
• Centre: With only stern thruster against the wind transverse speed SOG=2.11kn in COG 206°, i.e. downwind speed component 2.11*cos(26°)=1,89kn
• Right: Maximum side drift with bow thruster -100% to PT and stern thruster +40% to SB: wind transverse speed SOG=2.17kn in COG 132°, i.e. downwind speed component in 180° is 2.17*cos(25°)=1,45kn

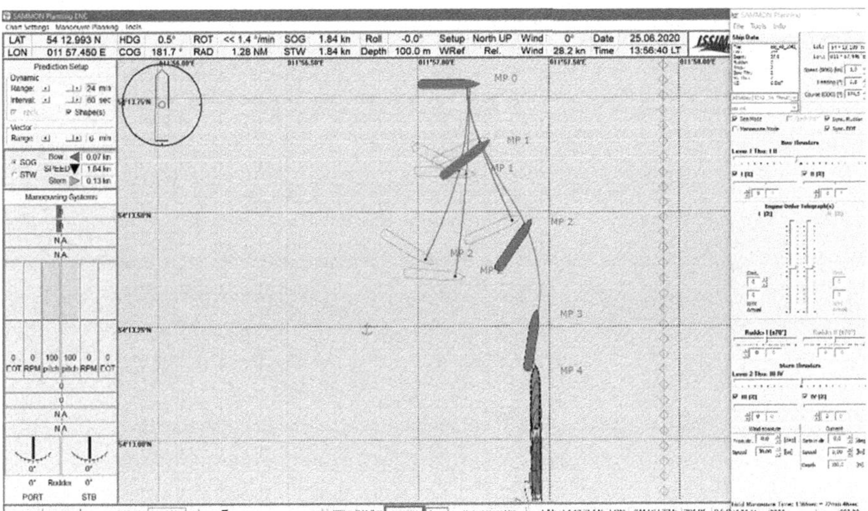

Fig. 1.4 Manoeuvre plan for turning the ship bow by thrusters into the wind until drifting astern

Rudder and Drift Angle for Balance on Straight Track Under Wind

For introduction, it is shown in Fig. 1.5 what rudder and drift angle are required to steer the ship under beam wind. In the SAMMON Manoevring Design & Planning tool, thc initial hcading and course can be easily adjusted in the top right window to get the ship's track between the fairway buoys.

For developing a mental model for the understanding of the wind effect, it is helpful to have an overview what rudder and drift angle are generally needed for different wind conditions and ship's speeds. In the planning module, this can be

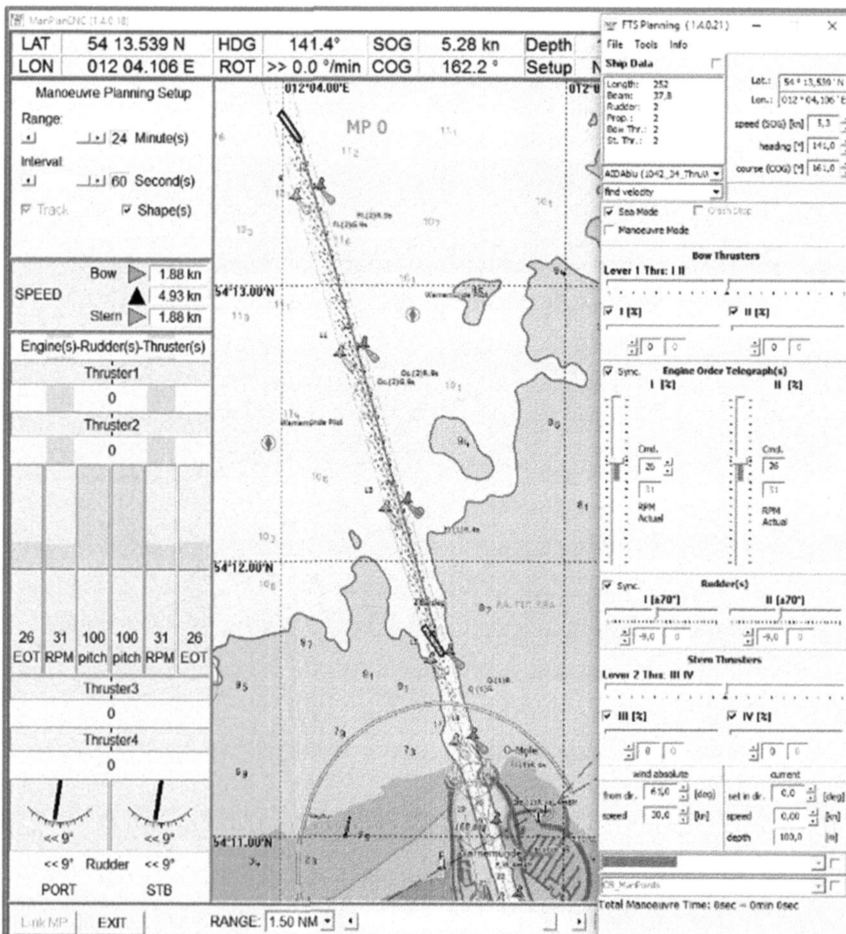

Fig. 1.5 Wind effect on straight motion in SAMMON Manoeuvring Design & Planning tool – approach to Rostock Port: AIDAblu enters the fairway under wind speed 30 kn from 60°. With EOT 26% (equivalent to 5.2 kn ship's speed), she needs 9° PT rudder; drift angle is 22°

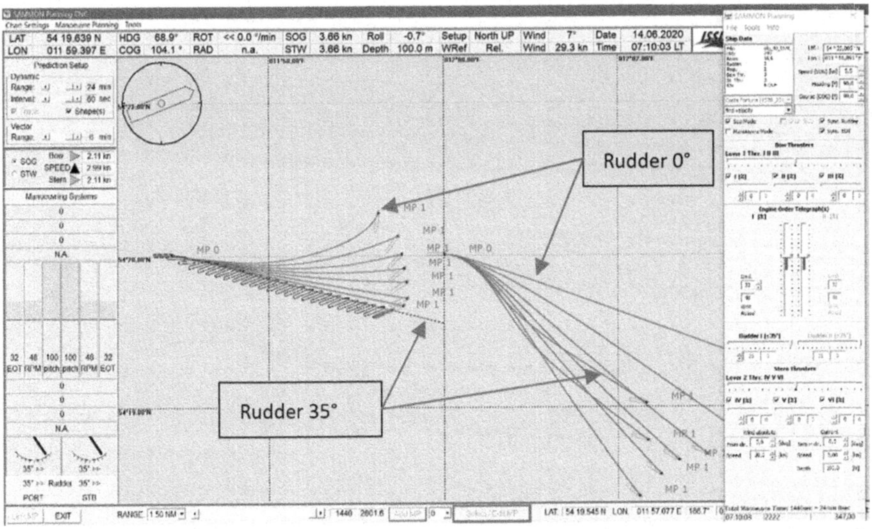

Fig. 1.6 Investigation of wind effect on straight motion in SAMMON Planning tool for two cruise ships with different superstructure for six different rudder angles 0°, 5°, 10°, 15°, 20°, 25°, 30° and 35° to starboard. Wind direction from North 0°, wind speed 30kn, initial ship speed about 6kn
• Left: Ship is windward turning (due to centre of wind area aft behind COG – Costa Fortuna).
• Right: Ship is leeward turning (due to centre of wind area fore of COG – AIDAblu)

investigated in trying out several wind encounters and wind to ship speeds in very short time. In Fig. 1.6 the result is shown for two different ships:

- Left: Ship Costa Fortuna has the centre of wind area aft behind COG and therefore windward turning tendency – with rudder amidships 0°, the ship turns into the wind; and with increasing starboard rudder, she finds equilibriums on different straight tracks out of the wind.
- Right: Ship AIDAblu is leeward turning due to centre of wind area fore of COG. With rudder amidships 0°, she finds an equilibrium on straight track out of the wind; for increasing starboard rudder with increasing COG, she turns further out of the wind. When the rudder exceeds 25°, then the rudder effectiveness is decreasing – for 30° and 35°, the ship cannot be brought out of the wind furthermore.

These simulations can be done as an individual training session or as a trainee group exercise, and the results are collected and entered into a diagram as in Fig. 1.7. In this diagram the rudder angles and drift angles are plotted versus the absolute wind encounter angle between wind direction and ship's course – it is zero for heading against the wind and 180° for stern wind. For given ship conditions, the wind impact changes with the encounter angle and the speed ratio V_R^2/V^2. For simplification and for better use of the data for route planning, the absolute wind speed V_A and the constant speed V_0 of the ship for a given EOT without wind will be used together with the absolute wind encounter angle (instead of the relative wind and encounter

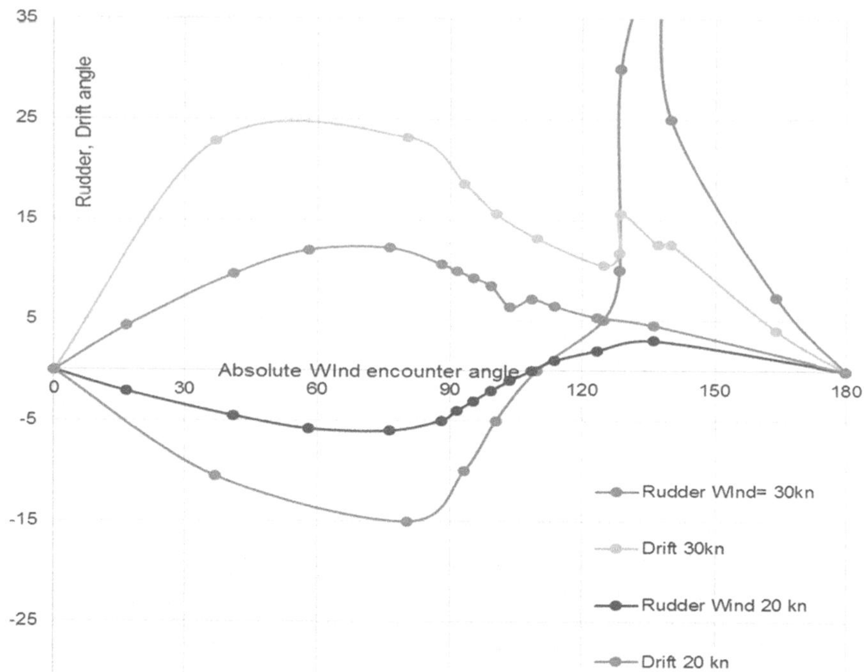

Fig. 1.7 Simulation results for balance of wind effects on straight track with constant speed for AIDAblu: equilibrium parameters rudder and drift angle versus absolute wind encounter angle for two wind speeds of 20 kn and 30 kn. (EOT 30%, according to ship speed 6.2kn)

angle). Figure 1.7 shows that the ship has a leeward turning tendency for wind from the bow to beam wind and a windward turning tendency for wind from the stern quarter. The maximum rudder angle is needed for an encounter angle of about 135°. In this course range, a change of rudder will not change the course. For the wind speed 30 kn, the required rudder angle would even exceed the maximum rudder angle of 35° because the turning moments due to wind and the unstable moment due to the drift angle point in the same direction – the ship could not be steered at this low speed. Such a diagram as in Fig. 1.7 is the basis for understanding the wind effect and for estimating the required rudder and drift angle beforehand.

Wind Impact on Turning Motion

The most important parameter for the wind impact is the "wind to ship speed ratio". For high ratios there can be situations where the ship can not be turned anymore without increasing the ship's speed or at least kicking the engine for increasing the rudder efficiency. In Fig. 1.8 several manoeuvres under different wind conditions are compared, and the efficiency of the FTS can be seen in the fact that all these simulations and the preparation of the figure took less than 15 min. The

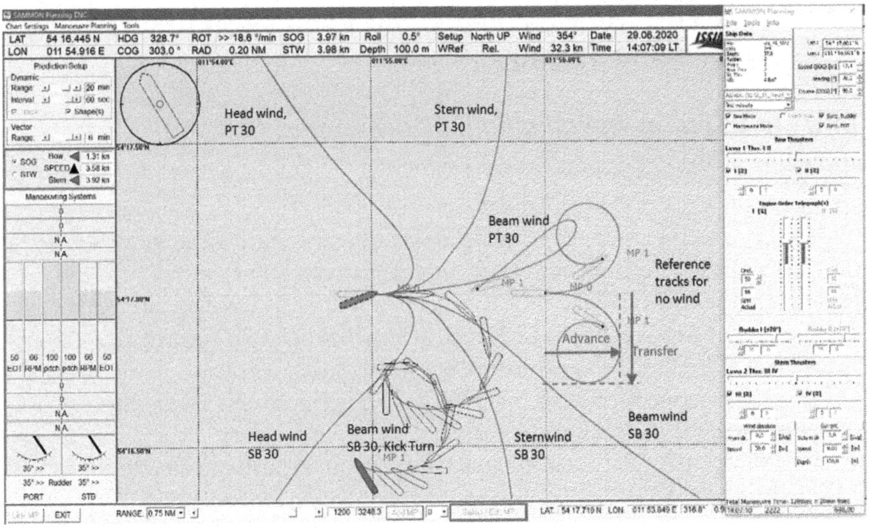

Fig. 1.8 Simulation results for turning circles' tracks (magenta curves) with hard rudder for cruise ship AIDAblu at slow speed (EOT 30%, i.e. 6.2kn without wind):
• Right side: Reference tracks for turning circles to SB and PT with no wind
• Left side: Tracks for same manoeuvres under 30kn wind with three different directions for head wind 90°, stern wind 270° and beam wind 90°
• Below: Additional track with shapes for kick turn from 30% to EOT 50% with beam wind from North 0°

manoeuvring tracks for hard rudder 35° turning circles with no wind on the right side serve as reference manoeuvres. All manoeuvres with wind are placed on the left side to be better visible. The turning tracks for head wind have smaller advance: due to the wind pressure, the ship reaches not so much out into the initial direction. This is the contrary under stern wind, where the wind pushes the ship in wind direction. However, the wind does not allow the ship to turn further than the equilibrium on straight track – i.e. there is no chance to identify a transfer distance.

However, the transfer is even not to be identified because the ship does not turn anymore. Specifically, under beam wind the ship's initial turning capability is reduced: because the ship needs a rudder angle of about −12° to port and a drift angle of about 20°, the rudder effectiveness is reduced: When the rudder is put to 35° starboard, then the inflow to rudder is: 20°drift+35°rudder=55° effective rudder inflow angle; therefore, the rudder is not very effective. Stalling effects reduce the rudder lift; the ship is not able to complete the turn to starboard. After course change of about 45°, the ship is on the straight track. This might be a very dangerous situation if the ship performed a collision avoidance manoeuvre as stand-on vessel to starboard under these conditions!

The turning to port side is also reduced at the beginning because from the -12° rudder port to hard rudder −35° port, there is only a small rudder change – so initially the ship starts to turn very slowly, and then the ship turns through the wind very fast – to go a straight track afterwards.

1.3 Use of Fast-Time Simulation for Planning of Manoeuvres for Simulator Training

Task Description: Introduction, Conventional Briefing and New Concept

During the exercise briefing, the navigational officer is introduced into the harbour area, the starting situation and the environmental conditions within this area on a conventional sea chart (Fig. 1.9). The objective is to bring the ship through the fairway channel of Rostock Port from North, to turn the ships and heading back through the channel to berth the ship with portside at the passenger pier.

In the conventional briefing, only these rough indications of the manoeuvring status can be used to develop a potential strategy for berthing the ship. In conventional berthing plans, only ship contours are used to be positioned in drawings with

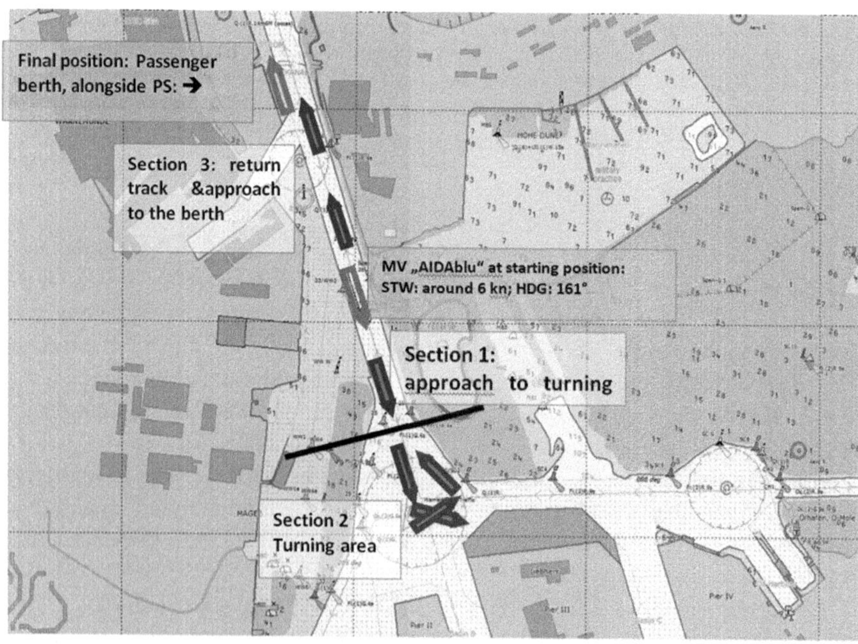

Fig. 1.9 Exercise area and environmental conditions in Port of Rostock for berthing scenario, divided into two sections for planning the manoeuvres and completed by guessing for desired positions as ship shapes only. The respective harbour area is divided into manoeuvring sections, which are following a specific aim:

• Section 1: Ship speed should be reduced until she is ready to be turned; SOG should be around 3 kn to be prepared for Section 2.

• Section 2: The ship should be turned and adjusted to go back in the fairway on opposite course to the final berth.

• Section 3: The ship should be stopped and berthed

Word or PowerPoint – The specific manoeuvres and settings of engine rudder and thrusters cannot be discussed in detail because specific manoeuvring characteristics can hardly be used for the specific situations. And real-time simulation is too much time-consuming. The FTS allows for new methods for individual exercise preparation with self-developed manoeuvring concepts:

- Drafting manoeuvring concept in more detail as manoeuvring plan with the SAMMON Manoeuvre Planning & Design Tool
- Optimisation of the concept by several planning trials with that tool
- Pre-training with the SAMMON Trial & Training Tool to try out the concept with real-time simulation on a laptop

Briefing by Means of the "SAMMON Manoeuvre Planning & Design Tool"

Basic Exercise with No Wind and Current With the new RAPIT, there is the chance for designing a manoeuvre plan as a detailed strategy with the specific settings at distinguished positions called manoeuvring points (MPs). In the following, the course of actions is described in a series of figures to make a full manoeuvring plan by means of the control actions at the MPs. This will be done first for easy conditions with no wind and current to explain the procedure of fast-time planning: In Fig. 1.10 the initial position MP0 is to be seen where the instructor has set the ship in the centre of the fairway. The ship has already been moved by the slider at

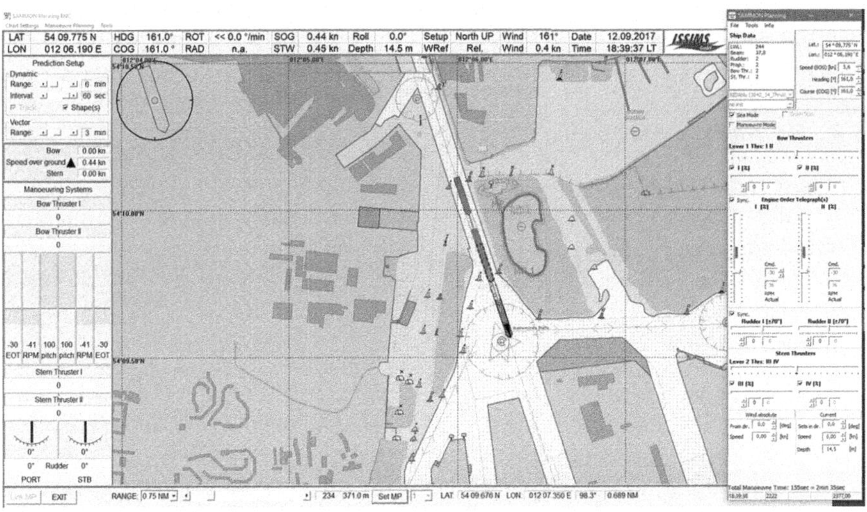

Fig. 1.10 Fast-time planning in sea chart: initial ship position at MP0 and prediction for the stopping manoeuvre at MP1: The prediction already shows that the ship reduces speed to the set handle positions EOT −30% astern

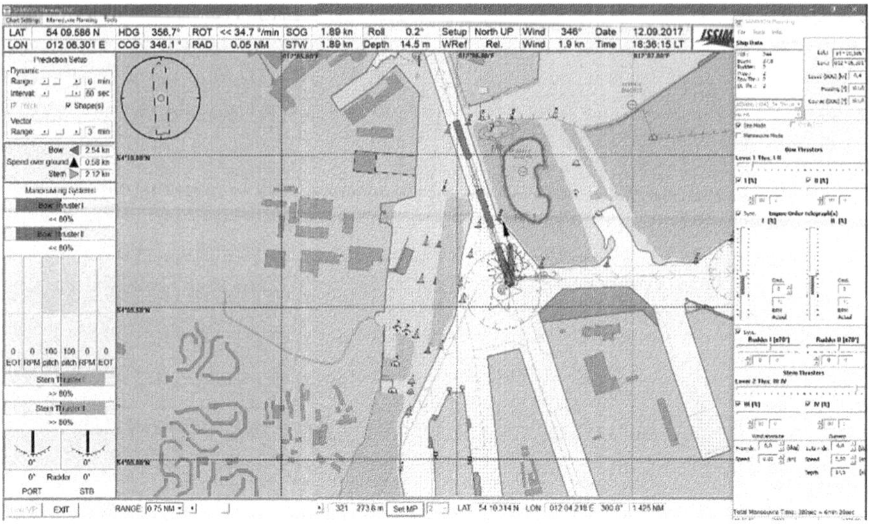

Fig. 1.11 Ship position at MP2 and prediction for the turning manoeuvre: The prediction shows that the ship is turning due to bow and stern thrusters set to 80%

the ENC bottom to set the next manoeuvring point MP1: there, the stopping manoeuvre is started with EOT−30%. The prediction already shows that the ship would lose speed according to the handle positions.

In Fig. 1.11 the ship is nearly stopped and turns by means of the thrusters: the contour is shifted to a position where the thrusters are stopped and the engines speed up to return to the fairway with opposite course. In Fig. 1.12 the vessel is brought close to the berth, and at MP5 the engines are reversed to reduce speed and to stop the ship at a position parallel to the berth to be shifted by thrusters to the pier from the next MP6. Afterwards the plan needs a further MP in order to reduce the transversal speed shortly before berthing.

Advanced Exercise with Strong Wind The full potential of the FTS can be seen for challenging weather conditions. In Fig. 1.13 the scenario is now to be solved for 25kn wind from 61°. The initial position is the same as in the previous example, but the first task for the trainee is to find the balance condition in the fairway: after some attempts, a drift angle of about 16° and rudder angle 3° was adjusted, and the ship contour was shifted to the buoys at the entrance of the fairway.

The next manoeuvring segment is for stopping and turning in Fig. 1.14. On the left side, it can be seen that in case the ship would be plainly stopped here as in the previous exercise, she would heavily drift with the wind. Therefore, the engines are split to support the turning by the starboard engine while the port engine goes astern.

In the final part of the manoeuvre (Fig. 1.15), the crucial segments are difficult because of the strong wind on the return track on opposite course: the ship enters the fairway now from south, and because of the strong wind from the bow, there is a

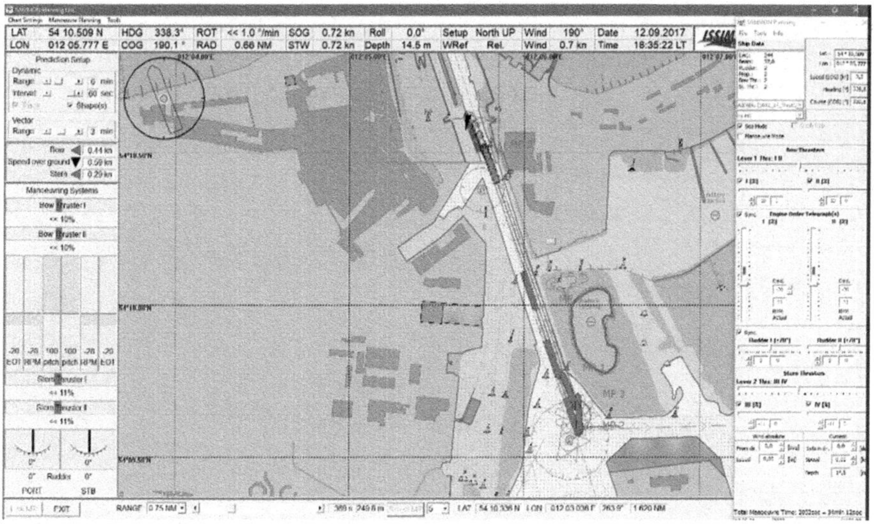

Fig. 1.12 Final part of the manoeuvring plan: the vessel is brought into a position parallel to the berth to be shifted by thrusters to the pier from the next MP6

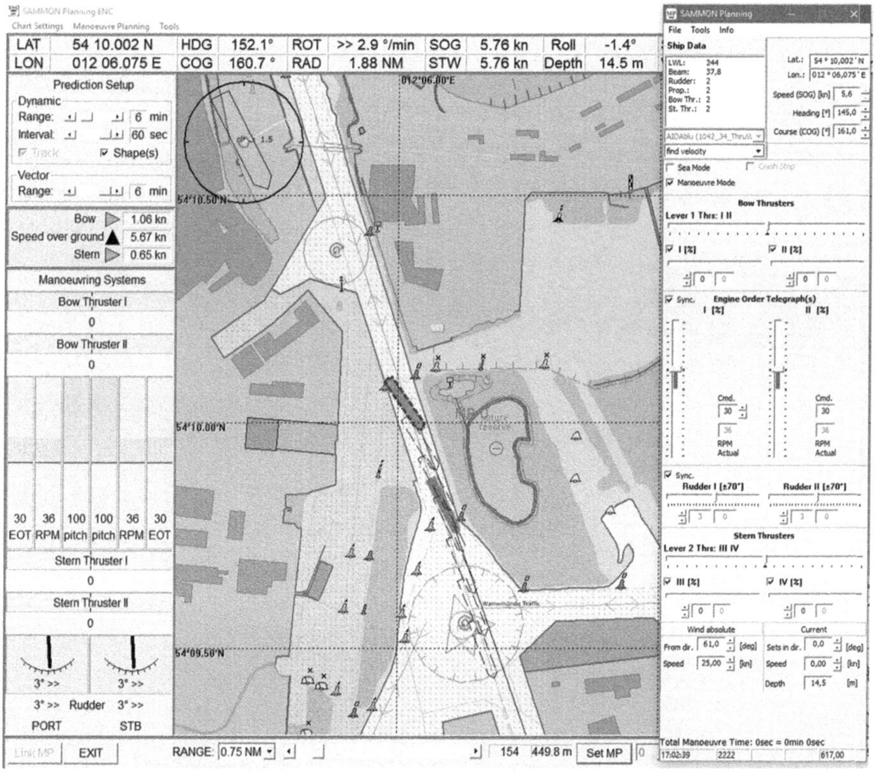

Fig. 1.13 Fast-time planning in sea chart under wind 25 kn from 61°: initial ship position at MP0 and prediction for future track under drift angle

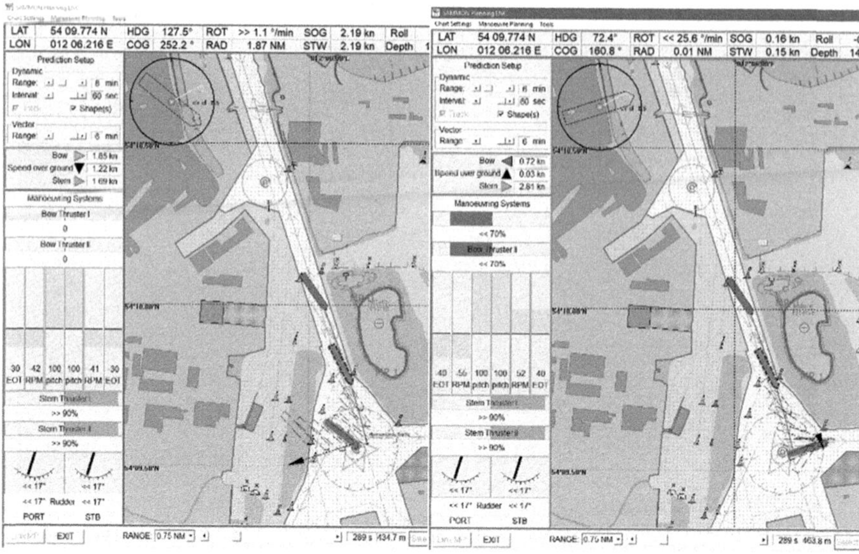

Fig. 1.14 Ship position at MP2 and prediction for the turning manoeuvre with two strategies. Left: turning only with thrusters (same concept as without wind in Fig. 1.11). Right: more powerful solution with split engines and rudder support

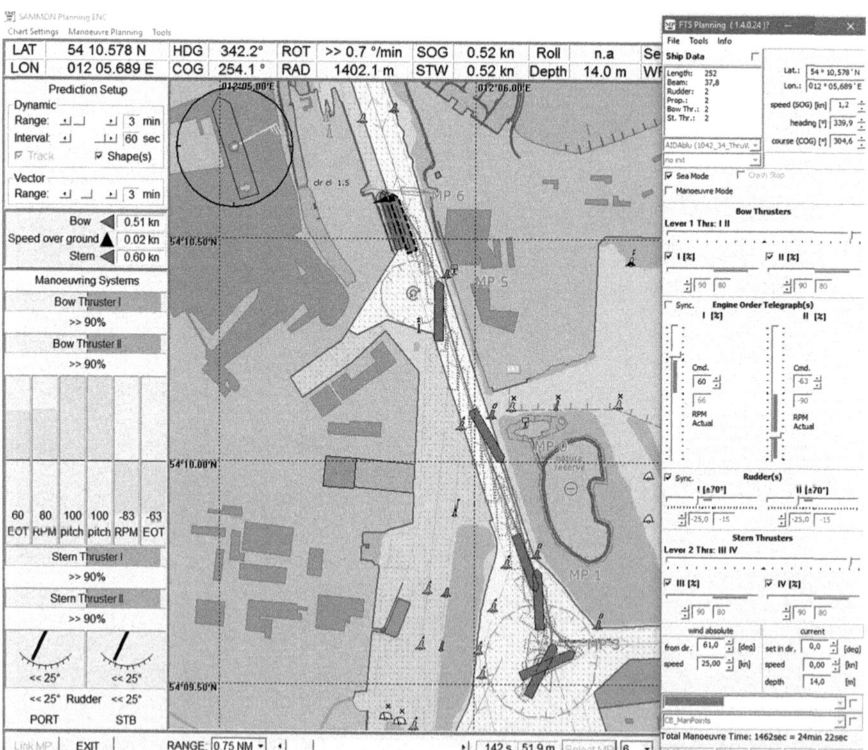

Fig. 1.15 Complete manoeuvring plan with final berthing manoeuvre

need to adjust heading, course and rudder. It is advantageous to split the engines because the rudder is more effective when one engine goes with more power. In addition, the ship is better prepared to stop because one engine is already going astern and does not need additional reversing time. On the right side of the figure, the stopping manoeuvre is to bring the ship into a position parallel to the berth. In the berthing phase, the thrusters and rudders are used with full power to counteract the wind effect for the final berthing; the approaching speed of the drift motion towards the pier is below 0.8 kn (for 30 kn, it would be over 1.5 kn).

1.4 Conclusions/Outlook

Fast-Time Manoeuvring Simulation has proven its benefits for both lecturing and training for improving ship handling knowledge and skills. In the paper were shown some examples for lecturing content on wind impact on ships where instead of theoretical explanations, simulated samples explain the ship behaviour and allow for preparing complete manoeuvring plans. The overall advantage of this planning procedure is that it is very fast. A student might need only minutes to prepare a plan, and then there is enough time to look for other options or other wind conditions and potentially limiting factors or elements. This is a contrast to the same manoeuvres in a full mission SHS which would take hours. For the future, the great potential will be investigated to integrate also fuel consumption and emissions into the simulation. Therefore, it could be used to increase efficiency and sustainability of manoeuvring – already in the education and training process and also on board [7, 8]. The majority of the participants in the ship handling courses expressed their opinion that the SAMMON Design & Planning Module could be used for preparing berthing plan on the ships. It is also possible to use the potential of RAPIT for various analyses (e.g. fairway layout, accidents) to find measures to make shipping safer.

Acknowledgements The research results presented in this paper were partly achieved in the following research projects:

- "EURO-ZA Capacity building in the field of maritime education" funded by the European Commission
- "Multi Media for Improvement of MET" (MultiSimMan), funded by the German Federal Ministry of Education and Research (BMBF) surveyed by Research Centre Juelich PTJ.The professional version of the SAMMON software tools has been further developed and maintained by the start-up company Innovative Ship Simulation and Maritime Systems GmbH (ISSIMS GmbH; www.issims-gmbh.com).

References

1. ISSIMS GmbH: SAMMON, *Simulation Augmented Manoeuvring Design & Monitoring System*. Version 1.4, https://www.issims-gmbh.com/yoomla/products/sammon
2. ISSIMS GmbH: SAMMON, *Simulation Augmented Manoeuvring Design & Monitoring System*. Version 1.4, https://www.issims-gmbh.com/yoomla/products/simopt
3. K. Benedict, M. Gluch, S. Fischer, M. Schaub, *Innovative Manoeuvring Support by Simulation Augmented Methods – On-Board and from the Shore*, in Conference of the International Maritime Lecturers Association IMLA 23, Durban /SA, 29 June to 3 July 2015
4. K. Benedict, M. Gluch, M. Kirchhoff, S. Fischer, M. Schaub, *Innovative Fast Time Simulation Tools for Briefing/Debriefing in Advanced Ship Handling Simulator Training at Aida Cruises Rostock*, in 19th International Navigation Simulator Lecturers' Conference – INSLC 19, Cape Town, SA, 5–8 September 2016
5. B. Günekbay, Application of Fast Time Simulation for Training Courses of Ship Officers. Master thesis at Department of Maritime Studies Rostock Warnemuende, Hochschule Wismar, University of Applied Sciences – Technology, Business and Design. Rostock, Germany May 2015
6. ISSIMS GmbH – Marine Prediction Technology Youtube channel, https://www.youtube.com/channel/UCR7yLtA5eqRUHNfQLXfgueA
7. M. Schaub, G. Finger, C.-M. Krüger, G. Tuschling, M. Baldauf, K. Benedict, *Innovative Simulation Method for Sustainable & Safe Operation*, in International Maritime Lecturers Association (IMLA) (Hg.): Modern Challenges in Maritime Education and Training, Batumi (2019)
8. M. Schaub, G. Finger, C. Krüger, G. Tuschling, M. Baldauf, K. Benedict, *Quantifying Fuel Consumption & Emission in Ship Handling Simulation for Sustainable and Safe Operation in Harbour Areas*, in IAMUC2019 (Hg.): Proceedings of the International Association of Maritime Universities (IAMU) Conference, Tokyo (2019)

Chapter 2
Artificial Intelligence as a Practical Approach to Represent Complex Dynamic Relationships in Maritime Navigation

Michèle Schaub, Knud Benedict, and Matthias Kirchhoff

2.1 Introduction

Artificial intelligence (AI) is often mentioned and discussed in relation to the disappearance of jobs. Despite that, AI is increasingly encountered in the maritime sector, either in the context of methods that aim to reduce fuel consumption or in discussions about unmanned ships and autonomous navigation. The latter leads to the assumption that artificial intelligence should replace natural human intelligence. Nevertheless, there are many tools basing on AI methods that support the navigational or technical officer. One essential basic element of AI is artificial neural networks (ANN). It is a data-based approach, which allows for mathematically capturing multidimensional relationships. Studying approaches that can reflect dynamic non-linear processes, the sample of diesel particle generation was chosen, where standard map representations or polynomial approaches quickly become impractical. Reasons for this include poor extrapolation properties or lack of manageability. ANN as universal approximators offer a possibility which, if sufficient training data is available, provides a good assignment of input data to output data, such as engine speed and consumption as input variables to particle emissions as output variables. The computing time needed to calculate the results is very small so it is possible to use this ANN method for fast-time simulation (FTS), e.g. in the SAMMON software system for rapid manoeuvring predictions. The example of ship particle emissions

M. Schaub (✉) · K. Benedict
Department of Maritime Studies Warnemuende/Institute ISSIMS, Hochschule Wismar, University of Applied Sciences – Technology, Business and Design, Wismar, Germany
e-mail: michele.schaub@hs-wismar.de; knud.benedict@hs-wismar.de

M. Kirchhoff
ISSIMS GmbH, Rostock, Germany
e-mail: mkirchhoff@issims-gmbh.com

S. Bauk, S. D. Ilčev (eds.), *The 1st International Conference on Maritime Education and Development*, https://doi.org/10.1007/978-3-030-64088-0_2

is used to demonstrate how methods of AI may potentially support people in competent decision-making for manoeuvring action. Having a better idea of the impact of intended manoeuvres will support navigational officers to take more easily decisions in the sense of sustainable shipping.

2.2 Artificial Neural Networks with External Dynamics

Definitions

Artificial Intelligence (AI) The term "Artificial Intelligence" goes back to the computer scientist John McCarthy [1], who coined the term in 1956 with the first AI conference. Philosophers are still divided on the concept of intelligence. In this article, AI is defined as a branch of computer science that uses various methods to try to equip a machine with capabilities that can be attributed to humans.

Artificial Neural Networks (ANN) One very often used method in AI is artificial neural networks. ANN were characterized first by McCulloch in the years 1940s [2] about one decade before the birth of AI. They are based on the attempt to describe human neuron structure and function mathematically.

External Dynamics In nature and technics, dynamic processes occur very frequently. Dynamics are characterized by the fact that an upcoming time step depends not only on its predecessor and/or any other actual impact but also on the previous time course. This leads to the fact that either models with internal memory are used or the input space must be multidimensionally spanned. The latter method is referred to as external dynamics [3]. It is a quasi-dynamic method, since it is actually based on a static model architecture and only extends the entrance space.

Model Architecture

The quasi-dynamic method introducing external dynamics into a static network architecture can make use of a multi-layer perceptron (MLP) network (Fig. 2.1). MLP networks are commonly used in AI tasks. They consist of an input layer, at least one hidden layer and an output layer.

The input data entering the input neurons are being normalized and forwarded to the hidden neurons. Before entering the neuron, the input is multiplied by the so-called synaptic weight. The hidden neurons consist of a summarizing function for all weighted inputs followed by the activation function whose orientation depends on the threshold value. Within the activation function, being a monotonically increasing function as, e.g. a hyperbolic tangent function, the summed-up

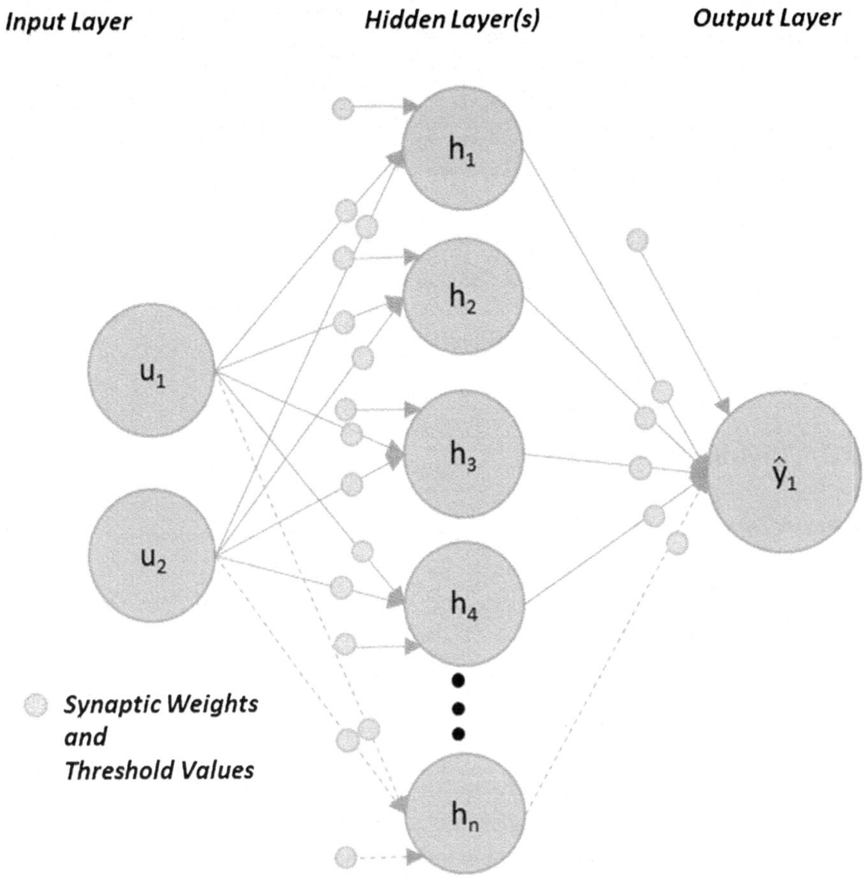

Fig. 2.1 Schematic structure of an artificial neural network (multi-layer perceptron)

value is assigned to a neuron output value. Following the link at the neuron's exit, the output value enters or another hidden neurons or the output neuron(s).

Training of the MLP Network

To adapt an ANN to a specific process, the synapse weights and the thresholds have to be tuned. The availability of large training data sets is the basis for well-trained ANN. The training is done by backpropagation algorithm [4]. It is based on a non-linear optimization procedure according to the gradient descent method. Due to the non-linear activation function, the model is non-linear in its parameters, i.e. in its synaptic weights and thresholds. The derivation of the cost function according to the ith weight of the jth hidden neuron is determined according to the chain rule (Eq. 2.1).

$$\frac{\partial C}{\partial w_{ij}} = \frac{\partial C}{\partial \hat{y}} \cdot \frac{\partial \hat{y}}{\partial h_j} \cdot \frac{\partial h_j}{\partial z_j} \cdot \frac{\partial z_j}{w_{ij}} \tag{2.1}$$

$\hat{y}$ is the estimated net output. h_j is the output of the jth neuron, whilst z_j is the argument of the activation function.

As mentioned above, the training data not only consists of one of each considered input type but also of their previous time course. The length of the previous data vector depends on the dynamic of the problem and the sampling rate.

2.3 Fields of Application

Theoretical vs. Experimental Modelling

If a theoretical, e.g. a physical, model meets the specified quality criteria, then this model should be preferred. However, there are still a lot of complex non-linear dynamic relationships which are not yet covered by theoretical models, e.g. the generation of particulate matters (PM) within combustion engines. Moreover, theoretical models need more time for calculation than experimental models do. On the other hand, experimental model takes a lot more time in setting a suitable model architecture and in training, i.e. parameterizing it. The need of training data implies that no process simulation can be performed during the design phase. Furthermore, the required amount of training data is much higher than for theoretical models.

The boundary conditions and quality criteria for the herein discussed example of PM generation are:

- The process is already running, collection of measurement data is possible.
- Short computation time in the range of a few milliseconds per time step is required.
- Calculation results should be within the variance of the measured training data
- Dynamics which had not been trained should be simulated with realistic tendencies.

The last criterion is the most sensitive one and the reason for the introducing sentence of this section. Even though ANN show very good interpolation and extrapolation behaviour among all experimental model types, there is no guarantee that it gives a realistic output in a totally new and unknown constellation of inputs. Theoretical models, in contrast, have given limits in which, e.g. physical laws determine the range of their outputs. This is where the "Natural Intelligence", the human brain, is needed – now and in future!

Manoeuvring Assistance Software SAMMON

The focus of the present studies lies on the implication of a new feature to an already existing manoeuvring assistance software for seagoing vessels, called SAMMON [5]. SAMMON stands for Simulation Augmented Manoeuvring Design & Monitoring System. It is the brand name of an assistance software developed by the research institute ISSIMS of the Wismar University of Applied Sciences in collaboration with the ISSIMS GmbH company.

SAMMON consists of a training, a planning and a manoeuvre online monitoring tool, all of them using a mathematical ship model in their kernel. The software is using a highly efficient and user-friendly fast-time simulation (FTS) technology called RAPIT. This "Rapid Advanced Prediction and Interface Technology" enables the prediction of the ship's path and time courses of the different velocities (longitudinal, transversal and rotational) over several minutes within only 1 s of computing time as well as their presentations in a user-oriented interface (Fig. 2.2).

Manoeuvring in confined waters with a lot of traffic is a challenge for each navigational officer. Not only do safety issues play a role, but time and, increasingly, environmental aspects are also coming to the fore. Recently, the calculation of fuel consumption had been integrated as a new feature into SAMMON software. Based on this new information and the available soft- and hardware interfaces, some more features regarding the pollutant emissions are targeted.

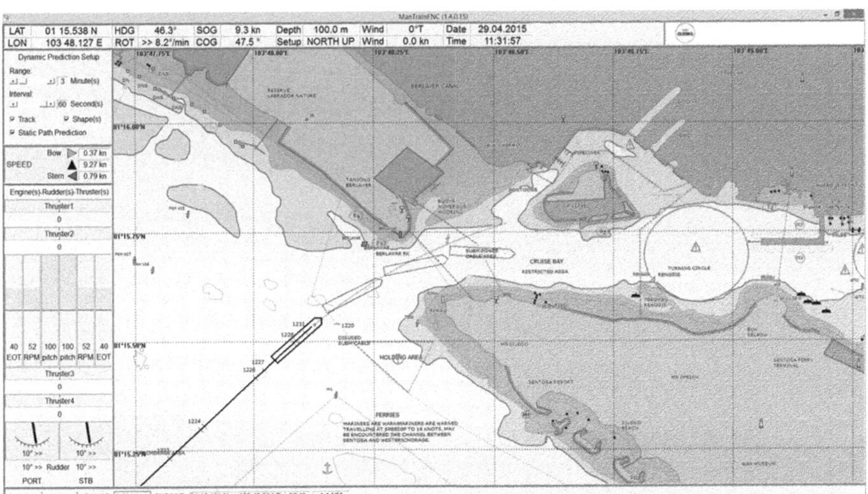

Fig. 2.2 Manoeuvring assistance software SAMMON for prediction and monitoring of future ship's path and velocities [6]

Integration of the MLP Network into SAMMON and SIMOPT

The ANN to predict particulate matters is fed by the inputs fuel consumption and current engine speed as well as their previous time courses. The experience as well as the cross-validation (Sect. 4.2) showed that previous data of the past 5 s are enough to get satisfying prediction results under the condition that sufficient training data are available. The input data for ship and engine dynamic to be fed into the ANN are taken from the SAMMON simulation process and its engine module.

The emission module is attached via a dynamic link library (dll) to the engine module and calculates pollutant emissions like PM by means of an ANN customized and trained for the specific propulsion engine.

Figure 2.3 shows the time history of a simulated load change undertaken with a virtual ship in the SIMOPT software [7]. SIMOPT (Simulation Optimization for Ship Models & Standard Manoeuvres) is compatible to SAMMON and allows for testing new software features and tuning as well as optimizing ship parameters and manoeuvres. In the present case, the time course for PM has been simulated whilst the ship's speed was increasing from 20% engine load up to 90%. In the period from 0 to 5 s, there is no information on PM yet, because the MLP network needs a lead time of 6 s to make a prediction from the past values.

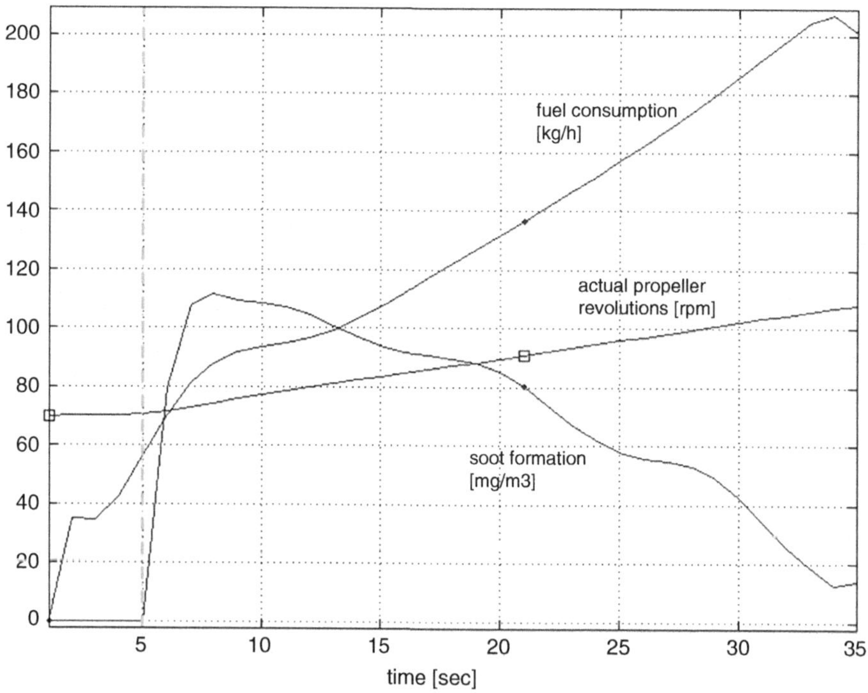

Fig. 2.3 Generation of PM due to a load increase. PM depend on the course of fuel consumption and the current engine speed resp. The propeller revolutions

2.4 Practical Implementation of the Method

Data Acquisition

Data acquisition on board of real vessels has already taken place several times but until now without the possibility of measuring particulate matters. Therefore, for the present studies, the engine at Hochschule Wismar Engine Lab was used as test bed (Table 2.1), equipped with the highly precise mobile PM sensor Pegasor Mi3 [8] which allows for measuring during transient engine operation.

The data acquisition was carried out in the time frame of 1 week. More than 200 reasonable load changes across the engine map have been measured considering generator and propeller modes: Generator mode means that the engine controller tries to keep the nominal engine speed constant. Load is changed by adapting the propeller blades of the controllable pitch propeller. Propeller mode is considered for ships having a fixed pitch propeller and therefore needs to adapt the engine speed in order to change the propeller torque.

Model Training and Validation

Due to, relatively seen, less data, a cross-validation approach was taken into account to find the best model structure. This means that 70% of the measurement data had been taken for training, whilst 30% were left to be used for validation. In this constellation, estimation and validation took place repeatedly whilst the MLP network was trained with a changing number of past input data and hidden neurons. After a whilst another 70% of the same complete data set was taken to estimate the parameters again and so on. At the end, best structure, i.e. the optimum number of past inputs and the optimum number of hidden neurons, was selected. The input dimension due to the MPL network with external dynamics increased from 2 (static case) up to 16 neurons in the input layer. The hidden layer consists of 20 neurons. Figure 2.4 shows a result where the combination of input and output data had not been used to train the network.

Table 2.1 Specification of test bed engine

Parameter	Test bed engine: MAN B&W 6L23/30
Type	Medium-speed 4-stroke marine diesel engine
Bore	225 mm
Stroke	300 mm
Rated output	1050 kW
Rated speed	900 min^{-1}
Compression ratio	13.5: 1
Fuel injection system	Unit injector system

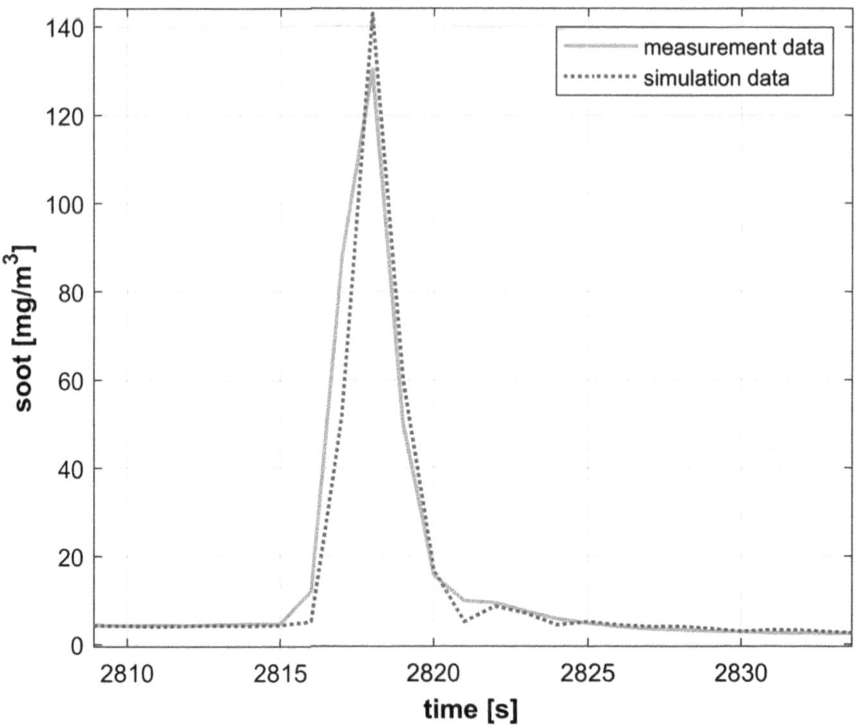

Fig. 2.4 Validation of the MLP network with a load increase resp. PM impact that had not been used for network training: solid line represents the measurement, whilst the dotted line is the simulation result

Most load changes and their influence on PM formation can be reproduced with the trained net. Exceptions exist in particular if a new setpoint is set during transient operation. Because these types of processes have not been trained, they are only simulated qualitatively, but not quantitatively correctly.

2.5 Application in Education

The SAMMON tools for planning, training and monitoring already form part of education and training at Hochschule Wismar, Department of Maritime Studies, Systems Engineering and Logistics. In the recently successfully completed national research project MEmBran (modelling of emissions and fuel consumption during ship manoeuvring), a potential analysis for the assistance software SAMMON was carried out. The result was very impressive: Without assistance software, the nautical officers navigated with much larger rudder positions and more frequent rudder movements. Many test persons manoeuvred with hardly any manoeuvring reserves.

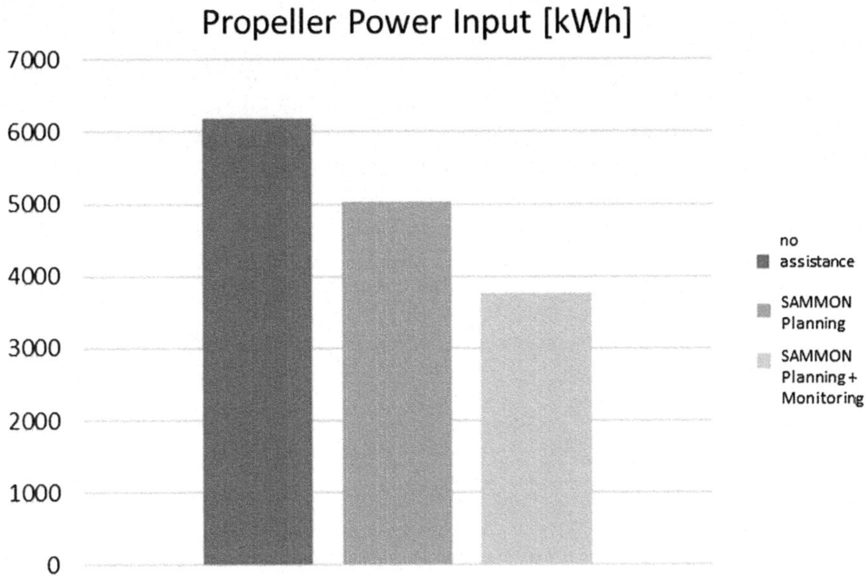

Fig. 2.5 Power input of the propeller as a mean value of all test scenarios depending on the use of an assistance software in advance (Planning) and/or during the manoeuvring itself (Monitoring)

The energy input into the water was obviously increased, and safety was partly at the limit. Improvements could already be observed by planning the manoeuvres in advance. Using planning software and online monitoring improved the performance even more (Fig. 2.5).

Since no real ship was available for the PM measurements, a simulation ship with a suitable propulsion engine had to be created for the use in the SAMMON software. This gives now the opportunity not only to teach about safety and time optimization during manoeuvres but also to consider fuel consumption and environmental aspects. As part of the ongoing ERASMUS+ project EURO-ZA between the partners from Europe and South Africa, the use of SAMMON software in teaching and further education is being promoted.

2.6 Summary and Outlook

In this article an efficient method was presented, which allows to visualize dynamic non-linear ship motion and engine processes under the condition of availability of numerous training data. Multi-layer perceptron neural networks are universal approximators and able to represent any given mathematical problem under the condition that the structure and the training data are chosen appropriately. In the present example of PM formation, a simulation could be carried out with the help of this data-based method. This could not have been carried out otherwise due to the lack of computing speed for fast decision-making processes and the absence of an adequate physical model.

The SAMMON assistance software equipped with this new feature can thus offer a new element in maritime education and training. Implementation on board an ocean-going vessel is theoretically possible, but would require the necessity of measuring PM and other relevant variables during manoeuvres.

Acknowledgements The approach described above is derived from a German research project (03SX423E), granted by the German Ministry of Economic Affairs and Energy. The herein presented software tools SAMMON and SIMOPT were provided by ISSIMS GmbH. Further acknowledgement goes to the ERASMUS+ project EURO-ZA, funded by the European Commission, which supports the international establishment of the described assistance software at maritime institutions.

References

1. E. Konrad, Zur Geschichte der Künstlichen Intelligenz in der Bundesrepublik Deutschland, in *Sozialgeschichte der Informatik*, ed. by D. Siefkes, P. Eulenhöfer, H. Stach, K. Städtler, (Studien zur Wissenschafts- und Technikforschung. Deutscher Universitätsverlag, Wiesbaden, 1998)
2. W.S. McCulloch, W. Pitts, A logical calculus of the ideas immanent in nervous activity. Bull. Math. Biophys. **4**, 115–133 (1943)
3. O. Nelles, *Nonlinear System identification, from Classical Approaches to Neural Networks and Fuzzy Models* (Springer, Berlin, 2001)
4. R. Isermann, S. Ernst, O. Nelles, *Identification with Dynamic Neural Networks – Architectures, Comparisons*, Applications, in IFAC Proceedings, vol. 11, pp. 947–972, 1997
5. ISSIMS GmbH: SAMMON, *Simulation Augmented Manoeuvring Design & Monitoring System*. Version 1.4, https://www.issims-gmbh.com/yoomla/products/sammon
6. M. Schaub, G. Finger, T. Riebe, F. Dahms, E. Hassel, M. Baldauf, *Data-Based Modelling of Ship Emissions and Fuel Oil Consumption for Transient Engine Operation*, in OCEANS Proceedings, 2019
7. ISSIMS GmbH: SAMMON, *Simulation Augmented Manoeuvring Design & Monitoring System*. Version 1.4, https://www.issims-gmbh.com/yoomla/products/simopt
8. https://pegasor.fi/site/attachments/Pegasor_Mi3_web.pdf (2020)

Chapter 3
Statistical Analysis of the Weather Conditions of the Main SSS Routes in the Western Mediterranean for Its Presentation and Use as a Participatory Site Web to Support Learning Processes

F. X. Martínez de Osés and Elisenda Ventura Jariod

3.1 Introduction

The short sea shipping (SSS) market is crucial for our sector, especially in the Mediterranean [1]. However, although the measures taken by the European Union to improve its use have been numerous, these have not been implemented quite successfully and few routes have their continuity assured; despite the benefits that this mode of transport offers, there are still many reluctances regarding its implementation that slow down its growth and damage its image [2, 3].

This paper presents the results obtained in the project linked to the research carried out by the Nautical Science and Engineering Department (DCEN) of the Barcelona School of Nautical Studies, for the optimization of routes in the western Mediterranean. Through the analysis of the weather conditions in the western Mediterranean sailing area and, in particular, the most common adverse weather conditions in the area, a study of the impact of these weather conditions is carried on in the context of sea route optimization and their importance for the short sea shipping routes in the area. Furthermore, this project also thrives to become a platform used by the students of the faculty as a didactic tool for the subject of meteorology.

F. X. Martínez de Osés (✉)
Department of Nautical Science and Engineering, Barcelona School of Nautical Studies, Technical University of Catalonia – BarcelonaTech, Barcelona, Spain
e-mail: fmartinez@cen.upc.edu

E. Ventura Jariod
Barcelona School of Nautical Studies, Technical University of Catalonia – BarcelonaTech, Barcelona, Catalonia, Spain
e-mail: elisenda.ventura@estudiant.upc.edu

S. Bauk, S. D. Ilčev (eds.), *The 1st International Conference on Maritime Education and Development*, https://doi.org/10.1007/978-3-030-64088-0_3

Many publications establishing which are the factors that condition the SSS in the Mediterranean and which impact have for the function of costs of this mode of transport [4, 5]; generally, these articles analyse its efficiency and sustainability [6, 7]. However, some knowledge on climatology and meteorology is key to understand the case that occupies us; the climatology of the Mediterranean has been amply studied by numerous authors [8–10] and continues being a very present element in the recent academic production [11, 12].

The area surveyed, the basin of the western Mediterranean, has also been studied in the last years, especially its characteristics [13–15]. Its geographic localization makes the western Mediterranean Sea one font of indispensable resources for Europe's south countries and the north of Africa.

Other authors have remarked [16], that it is necessary an analysis of the main variables that affect the SSS in the western Mediterranean to comprehend which is the impact that these characteristic meteorological conditions have on the behaviour of the ship, and for the function of costs, in order to determine the best course at any season of the year.

This study considers one of the principal variables that affect the ship while in navigation: the waves. These are especially important when considering the consumption of fuel that, as it has already been established in literature, is the first factor that will affect the variable cost of the SSS [15].

With the data collected through the web of *Puertos del Estado,* [1] the main areas of affectation have been defined, and the changes produced through the seasons have been studied. Once these results have been analysed, they have been contrasted comparing them with the results obtained with the SIMROUTE software[2] for one of the routes of SSS of the Mediterranean for each month of the year 2017. The data of this study have been compared with the known parameters of the western Mediterranean to extract the conclusions.

As the information of the meteorology and climatology available for the students does not contain, in most cases, the data used for the studies, it is often difficult for them to work with it. To tackle this problem and provide the students with appropriate materials to fully understand the contents of the subject, the results of this study have been presented in participatory web format for dissemination and use as a didactic tool by them; the routes studied will also be used in class as practical examples to explain the contents of the course in a clear and concise way.

---

[1] *Puertos del Estado* is the government agency responsible for the coordination and efficiency control of the Port Authorities of the Spanish Port System. On its official website (*http://www.puertos.es/en-us*), it can be found, among others, oceanographical information and the reports used in this analysis.

[2] The SIMROUTE is software designed for maritime route optimizations which carries out simulations of short and long distances. This software has been developed by a team of researchers at the Technical University of Catalonia (UPC-BarcelonaTech) in MATLAB language. The spatial scope of SIMROUTE is the Mediterranean Sea, fully or partially selected, depending on the longitudes and latitudes introduced by the user. In order to run the program, there are several parameters to introduce: port of departure, port of arrival, vessel speed and the wave field for the days of the voyage to simulate.

3.2 Materials and Methodology

The pool of data analysed for this survey, nicknamed SIMAR, is made of some temporal series of parameters for waves numerically modelled; they are, therefore, synthetic data.

This pool of data arises from the concatenation of the two big pools of data simulated by *Puertos del Estado*, SIMAR-44 and WANA, and offers an extensive temporal series, updated and generated with a time cadence for the period 1958–2017.

The temporal stretch, together with its regularity, is the main reason for which this pool of data has been chosen instead of the data collected by the buoys placed near the coast. Although the latter offers real data, its proximity to the coast and the disparity among periods studied make difficult to establish the pool of data to study. However, SIMAR data set, is a forecasting simulation data, but offers a wider period that is more suitable and useful for a statistical survey of the climatological variables.

The reports studied presented the statistical characterization of the waves, in the medium term; the reports were completed with a statistical description of the series of heights, periods and directions of the waves. Three types of statistics were included for the waves: height and period distributions, wave roses and height and wave direction distributions. This survey has centred especially at the distributions of height and period to find a relation between the seasonal climate and the behaviour of the waves. However, the direction of the waves has been observed in order to determine which is the most repetitive.

To analyse the data, the following methodology has been established:

1. Selection of the points SIMAR, distributed at the basin of the western Mediterranean, considering their geographic situation. Some of them, after being considered, have been disregarded for being too close to the coast.
2. Selection of the meteorological variables. This survey has considered the significant height (Hs) at % and the direction of the swell, since they are the parameters that most influence when navigating.
3. Sorting of the data according to seasons and zones established.
4. Tabulation and elaboration of seasonal graphs for each zone.
5. Calculus of the average distribution for season and zone.
6. Determination of the behaviour of the variables selected at the period studied.
7. Weighting of the data by seasons and zones to establish which challenges do they pose for the SSS lines.
8. Obtention of the differences of time and fuel for the route studied for the cases of the optimal route and the minimal distance route.
9. Comparative of the results obtained through the statistical analysis of the conditions with the interpolated wave fields generated with the SIMROUTE software for the days chosen.

The available data for each point studied, regarding its annual distribution along the period 1958–2017, have been considered in order to establish distinct zones that presented a similar behaviour. From there, the basin of the western Mediterranean has been divided into four zones: the area 1 (in green), the area 2 (in lilac), the area 3 (in yellow) and the coastal area (in blue) (see Fig. 3.1). This last area is formed by points that, due to their proximity with the coast, showed a characteristic distribution that could induce errors if considered together with the points of the other areas. Therefore, the areas studied are limited to the first three.

Although initially the identification of each one of the points resulted crucial for their classification, once established the zones, the points have been treated as a data set.

Finally, the program SIMROUTE has been run for the route Barcelona-Genova for the days 13 and 14 of each month of the year 2017. The analysis has considered the interpolated wave areas for the zone studied for each month, in order to check if the real data resembled the statistics.

The results have been compared among them and with the data of swell analysed for each season to establish similarities and/or dissidences.

3.3 Results

The following tables (Tables 3.1, 3.2, and 3.3) show the average distribution for each season at the defined areas of the western Mediterranean.

Studying the seasonal behaviour, it is observed that all three areas show a similar evolution; in the period comprised between December and February, the percentage of waves with a significant height of more than 3 m increases, surpassing the 5 m, while, in the period comprised from March to May, the data show that, although the

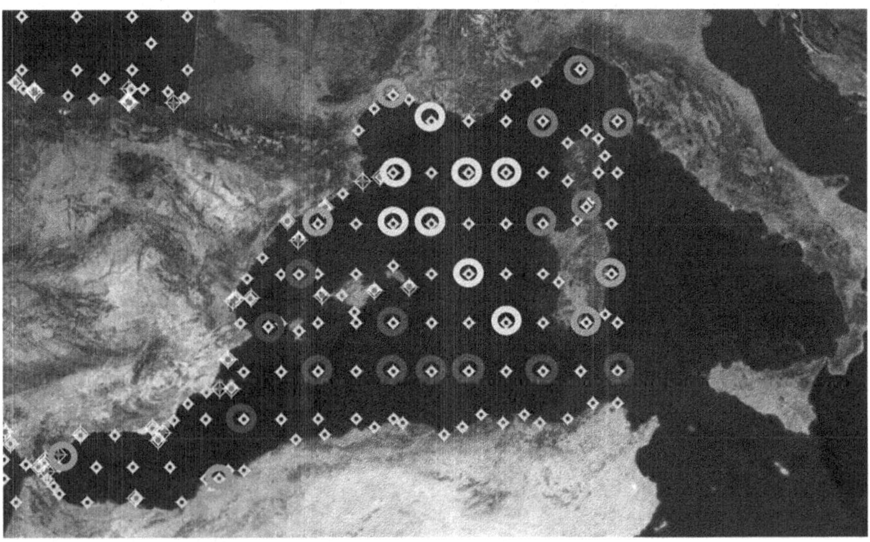

Fig. 3.1 Distribution of the points SIMAR studied

Table 3.1 Average seasonal distribution for the area 1

	Significative height (Hs) en %										
Period	<0,5	1	1,5	2	2,5	3	3,5	4	4,5	5	> 5
Dec.–Feb.	26,65	30,53	18,77	11,17	6,21	3,32	1,71	0,88	0,41	0,19	0,17
Mar.–May	35,62	32,73	16,31	8,05	3,86	1,85	0,85	0,39	0,18	0,08	0,07
Jun.– Aug.	54,32	28,48	10,12	4,33	1,75	0,64	0,24	0,07	0,02	0,01	0,00
Sept.–Nov.	39,24	30,34	15,30	7,96	3,88	1,85	0,80	0,35	0,16	0,06	0,05

Table 3.2 Average seasonal distribution for the area 2

	Significative height (Hs) en %										
Period	<0,5	1	1,5	2	2,5	3	3,5	4	4,5	5	> 5
Dec.–Feb.	15,45	27,23	21,50	14,65	9,05	5,38	3,15	1,82	0,96	0,51	0,39
Mar.–May	18,00	34,83	22,15	11,89	6,41	3,32	1,82	0,86	0,40	0,19	0,13
Jun.–Aug.	27,50	44,29	19,12	6,43	1,93	0,52	0,14	0,05	0,01	0,00	0,00
Sept.–Nov.	20,44	37,22	21,24	10,69	5,28	2,60	1,30	0,68	0,29	0,13	0,11

Table 3.3 Average seasonal distribution for the area 3

	Significative height (Hs) en %										
Period	<0,5	1	1,5	2	2,5	3	3,5	4	4,5	5	> 5
Dec.–Feb.	11,94	24,51	21,03	15,59	10,48	6,81	4,30	2,55	1,36	0,68	0,73
Mar.–May	19,54	31,17	20,50	12,06	7,27	4,32	2,54	1,32	0,65	0,32	0,32
Jun.–Aug.	33,78	32,76	15,12	7,58	3,96	1,92	0,73	0,27	0,09	0,03	0,01
Sept.–Nov.	21,25	29,99	20,17	12,51	7,41	4,12	2,20	1,17	0,57	0,27	0,33

distribution is similar, the incidence of waves with a superior height of 3 m diminishes roughly by 50%, compared with the previous season.

In all of them, in the period comprised by the months of June, July and August, the conditions of the sea improve notoriously, with a decrease of 40% of the significant height from 2 m, in comparison with the previous season. Despite this observation, in this same period, increases the incidence of a swell of less than 0,5 . Finally, the months of September, October and November show a new increase of the incidence of the swell from 2 m by 50%. However, although the changes of the significant height of the swell are similar across the three areas established, they do not have the same impact on them, since they present distinct characteristics.

It can be appreciated that in the area 1 there is a swell with a minor significant height arriving, in general, up to 1,5 m. The area 2, that has most of its points situated among the 36° and the 38° of latitude, presents a significant height of the swell minor than the area 3, but higher than the area 1. The area 3 stands up for a higher incidence of significant heights from 2,5 m to 5 m, at any epoch of the year, excluding the summer, where waves of 5 m result almost non-existent in all the areas.

Comparing these data with the interpolated wave area generated by the SIMROUTE for the days selected of each month, a similar behaviour can be observed. From the months of December to January, the swell represented has a

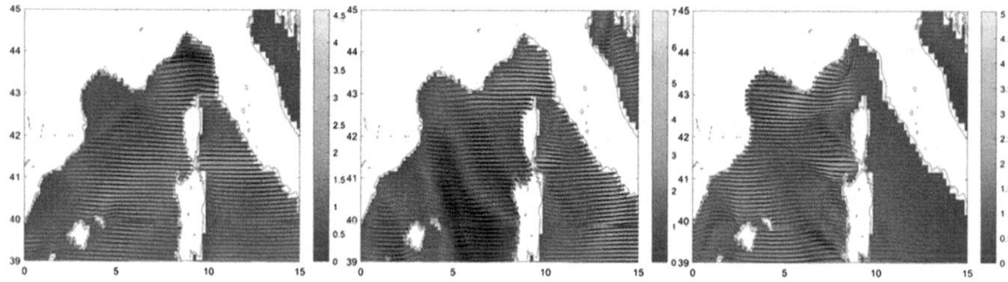

Fig. 3.2 December–January–February

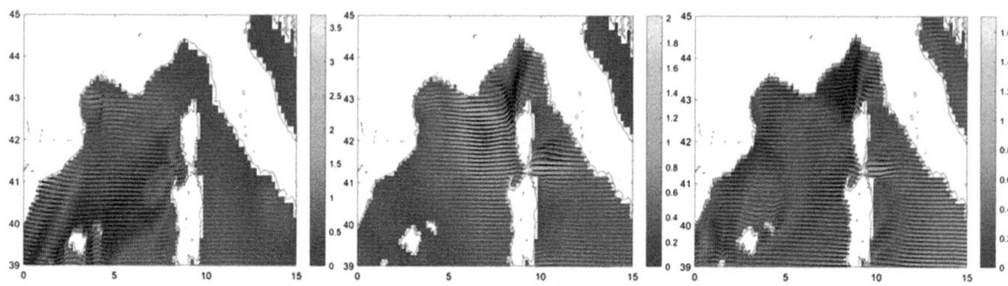

Fig. 3.3 March–April–May

height of 2–3 m, and, in some points, it arrives up to 5 m (Fig. 3.2). Conversely, for the months from March to May, the height of the waves diminishes gradually, never surpassing 3,5 m (Fig. 3.3). From June to August, the season that presents the best conditions, the swell does not surpass 2 m, with certain exceptions in some points close to Corse in July. Lastly, whereas September and October present some similar conditions compared with the period of March–May, November presents an increase at the height of the swell, especially in the area 2, that would correspond at a maritime storm, very common in this month of the year (Figs. 3.4 and 3.5).

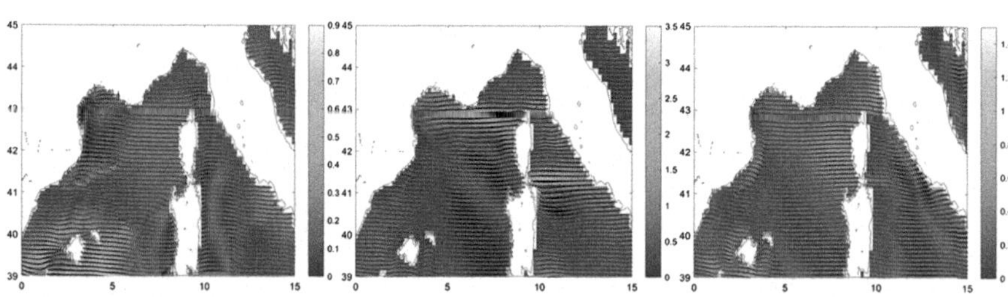

Fig. 3.4 June–July–August

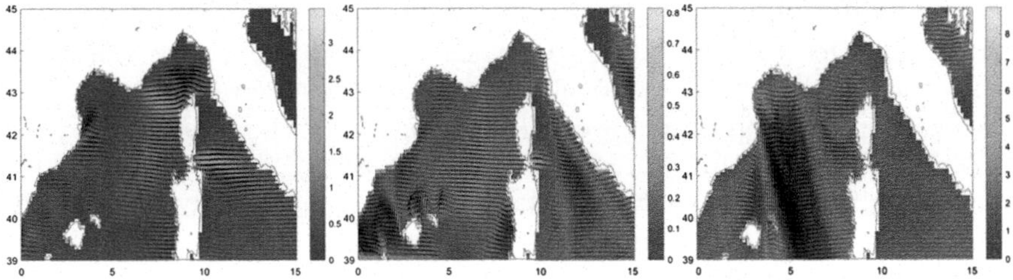

Fig. 3.5 September–October–November

3.4 Discussion

Although the Mediterranean Sea is weighted and studied as a whole, the fact is that it is formed by 12 seas, most of them known by their own name, such as the Ligurian Sea, the Tyrrhenian Sea, the Adriatic Sea, the Jonic Sea, the Aegean Sea and the Marmara Sea. Of all these seas, the Ligurian and the Tyrrhenian Sea are the main seas of interest for this survey since the points studied are mostly placed in their waters and since different areas show climatological differences (Fig. 3.6).

The area 1 is in the Ligurian and the Tyrrhenian sea. These seas are characterised for being warm, even in the winter, but with torrential rains in the autumn. The area 2, located close to the Moroccan and Algerian coasts, presents some conditions distinct of the Mediterranean Sea, but without the peculiarities of his neighbouring zones. This is the reason why its values in the table fall in the in-between range compared with the other two areas. The area 3 is situated in the Mediterranean Sea, next to the Gulf of Lion and the Balearic Sea. The prevailing winds in this area of

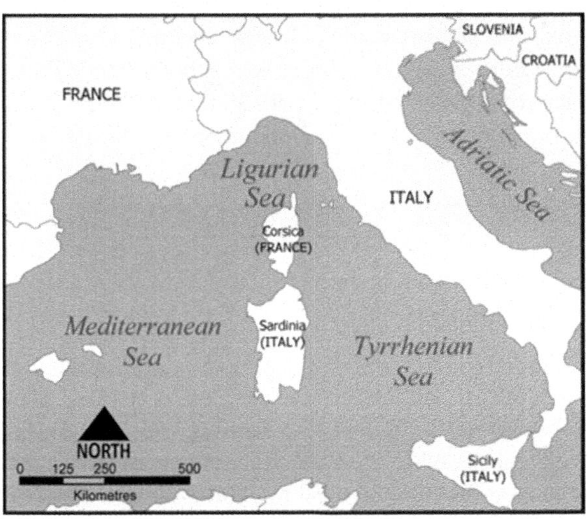

Fig. 3.6 Ligurian, Tyrrhenian and Mediterranean Sea

the golf are the north wind, called the Tramontana, and the Mistral, which are the unchaining of the most important storms of the western Mediterranean, with a swell that can exceed 8 m of height. A similar behaviour can be observed, although not as accused, in the interpolated wave field for the month of February (Fig. 3.2). Conversely, at the Balearic Sea, episodes of rain are usually coupled with the arrival of the Leveche wind and the Levant wind, both debited of humidity when passing for the Mediterranean; they also bring very intense precipitations in autumn and occasional floods (Figs. 3.4 and 3.5).

As it can be observed, the Mediterranean has a specific climate, with small peculiarities depending on the area of study. In general, it can be expected winters mildly cold, but humid and without rain, soft and rainy springs, warm and dry summers and soft autumns with a high potential of torrential rains.

Regarding the effect that these conditions have for the navigation, the most affected months are January, February and November. Poor weather conditions in January and February are associated, as it has been said before, with the Tramontana and the Mistral, as it can also be observed in the areas 2 and 3, where most of the swell came from the North and the West. On the contrary, the adverse weather conditions of November are caused by a north flux that provokes fanned strong winds from the north-west, due to the presence of an anticyclone in the Atlantic that is displaced to the north and the west of the Iberian Peninsula, and a low sited in the Gulf of Lion that worsened the navigational conditions in the area. Of the three months, November is the most affected by the meteorological conditions.

It is well-known that the optimisation of routes is an effective method to operate the ships, increasing the safety and reducing the time of navigation and the consumption of fuel. There have been a lot of studies to develop methods of effective calculus to compound the optimal route. In the case of this project, the SIMROUTE fulfils two purposes: on the one hand, it samples the optimised route and provides the differences of time, consumptions and emissions, in comparison with the minimal distance route, and, on the other hand, offers the interpolated wave fields for the dates chosen that can be used to compare the real conditions of the state of the sea with the statistical data gathered among the period 1958–2017. This comparison altogether with the knowledge on the climatology of the Mediterranean helps to represent the information visually and allows to detect meteorological anomalies in the area through all the seasons.

3.5 Result Dissemination

The Internet is ever-present in our contemporary society and recently has gained importance in the education field. Distance learning programmes and online courses are proliferating, thanks to new education tools and online platforms. In fact, nowadays, higher education institutions need to address the problem of preparing students for the current professional requirements, which are continually changing and which require a continuous update of their knowledge and skills [17].

Even when it has been argued [17] that online classes are neither practical nor effective and that they cannot, in any instance, substitute the traditional classroom environment, Internet learning platforms make education accessible to all. Virtual communities are created to provide information and offer new tools to the students, and they have been expanding steadily in the last decade; sharing knowledge by means of web-based interactions is a usual practice nowadays.

As a consequence of the digital revolution, and specially now, amid the COVID-19 pandemic, integrative and flexible forms of education are in great demand and represent an opportunity but also a challenge for the training professionals. The maritime sector is not an exception. In the early years, maritime education had been mainly focused on the vocational training of deck officers on board. However, as the maritime business evolved, a change in the maritime education curriculum started to be necessary.

To align maritime programmes with the needs of maritime stakeholders, maritime studies have become an interdisciplinary academic field. Today, MET institutions must comply with the Standard Training and Certificate Watch keeping (STCW) Convention and Code. However, it is also important to ensure that core subjects do not fade out under the integration of these new courses.

Over the last few years, the technological scenario has changed and communication between teachers and students and between students has become more fluid, through the platforms of each institution. These structured spaces make possible to propose learning activities, monitor each student and assess their performance while making the information about the centre available to the entire educational community.

An example of this is the Technical University of Catalonia, and by extension, the Barcelona School of Nautical Studies, which have enabled different platforms and dissemination tools. Among them there is the *UPCommons Portal* (Fig. 3.7), an open-access portal that promotes academic and scientific communication at the university. There you can find published articles, final degree and master's theses, exams and various teaching materials, among others. However, in the setting of our institution, the students only participate in these platforms when it is the time to publish their theses. Apart from that, our students do not publish anything else on any other platform.

With the hard believing that student outputs are the result of a process that encourages learning, this project has its results presented as a participatory web format for dissemination and use as a didactic tool by the students. The platform is designed by teachers to use the SIMROUTE program, a tool that is already designed to work on the contents of the subject of meteorology, along with the data collected in this study, to propose case study scenarios to the students. The results of these studies can be published on the website *El Mediterrani* (Fig. 3.8) (www.elmediterrani.cat) and are available for consultation.

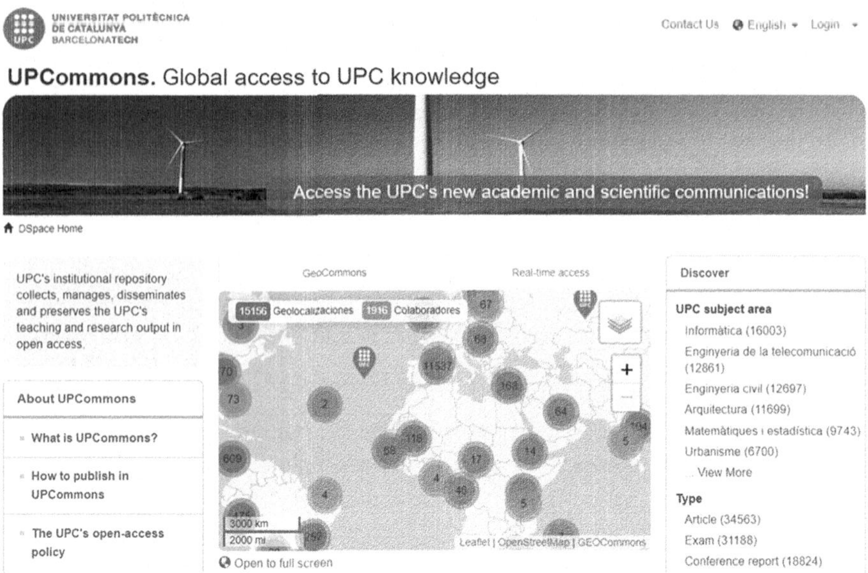

Fig. 3.7 *UPCommons* portal

Fig. 3.8 *El Mediterrani*, main page

3.6 Conclusions

The data studied in this project give a detailed picture of the climatology of the western Mediterranean. The latter has a very similar climate in all its extensions, although with small peculiarities. In general, mildly cold winters can be expected, but humid and without rain, gentle and rainy springs, hot and dry summers and soft autumns with a high risk of torrential rains. The climatological peculiarities of the western Mediterranean basin contribute directly to the generation of strong local winds, intense depressions, large floods and gusts.

The case study chosen is one of the short sea shipping routes in the Mediterranean, the Barcelona-Genoa line. As it has been said throughout the paper, short sea shipping is a crucial market in our sector. Its importance, as a more sustainable and environmentally friendly alternative, has been reiterated over the years, first with the publication of the White Paper in 2001 and then with the position taken by the EU in the face of the climate emergency. At this point, the use of this mode of transport is necessary, whether competitive or not, to address the problems posed by climate change. Full integration of this mode of transport with the logistics chain, together with the distribution of costs, is a key factor for a safe, sustainable and efficient maritime transport.

Despite the measures taken by the competent bodies, it has not been sufficient for a widespread implementation and, over the past decade, its growth has been derisory compared to that of the international maritime transport. However, it remains an indispensable tool in the fight against the climate emergency and is therefore defended in the context of the optimization of maritime routes and the advantages it entails.

The comparison between the statistical data and the interpolated wave fields of the Barcelona-Genoa route reinforces the image of the climatology of the western Mediterranean presented. Through the usage of the SIMROUTE software in conjunction with the database created, it is possible to compare the real conditions of the state of the sea with the period 1958–2017 which allows to detect meteorological anomalies in the area across all seasons. The presentation of the results in web format offers a useful platform for the teaching of the meteorology subject in our institution.

Academic publications in scientific journals are the proof of the generation of knowledge, a process that constitutes one of the main roles that the faculty must fulfil. The platform, presented in part with a similar format to that of a scientific journal, offers all those students in our faculty who are interested in publishing but have no experience a space where they can learn. At the same time, teachers can propose activities related to academic production, which could then be published. So, this platform will be used to:

- Promote academic production at the Barcelona School of Nautical Studies
- Provide a safe and relaxed learning environment
- Create another source of knowledge available for the students while allowing them to share their research

- Provide guidelines for students to start publishing in scientific journals
- Offer the professors a platform to encourage students

Through computer-supported collaborative learning, students will acquire life-long skills and they will participate in the activities of their institution.

References

1. M. Grosso, A.R. Lynce, A. Silla, G.K. Vaggelas, Short Sea Shipping, intermodality and parameters influencing pricing policies: the Mediterranean case. NETNOMICS Econ. Res. Electron. Netw. **11**(1), 47–67 (2010)
2. H.N. Psaraftis, *Challenges in European Short Sea Shipping*, in IMAM 2009 Conference, Istanbul, pp. 13–15, 2009
3. M. Douet, J.F. Cappuccilli, A review of Short Sea Shipping policy in the European Union. J. Transport Geogr. **19**(4), 968–976 (2011)
4. M. Castells, F.X.M. Oses, Studies of suitability on short sea shipping routes in SW Europe. J. Maritime Res. **3**(2), 43–52 (2006)
5. E. Musso, U. Marchese, Economics of short sea shipping, in *The Handbook of Maritime Economics and Business*, pp. 280–304, 2002
6. H.M. Hjelle, E. Fridell, When is short sea shipping environmentally competitive? Environ. Health-Emerg. Issues Pract., 3–20 (2012)
7. A. Christodoulou, J. Woxenius, *Sustainable Short Sea Shipping* (2019)
8. J. Martín Vide, J. Olcina Cantos, *Tiempos y climas mundiales: climatología a través de mapas del tiempo e imágenes de satélite*, pàg. 165–167 (1996)
9. A. Jansà Clar, *Meteorología y climatología mediterráneas* (1980)
10. N. Benítez, S. Campanario, V. Cuaresma, M. Digeuez, A. Morilla, *Clima Mediterráneo* (2012)
11. F. Pastor, J.A. Valiente, J.L. Palau, Sea surface temperature in the Mediterranean: trends and spatial patterns (1982–2016), in *Meteorology and Climatology of the Mediterranean and Black Seas*, (Birkhäuser, Cham, 2019), pp. 297–309
12. A. Iona, A. Theodorou, S. Sofianos, S. Watelet, C. Troupin, J.M. Beckers, Mediterranean Sea climatic indices: monitoring long-term variability and climate changes. Earth Syst. Sci. Data **10**(4), 1829–1842 (2018)
13. A.J. i Clar, Peculiaritats meteorològiques de la Mediterrània Occidental. Butlletí de les Societats Catalanes de Física, Química, Matemàtiques i Tecnologia, 91–97 (1995)
14. J.Q. Sala, E.M. Chiva, J.E. Barberá, La evolución de las precipitaciones en la cuenca occidental del mediterráneo: ¿tendencias o ciclos? Investigaciones Geográficas (Esp) **24**, 17–35 (2000)
15. J.I. Frigola Ferrer, *Variabilitat climàtica ràpida a la conca occidental del Mediterrani: registre sedimentològic* (2012)
16. F. Medda, L. Trujillo, Short-sea shipping: an analysis of its determinants. Maritime Policy Manag. **37**(3), 285–303 (2010)
17. G. Marzano, L. Ochoa-Siguencia, *Challenges of Web-Based Participatory Learning*, in Society. Integration. Education. Proceedings of the International Scientific Conference, vol. 2, pp. 458–467 (2017)

Chapter 4
Modeling of the Navigation Situation Monitoring Ship Control System

Igor Afonin, Vladimir Iskiv, Yuri Mickhayluck, Andrey Schekaturin, Ivan Skorik, and Elena Redkina

4.1 Introduction

At present, much attention is paid to the development of remote monitoring systems for the navigational situation and the state of ship control systems, since with the current level of development of telecommunication systems and the functionality of ship complexes and assemblies, the ability to remotely monitor the navigation situation and the condition of ship control systems can significantly increase the efficiency of solutions of vessels data transmission in practical tasks.

From the latest developments in this area, several of the most significant technical solutions can be distinguished.

At the present, there is well known monitring system for moving objects "Impulse-GLONASS/GPS: water transport" allows monitoring of small vessels only [1]. The monitoring system for moving objects is built on a modular basis.

Monitoring systems are known that make it possible to monitor the condition of ship units and to remotely control their operation.

In particular, Altair LLC offers a software and hardware complex for a ship monitoring system designed for automated monitoring of a ship's power plant and monitoring the location of river and mixed (river-sea) vessels [2].

Transas vessel traffic control systems are modern automated systems used to improve the safety of navigation and the safety of life at sea and protect the environment from possible negative consequences of shipping, as well as improve the efficiency of navigation and cargo transportation [3]. Ship traffic control systems provide users with a variety of navigation information, allow the identification and

I. Afonin · V. Iskiv · Y. Mickhayluck · A. Schekaturin · I. Skorik (✉) · E. Redkina
Sevastopol State University, Sevastopol, Russia
e-mail: ILAfonin@sevsu.ru; VMIskiv@sevsu.ru; YPMikhaylyuk@sevsu.ru; AASchekaturin@sevsu.ru; IVSkorik@sevsu.ru; EARedkina@sevsu.ru

S. Bauk, S. D. Ilčev (eds.), *The 1st International Conference on Maritime Education and Development*, https://doi.org/10.1007/978-3-030-64088-0_4

tracking of ships and other navigation objects in coastal waters, and also plan shipping in coastal waters.

In the field of electronic navigation, global improvements are also underway. For example, the International Maritime Organization is developing and implementing the concept of e-Navigation. The concept involves improving the safety of shipping all over the world through the active use of modern technologies, increasing the information support of ships and coastal services, and improving the exchange of navigation data between them [4].

In 2013, the Russian company ZAO KB NAVIS developed and certified in the marine and river shipping registers a shipborne GNSS GLONASS/GPS combined receiver-indicator type SN-5703 with a monitoring function [5]. A distinctive feature of the transceiver is the ability to build on its basis monitoring systems for sea and river vessels of various sizes and internal organization, suitable for a small number of special purpose ships, and integration into existing global coverage monitoring systems, both departmental and public companies.

The current system for monitoring the navigational situation and the state of the ship's control systems have, as a rule, a low-speed communication channel between the ship and coastal subsystems or are characterized by significant complexity and cost, as well as narrow specialization; therefore, the urgent task is to develop a system for remote monitoring of the navigational situation and condition of ship control systems having a high-speed data channel that allows streaming video transmission.

4.2 Onboard Subsystem of the Remote Monitoring System

The onboard subsystem should have the following elements:

- Shaper of group information signal, which converts heterogeneous signals from all systems of the vessel into a single group digital signal
- A modem that generates a signal with a given type and modulation parameters for transmission through the communication channel and demodulates the signal received from the communication channel
- A control computer that transmits the generated group information signal to the modem and receives a signal from the modem for transmission to the driver of the control signals

The block diagram of the onboard subsystem is shown in Fig. 4.1.

The group information signal generator is a set of adapters that converts digital signals into discrete or analog signals of the required amplitude and power for discrete or proportional (analog) control of actuators (relays, motors, etc.). The block diagram of this unit depends on the composition of the equipment and will be different for each vessel.

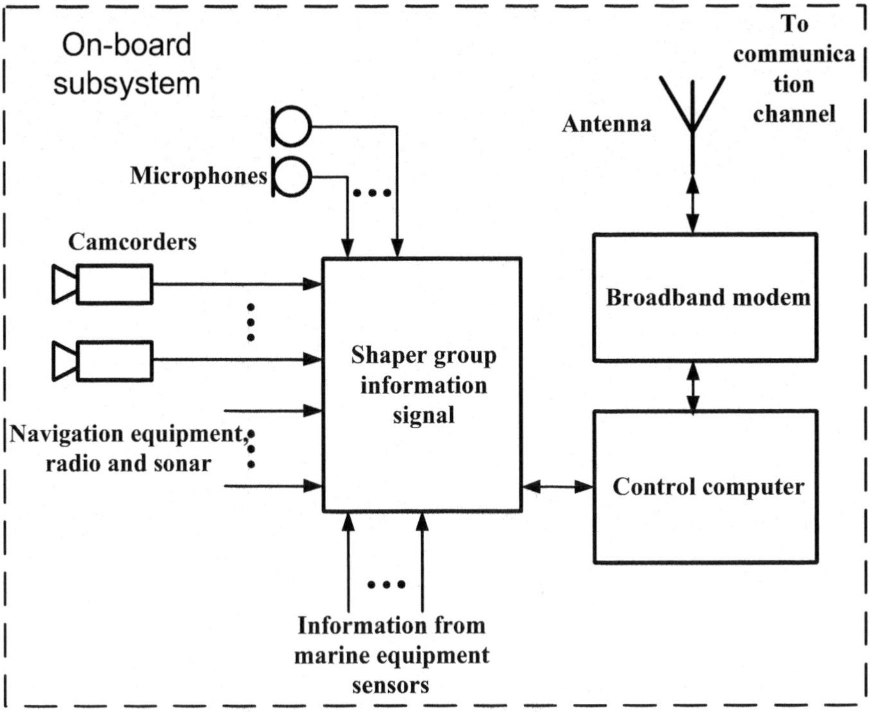

Fig. 4.1 Block diagram of the onboard subsystem

It is possible to use an integrated ship bridge. In this case, a remote control system can be used, most of the signals are already brought to the same interface and standard, and a specialized controller with an interface converter is used as a shaper of the group information signal.

A data recorder can be used as a shaper of a group information signal. In this case, most sensors are equipped with the necessary adapters, and their signals are grouped into a single digital stream, which is recorded by the recorder. Data recorders are equipped with the necessary network interfaces for connecting to control computers and modems.

The signals from sensors equipped with various interfaces in the interface converter are brought to a single digital form and then fed to the registration units and the group signal shaper.

The registration unit records the current information, saves it for a predetermined time interval, and allows you to search and access it upon request.

The group signal generator generates a signal of a uniform standard from heterogeneous digital signals, suitable for transmission, and sends this signal via the onboard transmitter to the coastal subsystem.

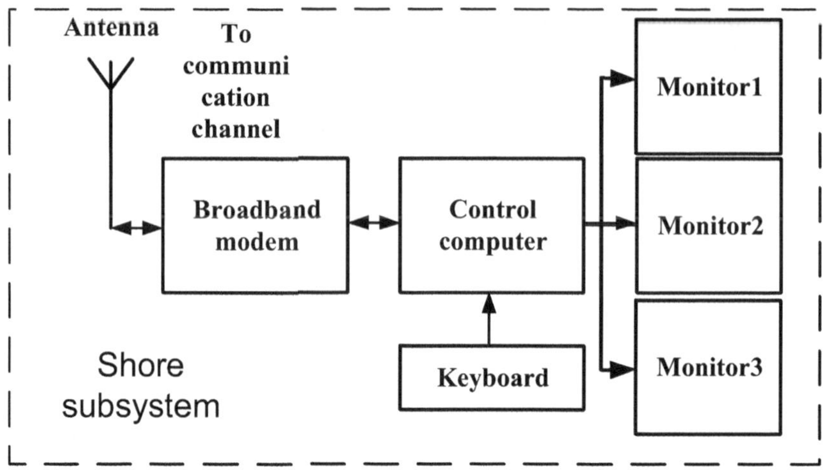

Fig. 4.2 Block diagram of the coastal subsystem

4.3 Coastal Subsystem of Remote Monitoring System

The block diagram of the coastal subsystem is shown in Fig. 4.2.

The subsystem uses several monitors, each of which displays its own type of information. The coastal receiver receives the radio signals transmitted by the onboard subsystem, detects them, and transmits to the group signal decoder. The group signal decoder extracts individual components from the general group signal – video signals, audio signals, and information signals from ship's sensors – and transmits them to display and playback devices.

The video information display device displays on the screens information from video cameras and monitors of a ship's locator, navigation, and other computers.

The audio information-reproducing device reproduces sounds perceived by the microphones of the onboard subsystem.

The sensor status indicator displays the status of sensors of various ship systems and devices.

4.4 Modeling the Operation of a System for Remote Monitoring of the Navigation Situation and the Status of Ship Control Systems

The simulation of the operation of the remote monitoring system of the navigational situation and the state of ship control systems was performed using the Cisco Packet Tracer software product. The main nodes of the network of the coastal and onboard system of the system are formed (see Fig. 4.3), and the IP addressing of the info-

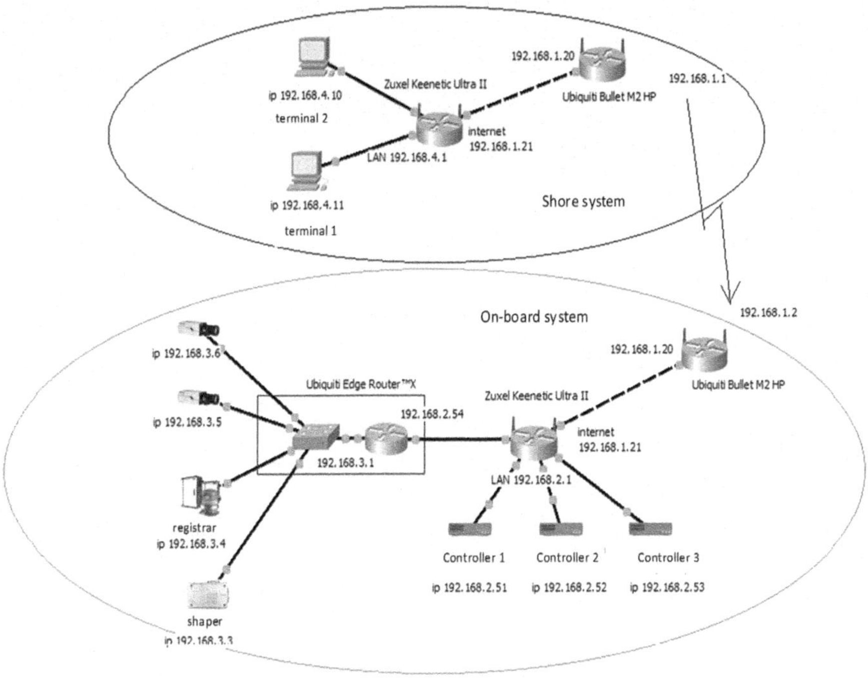

Fig. 4.3 Wireless network model of a system for remote monitoring of the navigation situation and the status of ship control systems

communication network of the remote control system is developed, which is presented in Table 4.1.

To simulate Ubiquiti Edge Router™ X in a network emulator, two network nodes are used: a switch and a router.

Settings of the Zuxel Keenetic Ultra II router of the coastal system: Internet Setup IP address 192.168.1.21, subnet mask 255.255.255.0, Gateway default 192.168.1.20; Network Setup IP address 192.168.4.1, subnet mask 255.255.255.0, security type WPA2 Personal, AES encryption.

Settings for the Zuxel Keenetic Ultra II router onboard system: Internet settings, IP address 192.168.1.21, subnet mask 255.255.255.0; LAN settings, IP address 192.168.2.1, subnet mask 255.255.255.0; wireless settings, channel 6, WPA2-PSK authentication, AES encryption.

Table 4.1 IP addressing of the network of the onboard and coastal complex of the remote control system

Shore system	
Ubiquiti Bullet M2 HP	Bridge 192.168.1.1 LAN gateway 192.168.1.20 (trunk)
Zuxel Keenetic Ultra II	Internet 192.168.1.21 (trunk) LAN gateway 192.168.4.1 (trunk)
Terminal 1	192.168.4.10
Terminal 2	192.168.4.11
Onboard system	
Ubiquiti Bullet M2 HP	Bridge 192.168.1.2 LAN gateway 192.168.1.20 (trunk)
Zuxel Keenetic Ultra II	Internet 192.168.1.21 (trunk) LAN gateway 192.168.2.1 (trunk)
Controller 1	192.168.2.51 (vlan 10)
Controller 2	192.168.2.52(vlan 10)
Controller 3	192.168.2.53(vlan 10)
Ubiquiti Edge Router™ X	LAN gateway 192.168.2.54 (trunk) LAN gateway 192.168.3.1 (trunk)
IP Webcamera 1	192.168.3.6 (vlan 11)
IP Webcamera 2	192.168.3.5 (vlan 11)
Registrar	192.168.3.4 (vlan 12)
Shaper	192.168.3.3 (vlan 12)

4.5 Stand Development and Experimental Research

The stand was developed using a Wi-Fi data channel. Onshore and onboard subsystems use MaxLink 01-VS-M15 Wi-Fi antennas, Ubiquiti Bullet M2 HP Wi-Fi adapters, and routers (Zuxel Keenetic Ultra II). Ubiquiti POE-24-24W-G are used as POE injectors for Bullet M2 HP.

The blocks for decoding, generating control signals, displaying video information, playing audio information, and indicating the status of sensors are implemented on the basis of computers and their peripheral devices.

In the onboard subsystem, the EMV-800HD video recorder was used. The signal converting unit also includes VGA-AV (VGA to VIDEO) video converters, which allow you to convert the usual standard VGA (D-sub) signal, which is used in almost all monitors of marine radar, navigation, sonar, and other systems, into a video signal Video standard PAL or NTSC to register this signal with the DVR.

The onboard subsystem also includes three STM32F746 Discovery controllers. These controllers are used to organize the management of onboard systems and systems, for which interface blocks are also provided that support CAN Bus and Ethernet.

The high-performance Ubiquiti Edge Router™ X router allows you to connect a number of peripheral devices and a signal conditioning unit to the onboard subsystem.

The image from the video cameras was transmitted from the onboard subsystem of the stand to the video cameras of the coastal subsystem and made it possible to track the movement of ships in the Streletskaya bay of Sevastopol. According to the measurement results at distances from 850 m to 3 km, the total data transfer rate was from 4.7 Mbit/s to 0.9 Mbit/s.

4.6 Conclusion

The developed system allows monitoring the navigational situation, including by transmitting video information from the ship, and the status of the small ship control systems at a distance of several kilometers from the coast subsystem when using a Wi-Fi data channel and thousands of kilometers when using satellite communication channels. The developed monitoring system can be part of the crewless ship management system. The system is simple and very flexible due to the inclusion of system elements as nodes of a local network and the use of programmable controllers, which allows you to change the composition and purpose of the system parts. The monitoring system provides a data transfer rate sufficient to transmit video data from cameras installed on the ship. The equipment of the monitoring system stand has a low cost.

References

1. The monitoring system for moving objects "mpulse-GLONASS/GPS: Water Transport" for the management of water transport, http://24glonass.ru/solutions/vodnyy-transport/vodnyy-transport
2. Control of sea and river transport, https://www.sirius.su/catalog/glonass-gps-trackery/otraslevye-resheniya/shipenergy
3. Ship traffic control systems, http://www.poseidon.su/products/suds.html
4. Ship monitoring systems in the concept of E-Navigation Speaker: Milyakov D.F. VII International Forum "Communications on the sea and river – 2011", https://officemagazine.ru/auto/50372
5. D.F. Milyakov, Testing SN-5703 for monitoring ships, https://www.korabel.ru/news/comments/aprobaciya_sn-5703_dlya_monitoringa_sudov.html

Chapter 5
The Future of Shipping: A Shore-Based Experience?

Tomke Julia Janßen, Michael Baldauf, Gisela Müller-Plath, and Momoko Kitada

5.1 Introduction

According to UNCTAD [1], international shipping industry is responsible for 80% of world trade. Rose [2] states that shipping moves 90% of all the good transported worldwide. Without shipping, intercontinental trade; the transportation of bulk, raw materials; and the import/export of affordable food and manufactured goods simply would not be possible.

Therefore, it is important to identify and address the challenges international shipping industry has to face in the next couple of years. These challenges include, among others, reduction of greenhouse gases, lack of suitably trained and experienced marine personnel, increased demands for safe operation and minimizing the number and consequences of accidents. In general, these challenges can be addressed by developing and integrating digitalization, automation and introducing automation technologies on board ships and ashore (e.g., Danish Maritime Authority [3]). These smart technologies represent great potentials, for example, to increase fuel savings by new means of propulsion. It also may enhance the safety for the crew by

T. J. Janßen · G. Müller-Plath
Technische Universität Berlin, Berlin, Germany
e-mail: gisela.mueller-plath@tu-berlin.de

M. Baldauf (✉)
Department of Maritime Studies Warnemuende/Institute ISSIMS, Hochschule Wismar, University of Applied Sciences - Technology, Business and Design, Wismar, Germany
e-mail: michael.baldauf@hs-wismar.de

M. Kitada
World Maritime University, Malmö, Sweden
e-mail: mk@wmu.se

S. Bauk, S. D. Ilčev (eds.), *The 1st International Conference on Maritime Education and Development*, https://doi.org/10.1007/978-3-030-64088-0_5

simply reducing the number of crew or even providing the opportunity to operate a ship unmanned, just monitored by an operator from ashore.

Despite all those expectations and outlook, it is necessary to have a look at the present maritime law and determine how all these new technologies, especially for autonomous or remotely controlled ships, can meet all the requirements. Another highly important issue is the training standards for marine personnel on board as well as ashore. The integration of remotely controlled or even autonomously navigating ships requires, among others, a shift of responsibilities and, obviously, a substantial change of skills of operators and other stakeholders involved, e.g., in Maritime Authorities and VTS centers [4].

This paper focuses on the implementation of smart operating systems and discusses their potential impact on legal, operational, and educational issues in the maritime sector on a very basic level. A draft research strategy will be suggested in order to identify maritime legal issues, issues regarding the standards of education and training for seafarers, and also to discuss possible impacts on future modes of operation of ships. Th research will take into account pilotage, traffic monitoring stations, and ship-shore/ship-ship communication respectively.

5.2 Maritime Transportation: The System's Perspective

Components and Actors

Each transportation system, basically, can be structured into five main components and can be applied to every transport mode (i.a., [5]) to describe processes, functions, relations, and dependencies. For providing common understanding, a descriptive model is developed below.

Transport Means Every transport mode has a specific means for moving goods from origin to destination, e.g., cars, lorries, trains, and airplanes. In the maritime system the transport means are ships of varying dimensions and types for numerous purposes. The basic division we need to take into account is between commercial and non-commercial ships.

Drivers Drivers, in principle, are the persons who are in charge for steering and controlling the transport means. In shipping, this is usually the captain with his team of navigating and technical officers and especially the respective Officer of the Watch (OOW) on the ship's navigational bridge. In certain cases, especially when navigating in challenging sea areas, a pilot is providing navigational assistance to support her or him.

Transport Ways Transport ways or paths are the highways, roads and streets, railways, air traffic corridors, etc. In the maritime transportation system, these are the open sea and coastal areas, as well as the river and other inland navigation waterways, canals to harbors and port terminals.

Traffic Organization and Administration Components for traffic organization and administration cover the activities of the dedicated and recognized international, regional, and national institutions responsible for developing, negotiating, and establishing the set of rules and regulations including also legal and technical standards. The system's component "Traffic organization and administration" is to represent the system's components providing overall legal frameworks for a globally harmonized implementation and enforcement of rules and standards. In the maritime transportation system, we consider the International Maritime Organization (IMO) as the main global body, accompanied by, among others, the International Association of Marine Aids to Navigation and Lighthouse Authorities (IALA), the International Chamber of Shipping (ICS), the International Association of Classification Societies (IACS), and other international but also intergovernmental, nongovernmental, and further maritime organizations. On a regional level, there are continental organizations like the European Maritime Safety Agency (EMSA) or Australian Maritime Safety Authority (AMSA) relevant actors under this subsystem. On a national level, the waterway and shipping administrations (like e.g. Danish or Swedish Maritime Administration) together with their related agencies and other bodies take action to ensure implementation and enforcement in their respective areas of jurisdiction.

Traffic Management A system component that is becoming more and more important in every transport mode is traffic management. Nowadays, every transportation system operates increasingly sophisticated traffic management systems reaching from provision of basic information up to advanced support to ensure safe and efficient flow of vessel traffic. This component covers sub-systems, e.g., Vessel Traffic Services (VTS) with its different services but also Pilotage, Search and Rescue, Maritime Assistance, Ice Patrol, and other commercial or noncommercial maritime services provided from ashore.

All components of the system are linked and interact with and affect each other. Overall aim of all the components' activities is to serve the safety and efficiency of maritime transportation.

Reviewing Ongoing Developments

Presently the maritime transportation system is undergoing substantial changes including the introduction of disruptive technologies requiring adequate action and modifications in the system and its components. Major examples for such substantial changes are the introduction of new technologies basing on substantial and sustainable digitalization and the development and implementation of the e-navigation concept.

Digitalization The fourth industrial revolution, also known as Industry 4.0, is, among others, driven by Internet, advanced technologies, and a network of smart systems and products. It includes technology of robotics, autonomous decision-making tools, and applications using augmented and virtual reality. The Industry 4.0 has already affected several domains connected to maritime logistics [6]. The concept is driven by digitalization and optimization of industrial processes that use the new emerging technology [7, 8].

Actors from all components of the maritime transportation system are involved in the digitalization process, ranging from ship and port management, waterway and port authorities, terminal operators, shipping companies up to truck and railroad companies. Sharing data between all actors is improving the traffic flow on the waterways, resulting in less waiting times for the ships and more efficient cargo operations in ports and hinterland transportation [8]. Moreover, Jahn and Saxe [9] are suggesting that only a network of interconnected ports can improve efficiency, reduce cost, increase reliability, and decrease waiting times for ships.

New technologies will not only improve the supply chain and maritime logistic processes but will improve the monitoring of traffic flows and cooperation between ship and shore as well. With new technologies and further developments in the marine surveillance sector, many shipping companies are already operating and monitoring their fleet with Fleet Operation Centres (FOC). Shore-based operators provide additional information and warnings to the ship. The operator may even be allowed to give recommendations or instructions [10].

e-Navigation e-Navigation is a concept developed by IMO and IALA. The concept is clearly focusing on the human operators on board the ships (e.g., OOW) and ashore (e.g., a VTS operator). It is aiming at "the harmonized collection, integration, exchange, presentation and analysis of marine information on board and ashore by electronic means to enhance berth to berth navigation and related services for safety and security at sea and protection of the marine environment." A Strategy Implementation Plan has been agreed by IMO at 94th session of MSC in 2014 and is set into force. There are many ongoing initiatives to implement the concept by testing new applications in e-Navigation test beds. The e-Navigation concept and its implementation are primarily basing on digitalization and consequent use of latest information and communication technologies [11, 12].

Digitalization and e-Navigation require not only adaptation of logistic processes but, moreover, further development of job profiles and existing training and education standards and schemes of maritime studies' syllabuses and course curricula, respectively.

MASS and Autonomous Shipping Maritime Autonomous Surface Ships (MASS) is the IMO's term addressing the recent developments heading for autonomously navigating and once maybe unmanned ships. Even though, autonomous or remotely controlled vehicles or ships are not a new idea. Autonomous or remotely controlled underwater vehicles are already common in marine geoscience for quite a time. They are in use for military, commercial, or scientific research. MASS is expected to be the next step in the transformation of the maritime transportation system. Digitalization and e-Navigation will contribute to increase the level of autonomy and will allow introducing unmanned and even autonomously navigating ships into the existing traffic system from the perspective of technical feasibility. Legal frameworks to provide the adequate legal background are under development. This new transport means will come along with further monitoring and remote operation facilities like shore control centers [13, 14].

5.3 Traffic Monitoring and Control: Today and Tomorrow

State of the Art

Today's traffic management is a result of the development of shore-based Vessel Traffic Services (VTS) [15]. Originally introduced, among others, to ensure safe approach to and from harbors under conditions of restricted visibility by providing radar-based navigational information from shore to ships, VTSs nowadays provide Information (IS), Navigational Assistance (NAS), and Traffic Organization Services (TOS). VTSs are part of SOLAS. According to IMO resolution A.857(20), VTS is defined "as a service implemented by a competent authority, designed to improve the safety and efficiency of vessel traffic and to protect the environment. The service should have the capability to interact with the traffic and to respond to traffic situations developing in a VTS area." This resolution, however, is under revision mainly to modernize and adapt resolution to common practice.

IMO's activities are strongly supported by IALA, who has undertaken huge efforts and has provided numerous guidelines and recommendation for efficient establishment and harmonized operation of VTS around the globe. VTS, clearly, is a risk management tool that functions like a control loop, as depicted below (Fig. 5.1).

Usually, a VTS works in national waters of a coastal state. It collects traffic and environmental data through its sensor infrastructure and interacts with vessel traffic by voice communication through VHF by sending out information at regular times or on demand, giving recommendation or advice to support onboard navigation processes and even may intervene by giving instructions, if needed, e.g., to contribute to avoidance of collisions and groundings. However, the final decision always remains with bridge team on board [16].

Besides VTS there are meanwhile further shore-based monitoring bodies established, namely, so-called Vessel Coordination Centers (VCC) as well as Fleet Operation Centers (FOC).

A sample of a VCC is the Hamburg Vessel Coordination Center (HVCC). It is operated by Hamburg port authority and coordinates ship movements inside the port terminals. Moreover, the HVCC also provides support to captains and shipping companies when they are leaving their last port of call and heading for Hamburg. The VCC provides passage plans taking into account, e.g., tidal conditions on river Elbe and expected traffic to organize smooth passage avoiding delays that may be caused, e.g., by meeting at fairway bottlenecks.

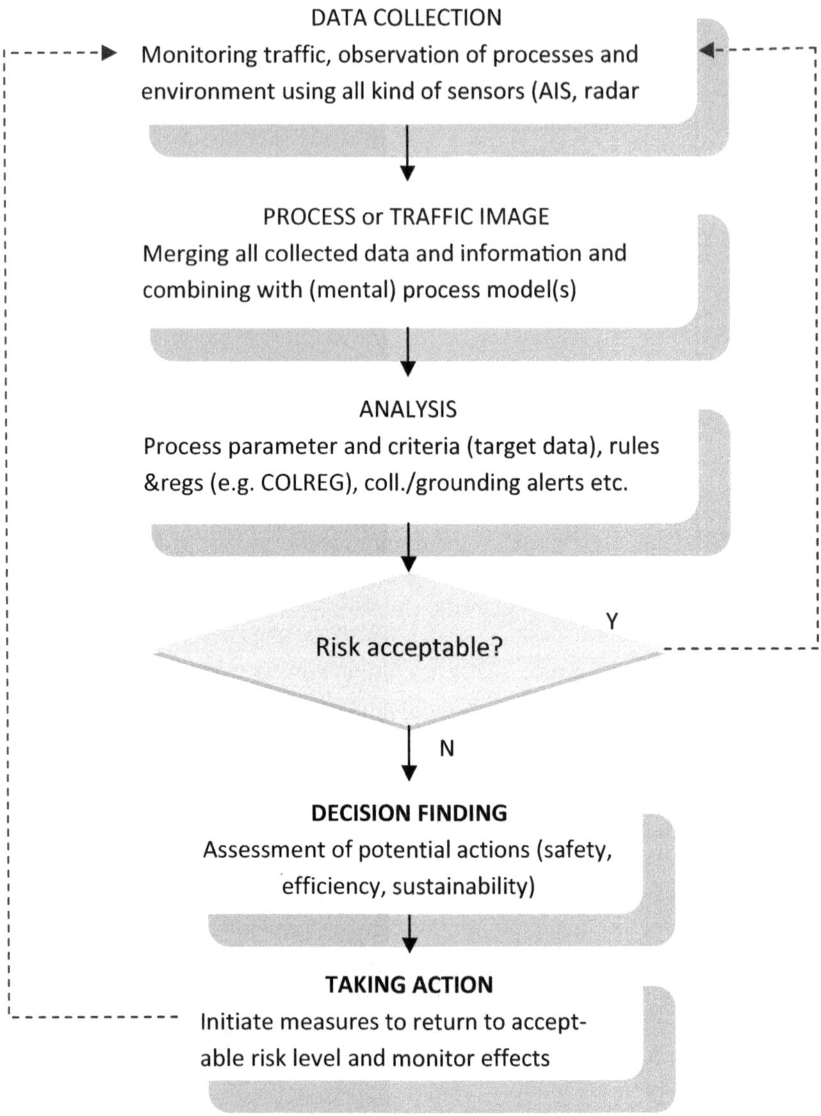

Fig. 5.1 Risk management by VTS presented as a control loop

An FOC is a company owned facility and monitors the safe progress of ships of its own fleet. FOCs use most modern data communication links and have installed same navigational tools and displays ashore as they are on board. FOCs act as an additional safety barrier and have continuous connection to the members of the bridge team. Shore-based operator in a FOC may even know the bridge team members on board.

Participant observations as part of field research are ongoing and include visits to VTS centers along the coasts of North and Baltic Sea as well as the fairways to ports. First preliminary observations and interviews indicate that, although the services are provided according to the same basis of operational procedures, there is a wide variety of different tasks and services are sometimes carried out differently. A main reason for this is the specific conditions of the monitored area in terms of the fairway conditions and the partly different specific rules. On the other hand, it is observed that services are provided with highest ambitions to take into account the specific characteristic of the concrete situation (type of ships, rules, environmental conditions, and much more). Similar to the rules of the "International Regulations for Preventing Collisions at Sea," it is valid that exact sequences of actions and measure a VTS operator shall give cannot be provided for each and every specific situation and challenge skill, knowledge, and experience of the operators. Similar results have been generated recently in empirical studies in other regions [17, 18] as well.

A Future's Scenario

Assuming that shipping will continuously develop like before the covid-19 times, where ship's dimensions were growing and the navigable space becomes more and more limited by other new users, e.g., renewable energy installations, fish farming, and others, the need for long-term coordination of traffic will further increase. The need for the protection of the marine environment will need to be supported by reducing emission of greenhouse gases to a minimum or even to zero.

On the other hand, composition of vessel traffic in future will include ships equipped with most sophisticated propulsion systems for high maneuverability; ships might sail with minimum crews but can be unmanned. Conventionally navigating ships will share navigable space with ships monitored by the company's FOC and may be even controlled from the remote SCC [13, 14]. Several research studies are ongoing to provide input for such future developments [19–21].

5.4 The Shore-Based Operator's Perspective

Besides the technical issues in shipbuilding, vessel traffic, and ship navigation, it is of utmost importance to consider changing the port's infrastructure as well. Moreover, yet importantly, there is obvious need to consider operational issues of how to organize the regimes and interventions of a VTS and to coordinate their measures with those of VCCs and FOCs respectively. These essential operational issues will also need to be taken into account for future training and education schemes.

The autonomous or remotely controlled ships have to "communicate" with the fairway's infrastructure such as dolphins and light buoys and port authorities to get the information they need. Communication is also necessary with VTS and pilots on conventional ships, e.g., to know if they are on the right track and if they have to slow down due to traffic congestion in the harbor and to get the information about their berth or jetty [9].

Transition from Navigators, Pilots to Shore-Based Operators

The progress of automation does not only change technical standards in the industry but has an influence on the human-machine interaction in the same way [22]. The automation progress in the shipping industry had many advantages, such as reducing the number of crew on board from 30–40 in the 1970s and 1980s to presently 10–20 and increasing the overall safety of ship operations [22]. Nevertheless, seafarers are trained to interact with the technology on board and make decisions based on their experience and evaluation of all elements [23]. In the process of navigation, the captain or OOW has an active role in maneuvering and steering the ship. The person in charge of navigation is aware of his surroundings and is still obliged to check the situation by finally looking out of the window and applying the rules of good seamanship (Fig. 5.2).

However, when reducing the number of crew on an autonomous ship to a minimum, for instance, four to five or even none, the ship is playing an active role in the navigation process. The autonomous ship navigation system, among others, needs to be programmed to detect and classify certain objects in its surrounding area according to a programmed algorithm [24]. The algorithm needs to be able to classify, prioritize, and assess certain information given by electronic sensors and signals from the onboard system or other ships in the area to navigate the ship like a human operator [25].

The same concept of information processing applies for the remote human operator. He needs to be able to get all the information, evaluate their meaning, and predict the outcome of the situation [23]. Nevertheless, most of the time, the operator (on board or ashore) will switch to a passive role, where he is just monitoring the Autonomous Navigation System.

Fig. 5.2 Human operators in the maritime transportation system (left, bridge team on board a container vessel; right, operators in VTS center Singapore). (Photos: author)

Need for In-Depth Studies to Adapt Maritime Training and Education

The potential development briefly described and discussed in the above paragraph highlights the urgent need for comprehensive research into the modernization and adaptation of the existing MET schemes. Necessary skills and knowledge of operators in the future maritime transportation system need to be reviewed and carefully determined. Until today, many organizations value the skills, knowledge, and experience of mariners when recruiting for positions ashore. Applicants for pilot's or VTS operator's positions shall have preferably a master's license and several years of experience as a captain or at least as a chief mate. This is similarly valid for training instructors and lecturers of cadets. However, institutions increasingly lacking appropriate applicants matching those requirements. Consequently, a fundamental question to answer is: will we have in the future experienced mariners, who will undergo the traditional training programmes to become a pilot or a VTS operator or do we need to rethink and to adapt and to change MET to completely new profiles of applicants, with less or even no seafaring experience at all? And, consequently, how to compensate missing seafaring experience?

If operators in the future just rely on displays of technical data and not question the outcome, it may result in major accidents, such as the collision between two ships in the fjord of Kiel in 2014. The navigating officers on both ships trusted the ECDIS data and did not realize that there was a GPS failure [26]. These failures are described as an "out-of-the-loop unfamiliarity" [22]. Due to the advanced technology on board ships, the navigating officers or captains are already taking over the passive role and just monitoring data given by the RADAR, ECDIS, AIS, or other bridge equipment. Over the time this can lead to a possible skill loss, and decisions in case of emergencies (collision, grounding, etc.) will be harder to make by the ship handler [22]. In respect to maritime safety, seafaring experience and good seamanship are considered to be most essential today. If this is true and shall remain in future, then we need to exactly know why seafaring time is needed and to what extend. If we cannot ensure sufficient practice times, then we need to ask how to compensate, e.g., by simulation-based training? Moreover, if we found, it is impossible to compensate, then we need to study what other skills and knowledge of the younger generation have potential to replace them.

To answer such questions, research studies are ongoing, applying qualitative and quantitative methods, in order to determine the need for a change in the skill set of marine personnel, handling and operating ships, as well as evaluating the possibility of unmanned autonomous or remotely controlled shipping within the coastal area of a coastal state. The focus will be laid on evaluating training standards for seafarers and communication processes (e.g., ship-ship and ship-shore) and proposing necessary changes. The research aim and objective are defined by applying the SMART (specific, measureable, attainable, relevant, and time-bound) rule [27].

5.5 Conclusion and Outlook

A holistic system's approach to study the existing maritime transportation as is has been carried out. It highlights the connections and interrelations of the single components of the maritime transportation system. The research developed a future scenario, specifically focusing on shore-based infrastructures for monitoring and controlling vessel traffic taking into account the introduction of potentially unmanned and autonomously navigating ships.

Basing on participant observation, expected changes of the existing regime for shore-based services are suggested and discussed. Main focus of the studies is laid on the shore-based perspective and potential changes in relation to the human operators in the loop for monitoring and controlling the traffic flow.

Basing on expected changes, driven by technological developments, the need for corresponding operational changes has been elaborated and the need for adaptation of the existing schemes for maritime education and training is indicated and briefly discussed.

Acknowledgments The authors wish to thank those VTS operators, pilots, and navigators who shared their opinions in interviews. They also thank for the opportunities to join watches for participant observations in VTS centers. Some of the research presented in this paper is carried out in the frame of the joint Polish-German project Collision Avoidance Domain-Method Used by Ships and aShore (CADMUSS) carried out under the MarTERA program, an ERA-NET Cofund scheme of Horizon 2020 of the European Commission.The project is funded by the German Federal Ministry for Economic Affairs and Energy and supervised by Project Management Jülich (PtJ).

References

1. UNCTAD, *Review of Maritime Transport* (New York/Geneva, United Nation Conference on Trade and Development., 2016)
2. G. Rose, *Ninety Percent of Everything: İnside Shipping, …* (Metropolitan Books/Henry Holt and Co, New York, 2013)
3. Danish Maritime Authority, *Analysis of Regulatory Barriers to the Use of Autonomous Ships – Final Report* (Korsør, Danish Maritime Authority, 2017)
4. European Commission, *Horizon 2020 – Work Programme 2018-2020: Smart, Green and İntegrated Transport* (Brussels, European Commission. Brüssel, 2019)
5. M. Kitada, M. Baldauf, A. Mannov, et al., Command of vessels in the era of digitalization, in *Advances in Human Factors, Business Management and Society*, AHFE 2018, Advances in Intelligent Systems and Computing, ed. by J. Kantola, S. Nazir, T. Barath, vol. 783, (Springer, Cham, 2019). https://doi.org/10.1007/978-3-319-94709-9_32
6. N. Taylor, C. Human, K. Kruger, A. Bekker, A. Basson, Comparison of digital twin development in manufacturing and maritime domains, in *Service Oriented, Holonic and Multi-agent Manufacturing Systems for Industry of the Future*, SOHOMA 2019. Studies in Computational Intelligence, ed. by T. Borangiu, D. Trentesaux, P. Leitão, A. Giret Boggino, V. Botti, vol. 853, (Springer, Cham, 2020). https://doi.org/10.1007/978-3-030-27477-1_12
7. D. Sarabia-Jácome, C.E. Palau, M. Esteve, F. Boronat, Seaport data space for ımproving logistic maritime operations. IEEE Access **8**, 4372–4382 (2020)
8. M. Kitada, P. Baum-Talmor, Maritime Digitization and Its Impact on Seafarers' Employment from a Career Perspective. Paper presented at International Association of Maritime Universities (IAMU) Conference, October 2019 Tokyo (2019)

9. S. Saxe, C. Jahn, *Digitalization of Seaports – First Ideas* (Fraunhofer Publica, Hamburg, 2017)
10. M. Baldauf, M. Kitada, R. Mehdi, D. Dalaklis, E-Navigation, digitalization and unmanned ships: challenges for future maritime education and training. In: 12th Annual International Technology, Education and Development Conference (INTED), Barcelona, Spain. https://doi.org/10.21125/inted.2018.2374 (2018)
11. J. Jeevan, K. Ramamoorthy, N.H.M. Salleh, et al., Implication of e-navigation on maritime transportation efficiency. WMU J. Marit. Aff. **19**, 73–94 (2020). https://doi.org/10.1007/s13437-020-00194-z
12. M. Baldauf, K. Benedict, C. Krüger, Potentials of e-navigation – enhanced support for collision avoidance. TransNav Int. J. Mar. Navig. Safety Sea Transport. **8**(4), 613–617 (2014). https://doi.org/10.12716/1001.08.04.18
13. S. McKinnon, Y. Man, M. Baldauf, Shore Control Centre, *MUNIN Deliverable D8.8. Final Report: Shore Control Centre, Technical Report, 24 pages* (Chalmers University of Technology, Gothenburg, 2015)
14. M. Baldauf, S. Fischer, M. Kitada, R.A. Mehdi, M.A. Al-Quhali, M. Fiorini, Merging conventionally navigating ships and MASS – merging VTS, FOC and SCC? TransNav Int. J. Mar. Navig. Safety Sea Transport. **13**(3), 495–501 (2019). https://doi.org/10.12716/1001.13.03.02
15. G. Praetorius, *Vessel Traffic Service (VTS): A Maritime Information Service or Traffic Control System? Understanding Everyday Performance and Resilience in a Socio-Technical System under Change* (Chalmers University of Technology, Gothenburg, 2014)
16. A. Khalil Bargash, The effect of vessel traffıc servıce on mınımızıng collısıons a case study: collısıon between rıckmers dubaı and walcon wızard. Int. J. Mech. Eng. (IJME) **4**(3), 57–64, Tamil Nadu (Ind), Roseville (USA), Adelaide (Aus) (2015)
17. J.T. Mansson, M. Lutzhoft, B. Brooks, Joint activity in the maritime traffic system: perceptions of ship masters, maritime pilots, tug masters, and vessel traffic service operators. J. Navig. **70**(3), 547–560 (2017). https://doi.org/10.1017/S0373463316000758
18. J.T. Mansson, M. Lutzhoft, B. Brooks, Balancing on the boundary: vessel traffic services in the maritime traffic system. Paper presented at ERGOSHIP 2016: shaping shipping for people April 2016, Melbourne 2016
19. K. Wróbel, M. Gil, J. Montewka, Identifying research directions of a remotely-controlled merchant ship by revisiting her system-theoretic safety control structure. Safety Sci. **129**, 104797 (2020)
20. K. Wróbel, J. Montewka, P. Kujala, Towards the assessment of potential ımpact of unmanned vessels on maritime transportation safety. Reliab. Engi. Syst. Safety **165**(September), 155–169 (2017). https://doi.org/10.1016/j.ress.2017.03.029
21. T. Relling, G. Praetorius, O.S. Hareide, A socio-technical perspective on the future Vessel Traffic Services. Necesse **4**(1), 112–129 (2019). https://doi.org/10.21339/2464-353x.4.1.1
22. B.D. Seppelt, J.D. Lee, Human factors and ergonomics in automation design, in *Handbook of Human Factors and Ergonomics*, ed. by G. Salvendy, 4th edn., (Wiley, Hoboken, 2012), pp. 1615–1642
23. T. Porathe, J. Prison, Y. Man, *Situation Awareness In Remote Control Centres For Unmanned Ships. Human Factors in Ship Design & Operation* (The Royal Institution of Naval Architects, London, 2014)
24. H.-C. Burmeister, *D8.6 Final Report Autonomous Bridge* (Fraunhofer CML: Center für Maritime Logistik und Dienstleistungen, Hamburg, 2015)
25. B.J. Vartdal, R. Skjong, A.L. St. Clair, *Group Technology & Research Paper, Position Paper: Remote-controlled and autonomous ships* (DNV GL Maritime, Hamburg, 2018) Retrieved September 29, 2019
26. BSU, *Untersuchungsbericht 276/14: Schwerer Seeunfall – Kollision in der Kieler Förde*, vol 2014 (Bundesstelle für Seeunfalluntersuchung – Federal Bureau of Maritime Casualty Investigation, Hamburg, 2014)
27. M. Saunders, P. Lewis, A. Thornhill, *Research methods for business students*, 5th edn. (Pearson Professional Limited, Harlow, 2009)

Chapter 6
UAV for Ecological Monitoring of Water Surface

Ivan Skorik, Maksim Durmanov, Semen Yankovskiy, Yuriy Gimpilevich, Yuriy Mikhaylyuk, Denis Nacharov, and Igor Shirokov

6.1 Introduction

The proposed device is relevant at present, since over the last years of the twenty-first century, the problem of pollution of the oceans has become one of the most important facing the world community. This is undoubtedly associated with an increase in the world's population, the development of technology, an increase in oil consumption and production, as well as an increase in the number of sea communications.

The main pollutants of water are petroleum products, as well as discharges of various chemical industrial wastes, such as nitrates, phosphates, insecticides, and herbicides. An oil film on the surface of the ocean disrupts its biological processes and causes oxygen deficiency, changing the composition of water. Oil and fuel oil deposited at the bottom give secondary pollution; in addition, oil causes irreversible damage to human health when it enters drinking water bodies and objects [1].

Over a hundred cases of oil spills over the past decade are known from open sources, including the accident on April 20, 2010, at the Deepwater Horizon oil platform in the Gulf of Mexico, which is the fifth largest accident involving an oil spill (Fig. 6.1).

Oil spill is a global problem, the elimination of the consequences of which can last several years. Contaminants are most quickly detected by remote sensing methods, which make it possible to inspect vast water surfaces in a relatively short period of time [2]. The latter include photographs taken from satellites, as well as aerial photography.

I. Skorik (✉) · M. Durmanov · S. Yankovskiy · Y. Gimpilevich · Y. Mikhaylyuk
D. Nacharov · I. Shirokov
Sevastopol State University, Sevastopol, Russia
e-mail: IVSkorik@sevsu.ru

S. Bauk, S. D. Ilčev (eds.), *The 1st International Conference on Maritime Education and Development*, https://doi.org/10.1007/978-3-030-64088-0_6

Fig. 6.1 The ship moves through the crude oil spill zone. Gulf of Mexico. April 28, 2010 year

However, these methods have disadvantages: the relatively small accuracy in determining the place and volume of pollution, as well as the inability to quickly investigate the composition of water directly in the place of pollution.

To analyze water in a contaminated area, it is necessary to send a ship with special laboratory equipment to this place, which can inadvertently provoke the spread of oil products on the surface of the water.

In view of the foregoing, the proposed device is aimed at solving the following problems:

- Determine the location of contamination with high accuracy.
- Determine the area of pollution.
- Determine the thickness of the layer and the composition of the liquid at the site of contamination.
- Minimize contact with contaminated liquids in order to minimize expansion of contamination.

6.2 Operation Principles

Video Recording System

To solve the presented problems, an unmanned aerial vehicle (UAV) was proposed, which works in the immediate vicinity of the water surface, allowing you to quickly add information about the location and extent of pollution, as well as take samples and analyze the contaminated liquid directly in the pollution zone.

The proposed UAV is equipped with a video system. Using the UAV video system, many tasks can be solved, including monitoring the space surrounding the vessel, detecting objects on the sea surface, and obtaining additional information when navigating in conditions of insufficient visibility. Moreover, the solution of such problems can be carried out both as a result of a visual analysis of a video signal by a human operator and using automatic analysis algorithms implemented by computational means.

The principle of operation of the UAV is as follows. Based on the received coordinates from the satellite about the location of the pollution, a vessel is sent to the pollution area, on board of which there is a UAV. Upon arrival, the UAV takes off from the takeoff and landing site of the vessel and is sent to the place of pollution.

A video system equipped with a UAV performs photo-video shooting of the sea surface. For the video system to work effectively, the optimal UAV flight altitude must be selected. The choice of height above the water surface depends on the state of the atmosphere, lighting, and the viewing angle of the camera lens.

Moving in the direction of the place of pollution, the UAV shoots and analyzes the resulting images until a contrast appears in the image or a change in the uniformity of the frame. This is due to the fact that the contaminated surface of the water has a lower reflectivity of light and has darker colors on the screen. The position on the border of the UAV pollution will occupy when the number of pixels of greater intensity and less will be the same. At this moment, the UAV remembers the coordinate of this place using the GPS receiver and continues to move in a predetermined direction, for example, to the north.When changing the ratio of the number of pixels of higher and lower intensity in the captured image, the flight direction is corrected to the right or left, depending on which pixels are larger. With significant deviations from the current direction, GPS coordinates are taken and stored in memory. These actions will continue until the UAV is at the starting point.

Thus, the UAV will fly around the border of the pollution and record the coordinates of the pollution spot. Using the obtained coordinates, you can build a polygon and calculate the coordinates of its center. In the center of the pollution spot, it is best to take a sample to analyze the composition and determine the thickness of the pollution layer. Knowing the thickness of the layer in the center of the spot and knowing the composition (density) of the pollution substance, we can calculate the approximate volume of a foreign substance located on the water surface.

Fluid Analysis Probe

Samples for analysis of pollution are taken using a special probe, which is attached to the bottom of the UAV on the cable and immersed in the test fluid. The probe itself is, for example, a cylindrical tube with an open bottom, into which liquid is collected during immersion. However, other physical options for the implementation of the probe (sensor) are possible: first (previously described) is a metal cylinder with a metal rod in the center, and second is two parallel metal plates located at a fixed distance from each other. Both sensor options can be used to determine the electrical capacitance or to determine the electrical conductivity of the medium, which depends on the layer thickness and chemical composition of the medium. The final design of the sensor will be determined in a practical way in the laboratory. When the sensor rises, it is emptied, after which the sensor is ready for further measurements.

For the sensor of electrical conductivity of the medium, a relatively constant UAV height above the sea surface should be maintained. It should be noted that maintaining a constant UAV height above the water surface is possible under the condition of moderate waves of the water surface; otherwise, the measurements will be impossible. This is ensured, thanks to the high-precision altimeter and the associated system of automatic control of the engines responsible for the climb, which the UAV is equipped with.

Flight Altitude Control

When conducting photo-video shooting and analysis of the composition of the water surface, it is required to provide the most accurate UAV height above the water surface. Maintaining the height with the greatest accuracy is possible in two ways: first, when analyzing the composition of the water surface, it is advisable to use an ultrasonic height sensor, which will allow you to maintain the height of the UAV position above the water surface during measurements, and second, when conducting video photos, surveillance should maintain a relatively constant UAV height above the sea surface due to the barometric sensor and GPS or GLONASS assist systems.

It should be noted that maintaining a constant UAV height above the surface of the water is possible subject to moderate waves, so that the UAV flight controller has the ability to maintain and stabilize the specified UAV height above the water surface in the event of destabilizing factors (wind, waves, and others).

Additional Solutions

In order to successfully solve the tasks with the help of the UAV video system, it is necessary to provide an acceptable image quality under any lighting conditions and weather conditions. This is especially important due to the fact that many well-known object detection algorithms, as a rule, show their effectiveness only in conditions of good lighting and a clear atmosphere. The simplest, and at the same time effective from the point of view of required computing resources, method of image quality correction is the element-by-element brightness conversion described by the functional dependence:

$$s = T(I)$$

where s is the brightness of the resulting image, $T(I)$ is the conversion function, and I is the brightness of the original image. Different kinds of conversion function $T(I)$ allow to achieve increase or decrease of exposure, change of contrast of images, allocation of separate ranges of brightness, threshold processing, and other effects. However, the choice of quantitative parameters of the conversion function $T(I)$ applied to a particular type of distorted image is often a nontrivial task and requires additional research. In addition, factors such as the speed of the current and the degree of water disturbance influence the efficiency of the tasks performed. Due to strong currents or winds, the stain will move along the surface of the water, which requires corrections to the coordinates being taken. In order to take this influence into account, it is necessary to use wind and current velocity sensors.

6.3 Conclusion

The proposed device is aimed at solving the urgent task of detecting pollution of the sea surface. For this purpose, it is proposed to use a UAV equipped with a video system for accurately determining the location of pollution and its boundaries and center, as well as a probe for taking samples of the medium under study and analyzing its composition. This device will help to significantly reduce the time and reduce costs associated with the elimination of the effects of pollution of the sea surface.

References

1. Oil spill: consequences and remedies. https://businessman.ru/razliv-nefteproduktov-posledstviya-i-metodyi-ustraneniya.html
2. V.I. Kozincev, M.L. Belov, V.A. Gorodnichev, O.A. Smirnova, U.V. Fedotov, Detection of oil pollution on the sea surface by a two-frequency remote laser method. Vesnik MSTU the name. N E. Bauman. Ser. «Instrumentation». 2006. №4. pp. 3–12

Chapter 7
Key Features of the Autonomous Underwater Vehicles for Marine Surveillance Missions

Sanja Bauk, Nexhat Kapidani, Jean-Philippe Boisgard, and Žarko Lukšić

7.1 Introduction

Two main types of unmanned underwater vehicles are (1) remotely operated underwater vehicles (ROV) and (2) autonomous underwater vehicles (AUVs). The ROVs are controlled from the surface by a wired connection. These can do many different tasks, but the wired connection restricts their maneuverability and capacities to reach remote areas. The AUVs navigate autonomously due to the dedicated navigation algorithms and collect surrounding information. Once launched, they collect data and come back to the surface after completion of their specific task. Since AUVs are not connected via wire to the ship or ground they have high maneuverability, powered by battery or fuel cell, they can reach remote locations, follow narrow composite pathways, avoiding at the same time human fatigue and reducing operation expenditures.

In this paper, emphasis will be put on AUVs. These are very versatile autonomous systems and can be used for numerous purposes as (1) *military* (surveillance, anti-submarine warfare, mine countermeasures, site inspection, inspection of

S. Bauk (✉)
Maritime Studies, Durban University of Technology, Durban, South Africa

Administration for Maritime Safety and Port Management of Montenegro, Podgorica, Montenegro

N. Kapidani · Ž. Lukšić
Administration for Maritime Safety and Port Management of Montenegro, Podgorica, Montenegro
e-mail: nexhat.kapidani@pomorstvo.me; zarko.luksic@pomorstvo.me

J.-P. Boisgard
ECA Robotics SAS, Lannion, France
e-mail: boisgard.jp@ecagroup.com

S. Bauk, S. D. Ilčev (eds.), *The 1st International Conference on Maritime Education and Development*, https://doi.org/10.1007/978-3-030-64088-0_7

vwreckage, payload delivery to ocean floor, search and rescue, aircraft crash investigation, i.e., black-box search and retrieval during the investigation, ocean exploration and bathymetric study, mapping of ocean floor, locating and retrieval of dumped illegal loads, etc.); (2) *scientific* (marine biology studies, close-up observations of aquatic life without disturbance, geological survey, archeological survey, underwater environmental monitoring in rivers and lakes, track oil spill and gas leakage, etc.); (3) *industry* (repair and maintenance, track and repair underwater cables and pipelines, underwater structure inspection, etc.); and (4) *others* (underwater video footage collection, fishing, entertainment, sports, tourism, etc.) [1].

Within the COMPASS2020 project, UAVs are used for environmental recognition, sensing, mapping, and localization [2]. More precisely, in the combat with narcotic smugglers, illicit narcotic bundles (of cannabis, for instance) might be dumped from the smugglers' speed vessels and police and/or military forces have to locate them and retrieve afterwards. Namely, once located, the positions of the crates and/or bags with narcotics are communicated to the authorities for their efficient retrieval. It is to be mentioned that COMPASS2020 can be treated as a kind of follow-up of the SWARM (Smart and Networking UnderWAter Robots in Cooperation Meshes) project, whose main objective was to develop an integrated platform for autonomous maritime and underwater operations including AUV, aerial autonomous vehicles (AAVs), ROVs, and unmanned surface vehicles (USV) performing dangerous missions [3, 4] in multimodal modes. Similar is with COMPASS2020 project, which has to integrate several AAVs as Zephyr, AR5 Life Ray Evolution, and AR3 Net Ray, operating at different altitudes and providing in such manner broader coverage at different resolution levels [5]; sea surface speed patrol vessel; and AUVs.

The rest of the paper is organized as follows: Sect. 7.2 contains key features of AUVs including their technical detail, shape, navigation, and communication principles; Sect. 7.3 provides short description of AUVs deployed within COMPASS2020 project; while Sect. 7.4 gives some conclusion remarks and directions for further research work in this domain.

7.2 Key Features of AUVs

The idea of designing and developing AUVs is not a new one. The first AUV or the self-propelled underwater research vehicle was developed by Murphy and Francois in 1957 in the Applied Physics Laboratory at the University of Washington (USA). This craft operated at 2–2.5 m/s up to a depth of 3600 m. In the 1970s few AUVs were developed in MIT and also in the Soviet Union [1]. These early underwater robots were heavy, expensive, and inefficient. Today's AUVs can have six degrees of freedom, travel faster than 20 m/s, accurately detect obstacles, and map ocean floor at depths of up to 6000 m. They are more sophisticated, less expensive, and consequently accessible for wider exploitation like fishing, sports, tourism, entertainment, etc. However, AUVs have yet to go a long way in terms of becoming fully autonomous, capable to explore deep and hazardous underwater habitats. In

Table 7.1 are given some basic data on actual AUVs' producers, applications, and key technical features.

Inspired mainly from submarines, AUVs are generally of torpedo shape. These are highly maneuverable and can exactly travel in complex pathways and access remote areas, without engendering human life. Some AUVs are of hydrofoil shape, while some other mimicked aquatic animals as snakes, turtles, beetles, and crabs. So-called fish robots are mostly popular among the biomimetic AUVs. As an example SoFi (Soft Robotic Fish) can be given (Fig. 7.1a, b). It is designed and developed

Table 7.1 Actual AUVs' types, producers, applications, and basic features [1, 5, 6]

Heading level	Applications	Dimensions	Depth
AE1000, Japan	Inspection of underwater telecommunication cables	2.3 m × 2.8 m × 0.7 m	1000 m
Maya AUV, NIO, Goa, India	Oceanography study	1.742 m, dia. 0.234 m	200 m
Theseus AUV, Canada	Under-ice bathymetric surveys	10.7 m, dia. 0.127 m	2000 m
Autosub 6000, AUVAC, USA	Scientific survey and mapping	5.50 m × 0.90 m × 0.90 m	6000 m
HUGIN, Kongsberg	Seabed mapping, pipeline inspection, mine reconnaissance	5.2–6.4 m, dia. 0.75 m	3000–4500 m
REMUS-6000, Kongsberg Maritime	Oceanography study, monitoring, surveillance, reconnaissance, etc.	3.96 m, dia. 0.71 m	6000 m
AUV-150, CMERI, India	Oceanography study, mapping, surveillance, reconnaissance, etc.	4.85 m, dia. 0.5 m	150 m
D. Allan B, MBARI, USA	Seafloor mapping	5.18 m, dia. 0.54 m	6000 m
SOTAB, Osaka University, Japan	Track of leakage from oil mines	3.0 m, dia. 0.27 m	200 m
AE 2000A, Japan	Under-ice survey	3.0 m × 0.7 m × 0.7 m	2000 m
Tri-TON 2, University of Tokyo, Japan	Estimate ore resources in underwater hydrothermal deposits	1.4 m × 0.7 m × 1.4 m	2000 m
SeaCat, Germany	Autonomous inspection of underwater structures	2.5 m × 0.58 m × 0.67 m	600 m
Bluefin21, 2016, General Dynamics, USA	Search and explore, oceanography, mine countermeasures	5 m, dia. 0.53 m	4500 m
A27, ECA Group, France	Enduring vehicle capable of carrying a large payload in atmospheric housing	4.7 m × 0.730 m	300 m
A9, ECA Group, France	High-resolution data acquisition, 3D data acquisition, shallow waters, harbor and coastal survey	2 m, dia. 0.23 m	300 m

a. b.

Fig. 7.1 SoFi UAV. (**a**) SoFi in searching elusive marine environment. (**b**) SoFi's key modules. (Source: Web)

by the team from MIT's Computer Science and Artificial Intelligence Laboratory. SoFi is made of silicone rubber and enables closer study of aquatic life. In fact, it gets closer to marine life than humans can get on their own [7, 8].

The AUVs are often equipped with different acoustic sensors like side scan sensors, forward-looking sensor, multi-beam echo sounder, etc. They are usually of modular structure, containing propulsion, sensing, controlling, navigation, communication, and other modules. These modules can be easily and quickly replaced in case of malfunction and/or for the purpose of different missions. The AUVs usually produce low level of noise (or no noise at all in some cases) and consequently don't disturb aquatic ecosystem.

Navigation Principles

The AUVs navigate underwater autonomously based on predefined plan. Localization is of key importance in navigation, since it enables AUV to follow the predefined path precisely and reach the final destination. As Global Positioning System (GPS) does not function underwater and high-frequency radio signals' propagation in the underwater environment is suppressed, localization and navigation are very challenging for AUVs. When using GPS, AUV has to resurface in intervals. In general, methods of AUVs' navigation can be broadly divided into inertial, acoustic, and geophysical.

Inertial Navigation An embedded inertial navigation system (INS) is a navigation device, based on submarine of world war, that uses a computer, motion sensors (accelerometers), and rotation sensors (gyroscopes) to continuously estimate by dead reckoning the position, the orientation, and the velocity (direction and speed of movement) of a moving object (here AUV) without the need for external references [9]. INS is usually used for small, inexpensive UAV, since over time it can accumulate errors generated by accelerometer and gyroscope [1]. A regular recognition is needed to compensate the drifting.

Acoustic Navigation When it comes to acoustic navigation, the range is estimated from the time of travel of the acoustic signal from AUV to the external transducers (devices which generate and receive sound waves) and backward. The AUV position is known in real time, thanks to calculating on the principle of triangulation. Three different types of acoustic navigation are briefly described below:

(a) In the case of ultra-short baseline (USBL), the AUV is positioned relative to a surface vehicle fitted with an array of acoustic transducers (Fig. 7.2b). Relative distance is calculated from the time of travel of the acoustic signal and direction from the phase difference of the signal received by different transducers. Here the transducers are placed close to one another, and major disadvantage is precise range detection.
(b) In the case of short baseline (SBL), the transducers are placed in front and back of the surface vessel (Fig. 7.2b). Therefore the baseline is limited to the length of the vessel, which limits the positional accuracy of the AUV.
(c) In the case of long baseline (LBL), the transducers are widely placed over the mission area on the seabed (Fig. 7.2c). Localization is done by triangulating the range estimated by acoustic transducers. The major limitation is the huge cost and time involved in placing the transducers on the seabed [1].

Geophysical Navigation When it comes to geophysical navigation, external environmental features arc used as landmarks for positioning. Optical and sonar are two main modes of geophysical navigation. Simultaneous localization and mapping is predominantly used for it [1]. In the case of optical navigation, monocular or stereo cameras can be used to take images of the underwater environment, and features extracted from the images can be used for simultaneous localization and mapping (SLAM). In such setting, different visual odometry[1] techniques are used. On another side, high-power SOund NAvigation and Ranging (SONAR) is a device for detecting and locating objects especially underwater by means of sound waves sent out to

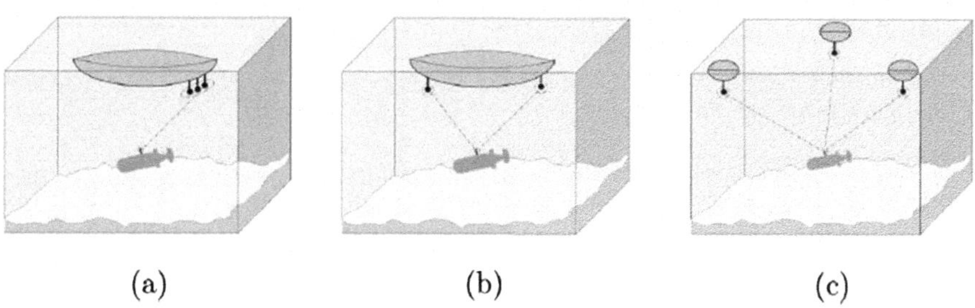

Fig. 7.2 Principles of acoustic navigation. (Source: [1])

[1]Odometry is the use of data from motion sensors to estimate change in position of an object (here AUV) over time. It is used in robotics by some robots (for AUV, too) to estimate their position relative to a starting location. The word odometry is composed from the Greek words *odos* (meaning "route") and *metron* (meaning "measure").

be reflected by the objects.[2] Sonar is a commonly used technique for communication, detection of objects, and navigation by using sound propagation. A comprehensible description of sonar can be found in [10], and it states: "When AUV[3] is used to map the topography of the ocean's floor, it sends out sound pulses, often referred to as pings, towards the bottom of the ocean within its vicinity. As these sound pulses travel downwards they will encounter physical features such as hills, valleys, rock, etc. These sound pulses are subsequently reflected back up towards the AUV, having been modified by the objects along their path. These reflected pulses are often called echoes. Receivers on the AUV that detect these echoes can then reconstruct the topology of the region from which the echoes bounced off". Sonar is very like the echo sounder. The difference is that the sound beam can be steered in the desired directions and present images of the bottom topography on suitable display. Synthetic Aperture Sonar (SAS) is a relatively new principle in hydroacoustics. Together with advanced image processing, the method can produce very detailed images of seabed and objects. It operates in such way that one moves sonar along a line and therefore illustrates stationary objects from several directions. The transmitting antenna's synthetic aperture in relation to the object will then be the length the sonar has moved (Fig. 7.3). These systems that are now in use can give resolution of 2 × 1 cm, which is typically ten times better than what ordinary sonars can give. Kongsberg produces an SAS (HISAS-1030), which is used at AUV Hugin (Fig. 7.4). It works at 70–100 kHz and can be delivered with Focus software [11].

Some AUVs should have ability to carry out long-distance missions fully autonomously and without supervision from surface ship. Combined with inertial navigation, the use of one or several transponders on the seabed is an accurate and cost-effective approach towards achieving this [12, 13]. An extensive description of actual advanced AUVs' propulsion solutions, control systems and their key components, state estimation methods, and path planning models and techniques along with object detection and obstacle avoidance can be found in [1, 14].

Communication Principles

Underwater wireless communications are implemented using communication systems based on acoustic, radio frequency, and optical (light and laser) waves. Underwater acoustic wireless communications have been one of the most used technologies since they can provide connection over rather long distances (Fig. 7.5). However, acoustic waves have many drawbacks as scattering, high delay due to the low propagation speeds, high attenuation, and low bandwidth. Additionally, acous-

[2] Merriam-Webster Dictionary. Retrieved from: https://www.merriam-webster.com/dictionary/sonar (last access: 10th April 2020).

[3] Originally "submarine"

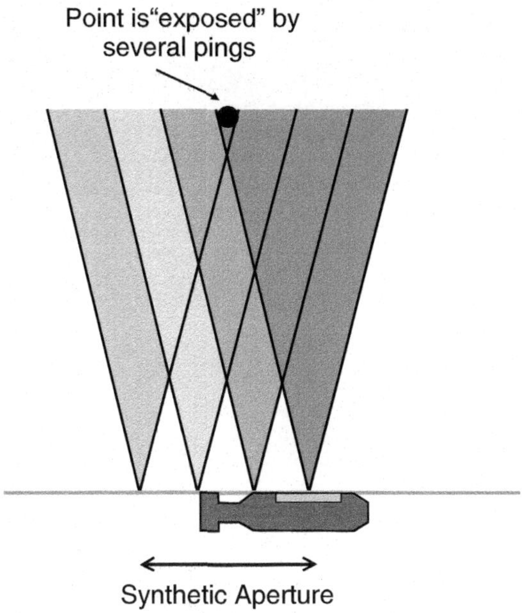

Fig. 7.3 Principle of synthetic aperture sonar. (Source: [11, pp. 4–40])

tic signals generated by communication systems and sonar devices have harmful impact on the underwater mammals and fishes. Therefore, research has been carried out in the past to use low-frequency radio waves (30–300 Hz). These waves have numerous disadvantages like high attenuation, low data rate, adverse effect of shallow areas, long antennas, etc. For worldwide communications with submarines, e.g., for depths up to a few 10 m, very-low-frequency (VLF) transmitters from 10 to 30 kHz are used. In oppose to acoustic and radio waves, optical waves can provide high-speed underwater optical communications at low latencies, thanks to high propagation speed and high data rate in return for a limited communication range (tens of meters). In Table 7.2 are given key features of underwater acoustic, radio, and optical wireless communications.

All three considered underwater wireless communication modes have certain advantages and disadvantages dependent on various underwater conditions. The subject remains open, and further research is necessary for conceiving and implementing more practicable and accurate communication, networking, and localization schemes [15].

7.3 The AUVs Deployed in COMPASS2020 Project

One of the COMPASS2020 project main goals is to develop an integral platform for efficient simultaneous deployment of unmanned aerial vehicles (Zephyr, AR5 Life Ray Evolution, AR3 Net Ray), offshore patrol vessel (OPV), and underwater auton-

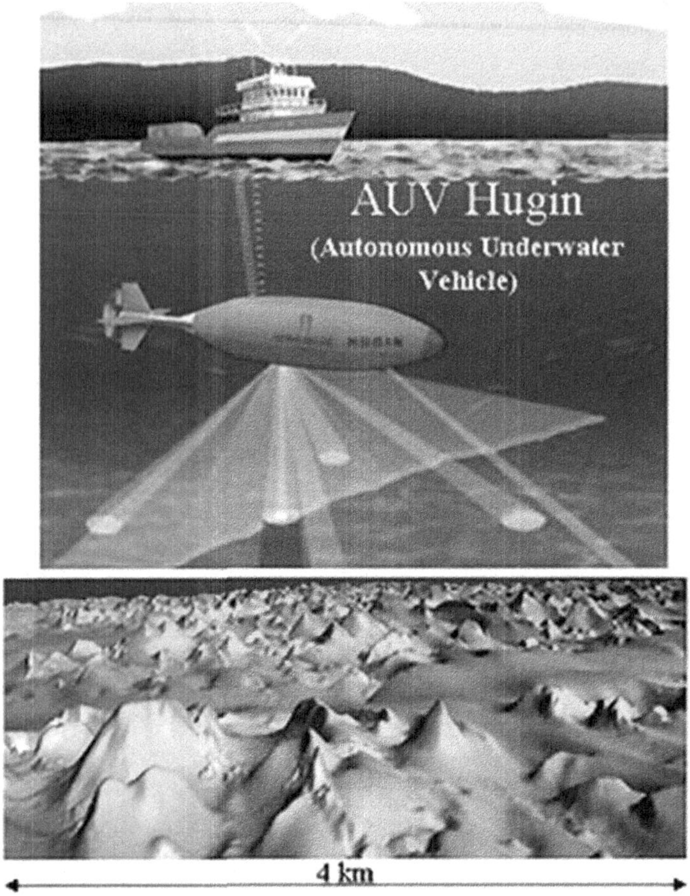

Fig. 7.4 AUV Hugin. Image from 1000 m depth. The tops that are shown on the map are 30–40 m high. (Source: [11, pp. 4–40])

omous vehicles (A27 and A9) for preventing and combating trafficking of narcotics over European borders. To illustrate the impact of narcotics in Europe, it is estimated that every year approximately 125 tonnes of cocaine are consumed. The majority of it comes from Latin America to Europe on transatlantic routes. However, in recent years transhipments from large vessels to various forms of transport (including leisure sailing vessels, fishing vessels, merchant vessels, and fast speedboats) have started occurring in sea waters along Northern and Eastern African coast. These new forms of transport target mainly Spain and Portugal as points of entry, while the most traditional forms target Belgium and the Netherlands (major European shipping ports) [16]. When it comes to interception of narcotics smugglers, the OPV, the Zephyr, and the AUV are in action in the border area. The Naval Group mission system (MS) is running on board the OPV, and it is always connected with its replica at marine operations center (MOC) ashore. Zephyr is

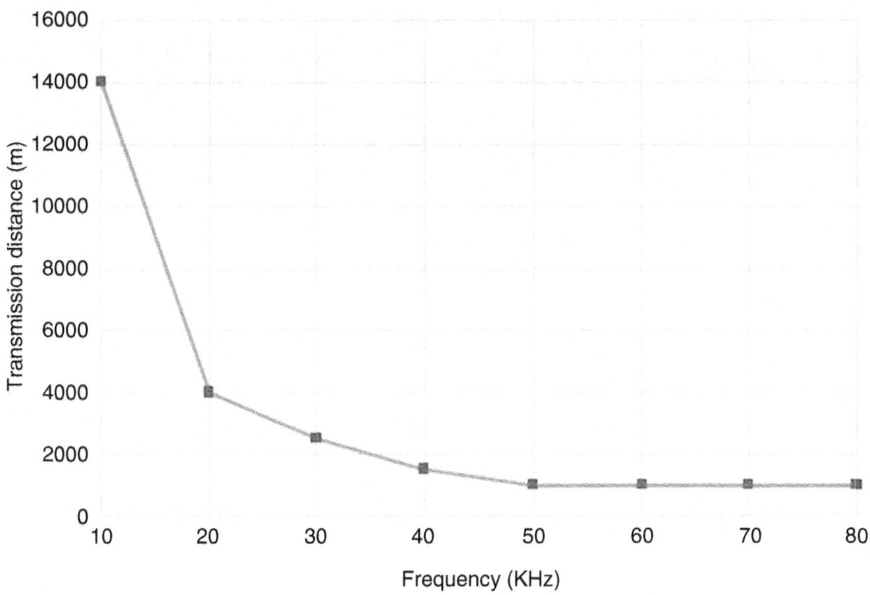

Fig. 7.5 Average achievable transmission distance by various commercial acoustic modems. (Source: [15, p. 2])

Table 7.2 Underwater acoustic, radio, and optical communications features [15, p. 3]

Feature	Acoustic	Radio	Optical
Range	<20 km	<100 m	100–200 m
Attenuation factors	Conductivity	Conductivity and frequency	Distance vs. inherent optical properties
Speed	1500 m/s	2.25 × 108 m/s	2.25 × 108 m/s
Power	~10 W	~100 W	~1 W
Cost	High	High	Low
Data rate	<10 Kbps	<0.1 Gbps	<10 Gbps
Antenna size	0.1 m	0.5 m	0.1 m
Latency	High	Moderate	Low

launched from MOC, and it has to collect an overall picture of the area that is being surveyed. In addition, the AUVs should be previously deployed from the OPV into a strategic location that is coincident to the traffickers' typical routes. The AUVs are programmed to follow specific trajectories in the area of interest, navigating underwater at low depth in order to remain undetected from the smugglers and at the same time staying closely enough to the surface in order to optimize the possibility of detecting the target. The low noise AUVs carry wide bandwidth streamer of hydrophones[4] that enable detecting speedboats. After detection of the target, the

[4] A hydrophone is a microphone designed to be used underwater for listening to underwater sound.

AUVs can communicate to the Zephyr, thanks to the OPV which is used as a communication relay in the system. The Zephyr sends automatically an alert to MS on board OPV and its replica in the MOC. Once the MOC receives the alert, the officers proceed with the deployment of an AR-5 platform [5]. The AR-5 has to come close to the vessel and acquire more detailed information about it. In accordance to this information, the officer on board OPV can decide how to intercept the threat and act efficiently. If the smugglers try to get rid of the cargo, UAV and AUVs have the capacity of searching for it by making use of sonars[5] [2]. In the following two subsections, short descriptions and some basic data on the AUVs employed within COMPASS2020 will be given.

The A27 AUV

The A27 is a development of ASEMAR of ECA Group Autonomous Underwater Vehicle family (Fig. 7.6). Big Size AUV with long endurance and high payload capability can be used for both defense and commercial purposes. It performs autonomous missions up to 300 m depth and is easily transportable by plane for overseas missions. Due to its large endurance, very high area coverage rate (2 km^2/h), and payload capacity, it is able to host high-performance payloads according to the mission's requirements: Synthetic Aperture Sonar (SAS), video, forward-looking sonar (FLS), multi-beam echo sounder, and others [17]. For navigation it uses Inertial Navigation System (INS), Doppler Velocity Log (DVL), military global navigation satellite system (GNSS), and Global Positioning System (GPS) periodically, after resurfacing. It can communicate via Wi-Fi, Ethernet, Iridium, and/or acoustic wireless communication channel. Its average speed is 3–5 knots (and max 6 knots). Key payloads are sonar and conductivity, temperature, and depth (CTD) sonde. It withstands harsh environmental conditions and offers a greater stability when encountering heavy turbulence from waves. The high degree of stability enables this AUV to capture high-resolution images. The information obtained by the platform is post-processed in the command center [18].

Most hydrophones are based on a piezoelectric transducer that generates an electric potential when subjected to a pressure change, such as a sound wave (in this case caused by narcotic smugglers' vessel/the author's comment) [Source: *Wikipedia*].

[5] Side-scan sonar is a category of sonar system that is used to efficiently create an image of large areas of the seafloor. Side-scan sonar imagery is also a commonly used tool to detect debris items and other obstructions on the seafloor that may be hazardous to shipping or to seafloor installations by the oil and gas industry. In addition, the status of pipelines and cables on the seafloor can be investigated using side-scan sonar (in this case it's used for detecting abandoned bundles with narcotics/the author's comment) [Source: *Wikipedia*].

Fig. 7.6 The A27 AUV. (Source: COMPASS2020 internal documentation)

The A9-E AUV

The A9-E AUV is the configuration of ECA Group for environmental monitoring (Fig. 7.7). In addition to the seabed image acquisition, it can record bathymetric data as well as environmental information such as water turbidity, conductivity, temperature, fluorescence, dissolved oxygen, and/or pH. Mission planning and monitoring are done through user-friendly software which allows operator to follow the vehicle at any time during its mission. This underwater drone has been designed to meet STANAG 1364 requirement, i.e., its acoustic and magnetic signatures are minimized in order not to trigger any underwater mines when doing the mine warfare survey. As part of early trials for the SWARM project, ECA Group's A9 AUV fitted with the interferometer side-looking sonar demonstrated ability to conduct surveys in a shallow water environment of 50m depth. It uses a phase differencing bathymetric sonar that increases area coverage by close to 200% over conventional multi-beam echo sounders in shallow water [19]. For navigation it uses INS, DVL, and GPS and for communication purposes radio (UHF), Wi-Fi, Ethernet, and the acoustic wireless communications. Its payload consists of, but it is not limited to, Interferometer Side Scan (ISS) sonar, video, CTD, environmental sensors (turbidity, pH, fluorescent Dissolved Organic Matter (fDOM)/wastewater discharge), etc. [20].

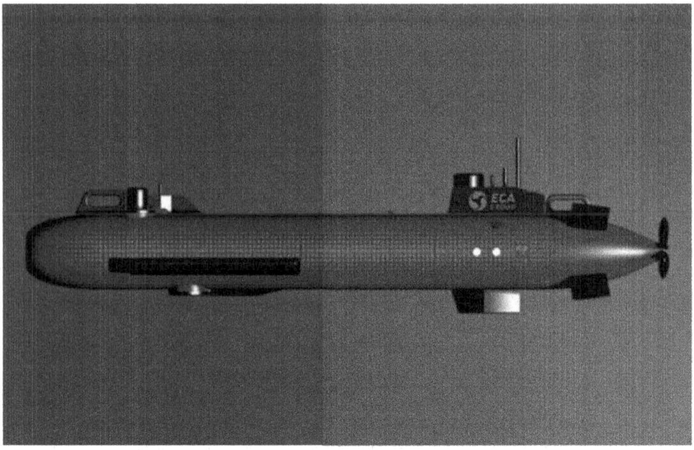

Fig. 7.7 The A9-E AUV. (Source: ECA Group Web site)

7.4 Conclusion

The paper gives an overview of some key determinants of AUV. These systems have a variety of military, scientific, industrial, and other applications. They can be very complex and expensive, but there are some that are available for educational and recreational purposes. It is a common opinion in the literature that these systems have been developed to a very high standard, but that there is plenty of room for further research and improvement when it comes to better adaptive control techniques using neuro-fuzzy techniques, more accurate localizing using improved INS nonlinear Kalman filters, cooperative localization (swarm intelligence), artificial intelligence vision and object detection, odometry, underwater wireless communications, high-density battery power supply, energy harvesting methods, etc. By improving all these dimensions, AUVs will become fully autonomous long-range underwater robots, capable to explore the deepest, inapproachable, and harsh corners of the seabed. The underwater autonomous vehicles, considered within COMPASS2020, have to be integrated into the complex system composed of autonomous aerial vehicles (Zephyr, AR5, and AR3) and sea surface vehicle (OPV), including Naval Group mission system (MS) on board OPV and shore-based marine operations center (MOC). At the moment, the experts' team within the project is designing algorithms for seamless data acquisition, analysis, storage, and presentation. This research work is based on the experts' knowledge, skills, and experiences acquired through several realistic case studies and recent test beds in European seas. Following research work should target harmonizing actions of all involved manned and unmanned vehicles and optimizing relevant data/information flow schemes. In parallel, improving bidirectional communication links between all involved parties in the case of emergency is to be further explored.

Acknowledgment This work has been partially funded by the European Commission Research and Innovation Program Horizon 2020, COMPASS2020 Project (Grant Agreement No: 833650).

References

1. A. Sahoo, S.K. Dwivedy, P.S. Robi, Advancement in the field of autonomous underwater vehicle. Ocean Eng. **181**, 145–160 (2019)
2. S. Bauk, N. Kapidani, Ž. Lukšić, F. Rodrigues, L. Sousa, Autonomous marine vehicles in sea surveillance as one of the COMPASS2020 project concerns. J Phys. Conf. Ser. **1357** (2020) https://iopscience.iop.org/article/10.1088/1742-6596/1357/1/012045
3. I. Leblond, S. Tauvry, M. Pino, Sonar image registration for swarm AUVs navigation: results from SWARM project. J. Comput. Sci. **36**, 101021 (2019)
4. E. Zereik et al., Challenges and future trends in marine robotics. Ann. Rev. Control **46**, 350–368 (2018)
5. S. Bauk, N. Kapidani, Ž. Lukšić, F. Rodrigues, L. Sousa, Review of Unmanned Aerial Systems for the Use as Maritime Surveillance Assets, Proceedings of the 24th International Conference on Information Technology (IT), Zabljak, Montenegro, 18–22 February 2020. https://doi.org/10.1109/IT48810.2020.9070718
6. Kongsberg, *HUGIN Autonomous Underwater Vehicle (AUV)*. Retrieved from: https://www.kongsberg.com/globalassets/maritime/km-products/product-documents/hugin-product-specification. Last access 11 Apr 2020
7. MIT News, *Soft robotic fish swims alongside real ones in coral reefs*, published by Adam Conner-Simons (CSAIL) on March 21, 2018. Retrieved from: http://news.mit.edu/2018/soft-robotic-fish-swims-alongside-real-ones-coral-reefs-0321. Last access 10 Apr 2020
8. U.K. Verfuss et al., A review of unmanned vehicles for the detection and monitoring of marine fauna. Maritime Pollut Bull **140**, 17–29 (2019)
9. Seminar on inertial navigation systems Tampere University of Technology, *Basic Principles of Inertial Navigation*. Retrieved from: http://www.aerostudents.com/courses/avionics/InertialNavigationSystems.pdf. Last access 10 Apr 2020
10. Study.com, *What is Sonar? – Definition, Uses & Examples*. Retrieved from: https://study.com/academy/lesson/what-is-sonar-definition-uses-examples.html. Last access 10 Apr 2020
11. N. Kjerstad, *Electronic and Acoustic Navigation Systems for Maritime Studies*, 1st edn. (NTNU Norwegian University of Science and Technology, Alesund, 2016)
12. Kongsberg, *Autonomous Underwater Vehicle – AUV – The HUGIN Family*. Retrieved from: https://www.kongsberg.com/globalassets/maritime/km-products/product-documents/hugin-family-of-auvs. Last access 11 Apr 2020
13. Ø. Hegrenæs, K. Gade, O.K. Hagen, P.E. Hagen, Underwater Transponder Positioning and Navigation of Autonomous Underwater Vehicles, *Proceedings of the MTS/IEEE oceans conference and exhibition*, BILOXI, 2009. Retrieved from: https://www.navlab.net/Publications/Underwater_Transponder_Positioning_and_Navigation_of_Autonomous_Underwater_Vehicles.pdf. Last access 11 Apr 2020
14. C. Yu et al., Onboard system of hybrid underwater robotic vehicles: ıntegrated software architecture and control algorithm. Ocean Eng **187**, 106121 (2019)
15. N. Saed et al., Underwater optical wireless communications, networking, and localizing: a survey. Ad Hoc Networks **94**, 101935 (2019)
16. COMPASS2020 – Coordination of Maritime assets for Persistent And Systematic Surveillance, Project internal documentation (Boosting the effectiveness of the Security Union, H2020-SU-SEC-2018-2019-2020), pp. 1–70

17. Naval Technology, *A18-M Autonomous Underwater Vehicle*. Retrieved from: https://www.naval-technology.com/projects/a18-m-autonomous-underwater-vehicle/. Last access9 Apr 2020
18. ECA Group, *A18-M/AUV/Autonomous Underwater Vehicle*, Data Sheet. Retrieved from: https://www.ecagroup.com/en/solutions/a18-m-auv-autonomous-underwater-vehicle. Last access 12 Apr 2020
19. ECA Group, *A9-E/AUV/Autonomous Underwater Vehicle*, Data Sheet. Retrieved from: https://www.ecagroup.com/en/solutions/a9-e-auv-autonomous-underwater-vehicle. Last access 12 Apr 2020
20. Shephard News, *Unmanned vehicles - A9-E AUV demos side-looking sonar*, November 2016. Retrieved from: https://www.shephardmedia.com/news/uv-online/a9-auv-demos-side-looking-sonar/. Last access 9 Apr 2020

Chapter 8
Low-Cost Unmanned Surface Vehicle for Autonomous Bathymetric Surveillance

Vladimir Nikishin, Maxim Durmanov, and Ivan Skorik

8.1 Introduction

The main goal of current development is to design complex solution including unmanned surface vehicle (USV) and custom autopilot firmware and software for real-time depth visualization. In the beginning of development, we have pointed next requirements for the USV:

- Easily accessible low-cost hardware parts
- Easy USV transportation
- Monohull
- Free software libraries and IDE
- Custom autopilot
- Custom base ground station
- Propulsion high reliability
- Maneuverability
- Minimal continuous USV work over 1 h
- Custom software for collected data post-processing

For suitable USV transfering, we have determined the longest dimension is 1 meter, which is dependent on average size of car trunk. Monohull shape was chosen because of simple manufacturing and because USV oriented to work in calm waters such as lakes and internal bays where the high waves are rare.

V. Nikishin · M. Durmanov · I. Skorik (✉)
Sevastopol State University, Sevastopol, Russia
e-mail: IVSkorik@sevsu.ru

S. Bauk, S. D. Ilčev (eds.), *The 1st International Conference on Maritime Education and Development*, https://doi.org/10.1007/978-3-030-64088-0_8

8.2 USV Main Parts

Hull

USV was made of two main materials – fiberglass and styrofoam as hull base. The hull base was CNC-machined and glued together in one part (Fig. 8.1).

Then styrofoam was covered by three layers of fiberglass (Fig. 8.2).

For minimizing interaction of electronic parts with water, they are placed over the top in separated covered box mounted with PVC pipes glued to the USV hull. Placing antennas as high as possible also gives advantages for reliable connection with ground station and easy access to electronic; it worthwhile because of magnetometer that is far enough from engine containing strong magnets. Final results of the hull are shown in Fig. 8.3.

Engine and Propulsion System

We have selected electrical propulsion system because modern brushless and brushed motors have enough power and they have small size suitable for use in USVs. Electrical propulsion system allows fast changing battery, and it does not pollute water by any hazardous materials or exhaust gases. Experimental USV uses 3S (12 volts) battery and 4074-sized brushless motor with 2000 kV and total power over 1 kW. Motor is driven by electronic speed controller (ESC). Because ESC dissipates a lot of heat, it requires cooling; cooling is produced by water injected by water jet.

For steering and movement of USV, we used water jet. Water jet has advantages in safety and maneuverability. Safety reason is very important if USV works near the beaches with swimming people. Because water jet rotor is located inside tube, it is almost impossible to injure someone by rotating blades. Open propellers are more dangerous; also, there are big chances that floating garbage, seaweed, or fish lines could wound up on the shaft and propeller blades. Even if water jet has no safety lattice, there are always problem of floating garbage and seaweed (Fig. 8.4).

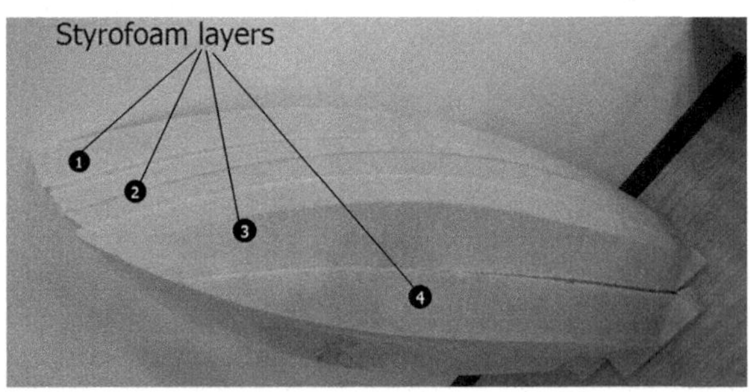

Fig. 8.1 CNC-machined styrofoam hull base

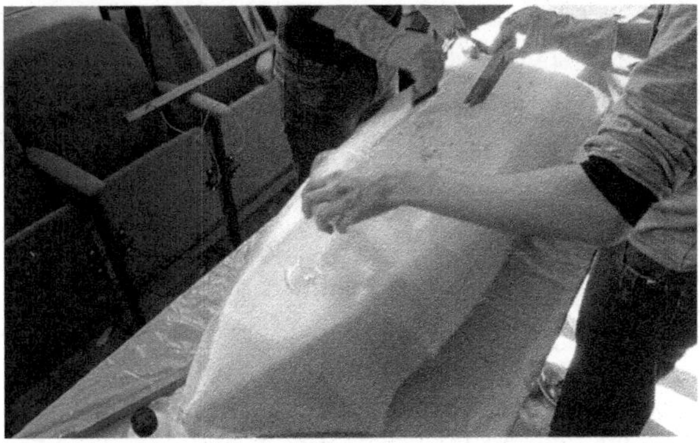

Fig. 8.2 Hull covering by fiberglass

Fig. 8.4 Seaweed in water jet

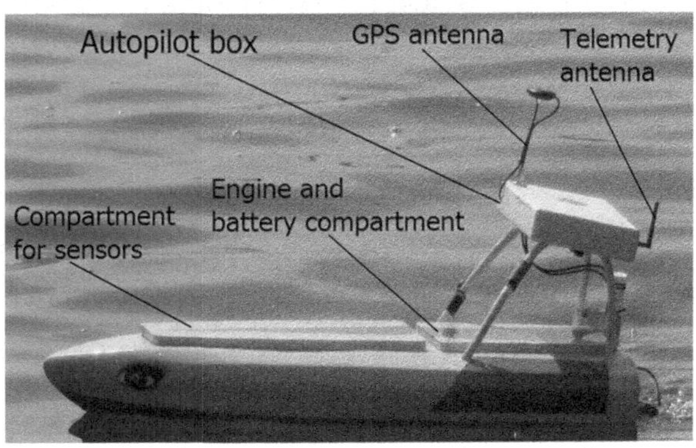

Fig. 8.3 Unmanned surface vehicle

For steering of USV was used rotating nozzle. Nozzle has rotating angle near 30° left and right. This allows to get very good maneuverability.

Autopilot Hardware

The main purpose of autopilot is to control USV by rotating nozzle to steer it at given route point. This requires sensors (gyro, accelerometer, and magnetometer) integration with microcontroller (MCU). We selected Atmega2560 on Arduino Mega board [1]. This MCU has enough interfaces for connecting with sensors and does not require high programming skill. For autonomous movement of USV, main measurements are current heading and current location. Heading is measured by MEMS magnetometer connected to MCU by I2C interface. Magnetometer requires calibration to minimize hard and soft iron distortion. For measurement, location of USV uses GPS module U-Blox connected to MCU by UART interface. For controlling nozzle and motor, ESC uses PWM signal generated by internal hardware timer. We have designed PCB sandwich board inserted in MCU board (Fig. 8.5).

Autopilot Software

Firmware was built using free Atmel Studio IDE. Firmware includes functions for azimuth and distance calculation to given point and converting controlling commands to PWM signal for servo and water jet nozzle.

Fig. 8.5 Autopilot

Sonar

The most useful part of USV development is integration with measurement devices. Current development is targeted to bathymetry. Selection of sonar depends on many reasons as precision, resolution, and single- or multi-beam ray. However, in the case of USV, the most important is connection with MCU – sonar has to be equipped with suitable interface; in most cases this is RS-232 with NMEA standard messages [2]. We choose Garmin Echomap 42 [3]; this sonar has 5 volts NMEA interface, but its signal is inverted. Because Atmega2560 does not have an option to invert UART signal, we have to add an inverter before signal goes to MCU using NPN transistor (Fig. 8.6).

Placing sonar in internal compartment protects it from water sprays (Fig. 8.7).

Fig. 8.6 NMEA inverter

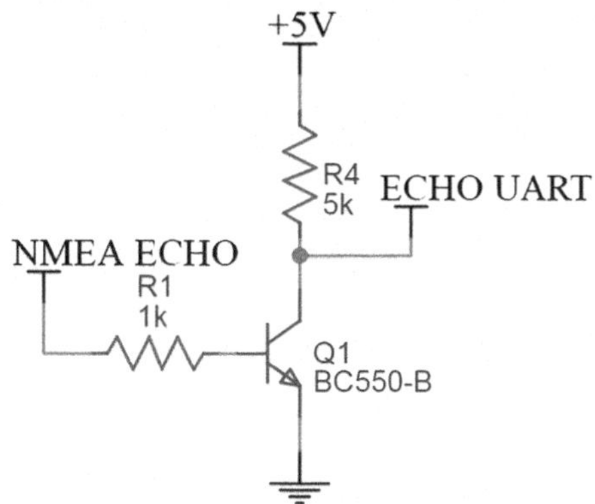

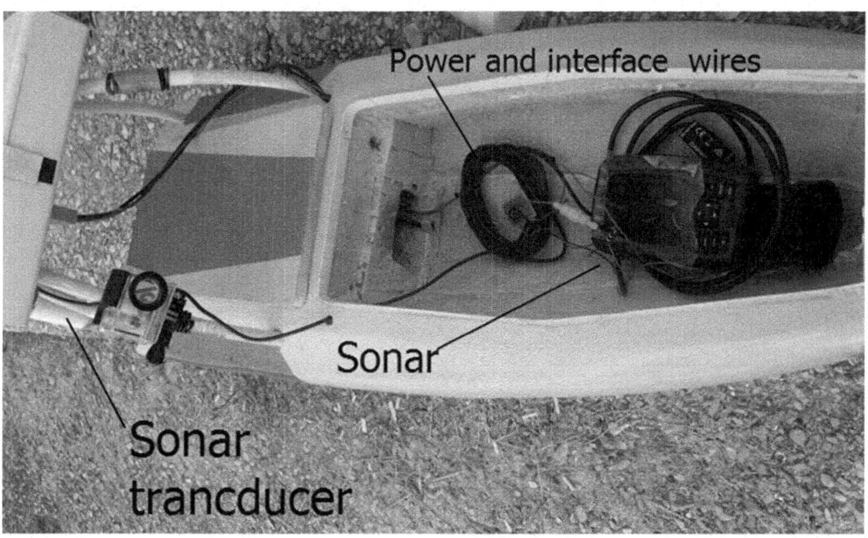

Fig. 8.7 Placing sonar in USV

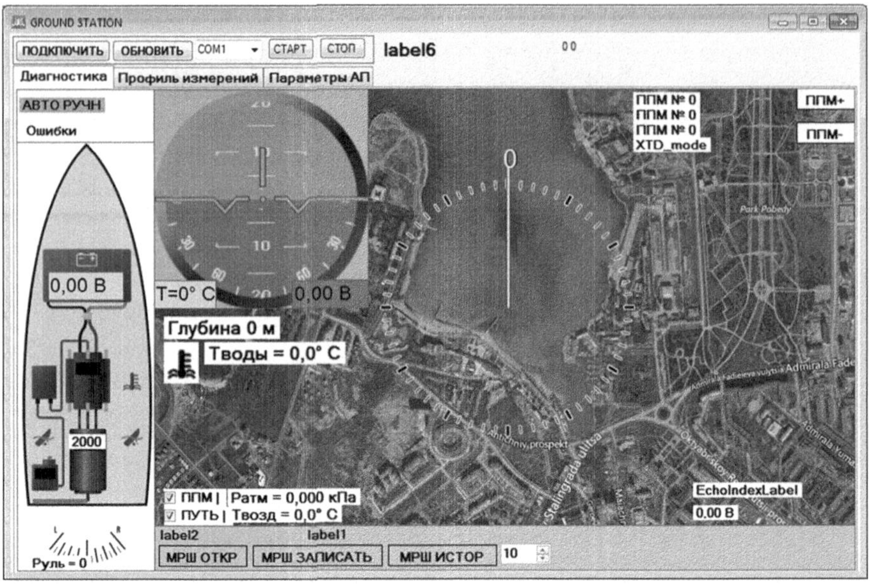

Fig. 8.8 Ground station GUI

Ground Station Software

Ground station (GS) visualizes USV data in real time. Graphical user interface (GUI) of GS is shown in Fig. 8.8. GS was built in free Microsoft Visual Studio C# using free GMap.Net library for map positioning of USV. GS receive data from USV using wireless 915 MHz telemetry modules. The received telemetry stores on PC in text document file for further post-processing. Post-processing could be done with software as Golden Surfer or ReefMaster, but we have designed our custom software.

8.3 Post-processing Software Development

Statement of the Bathymetric Data Visualization Problem

It is proposed to visualize bathymetric data using a heat map. To do this, select a color palette, and fill with color each area with the same depth. The initial data for building a heat map are:

- Geographical coordinates of points (latitude and longitude) and depth values measured by the sonar at these points

- Geographical coordinates of the boundary points of the test site, to create the boundaries of the displayed heat map
- Color palette (range and number of color shades)

To solve this problem, there are special services. For example, the HeatmapTool [4] online service allows you to use various color palettes; control the radius, zoom, and opacity of heat spots; and update the map in real time. However, this service is paid and designed to display statistical data in the selected region.

Heat Map Algorithm

The problem of building a heat map is not only to fill on the map the individual points at which depth measurements were made but to fill the entire area with the color. This problem is solved using the inverse distance weighting (IDW) [5] method, which consists in determining the weight of each point by interpolating neighboring points with a known weight at a given point.

The heat map algorithm consists of the following steps:

1. Converting the geographical coordinates of the points to the x and y coordinates of the Cartesian coordinates
2. Formation of the border of the polygon for cropping the resulting image of the heat map
3. Implementation of the IDW method for constructing a heat map image
4. Converting points in Cartesian coordinates to geographical coordinates and combining a heat map with a geographical map of the area

Geographic coordinates and depth measurements are taken from the sonar and saved in a text file. The geographic coordinates of the boundary points are taken in advance from the terrain map and saved in a file. The color palette is chosen arbitrarily; usually warmer tones correspond to smaller depths and colder ones to larger ones. The algorithm is implemented using developed software. An example of building a depth map for an artificial reservoir in Inkerman (Sevastopol) is shown in Fig. 8.9 and the coastal waters of Holland Bay (Sevastopol) is shown in Fig. 8.10.

Fig. 8.9 Heat map of Inkerman (Sevastopol)

Fig. 8.10 Heat map of Holland Bay (Sevastopol)

8.4 Conclusion

Developed USV shown in the paper has simple design which can be reproduced with minimal cost. In summary, the price of bill of materials without sonar is less than 500$. The main limitation of our USV is it works only in calm waters without strong wind and waves. Small draft of USV less than 3 inches allows mapping the depth of lakes or bays very close to shoreline; this is almost impossible with big boats with crew on board.

Acknowledgment The reported study was funded by the internal grant of Sevastopol State University, project number 516/06-31.

References

1. Arduino Mega official site, https://store.arduino.cc/arduino-mega-2560-rev3
2. National Marine Electronics Association, https://www.nmea.org
3. Garmin Echomap sonar, https://buy.garmin.com/en-US/US/p/592785
4. HeatmapTool.com, https://www.heatmaptool.com
5. GisGeography. Inverse Distance Weighting (IDW) Interpolation, https://gisgeography.com/inverse-distance-weighting-idw-interpolation

Chapter 9
Introduction to the Global Ship Tracking System via Mobile Satellite Constellations

Ilcev Marijan, Oludayo Olugbara, and Stojče Dimov Ilčev

9.1 Introduction

A major goal is proposed near-universal use of GNSS of the US GPS and Russian GLONASS infrastructures integrated with Satellite Mobile Communication Systems, whose very small units will be able to improve ship tracking, collision avoidance, and positioning facilities providing reliable Position, Velocity, and Time (PVT) data. The new augmenting system of GNSS is also proposed and projected to enhance Traffic Control Management (TCM) for merchant ships including for enhanced safety and security. As a result of these efforts, new tracking techniques have been projected and developed to utilize Communication, Navigation, and Surveillance (CNS) solutions and services for enhanced Ship Traffic Control (STC) and management for improved safety and security in commercial maritime transportation.

In order to meet the requirements for better CNS solutions of ocean sailing, approaching to anchorages, and inside of seaports, it is also proposed the development of Global Ship Tracking (GST) including Satellite Automatic Identification System (S-AIS) solutions. Thus, these new technologies will cover the entire African continent and the rest of the world for ocean and coastal navigation and can improve tracking and determination of all types of ships. The new GST and other existing solutions for determination will improve the basic GPS and GLONASS facilities and allow these GNSS networks to be utilized with satellite transceivers as a primary means of tracking of ships and all land vehicle movements in the seaports via Maritime Traffic Control Centres (TCC). Proposed satellite tracking systems may be used in all possible applications for determination of spatial coordinates

I. Marijan · O. Olugbara · S. D. Ilčev (✉)
Durban University of Technology (DUT), Durban, South Africa
e-mail: ilcevm@dut.ac.za; oludayoo@dut.ac.za; ilcev@dut.ac.za

S. Bauk, S. D. Ilčev (eds.), *The 1st International Conference on Maritime Education and Development*, https://doi.org/10.1007/978-3-030-64088-0_9

such as position, speed, and navigational status of target objects, which via GNSS equipment may provide PTV data for maritime and all transportation applications.

These systems mainly are necessary for improved collision avoidance of ships especially in areas with heavy traffic moving such as sea channels, in approaching anchorages, and inside of seaports. Thus, using stand-alone GPS or GLONASS data, these systems can provide speed and position of ships only. However, integrated with GNSS units and some sensors, these systems are also capable of controlling the main parameters of ocean vessels such as continuous control of position, mileage, fuel consumption and at a certain time to transmit the collected data of these and other CNS parameters. The position data can be used in case of ships' grounding and hijacking or emergency abandoned ships and for eventual Search and Rescue (SAR) actions. Appearance and implementation of these systems may provide very important contributions in enhanced safety and security for all types of ships and according to the International Maritime Organization (IMO) to improve distress and SAR operations.

The satellite tracking systems are working without man intervention in order to prevent human errors. Continuous increasing of ships transport is augmenting the necessity per explores the emerging safety and security solutions for maritime CNS systems and especially for tracking and data messaging facilities. These new tracking techniques consider the consequences of a global satellite communication framework supporting asynchronous messaging of navigation data that can be used to enhance the basic GST and AIS capability. In that manner, the analyzed modern satellite GST application can be pursued within the standardization process or independently developed with attention to compatibility with existing radio systems.

The GAT system is proposed by the authors of this paper, which is integrating GPS receiver (Rx) and Iridium or Inmarsat satellite transceivers with antennas installed on board ships. Thus, the ship tracking unit has to be installed discrete onboard ships secret location, which solution has to protect accidental or forced its shutdown [1, 2].

9.2 Development of GST Networks and Equipment

The scenario of Satellite Asset Tracking (SAT) is system employing the GNSS subsystem of US GPS and Russian GLONASS to provide free of charge position (PTV) data to different users at sea, on the ground, and in the air. This PVT data can receive ships, land vehicles and aircrafts via onboard GPS or GLONASS receivers (Rx) and used in navigation purpose, which Inmarsat network is depicted in Fig. 9.1. If GPS or GLONASS Rx is integrated with Satellite Transceiver, Rx, and Transmitter (Tx) in an integrated satellite unit with both antennas known as SAT, it will be possible to provide frequently transmission of PTV data via Geostationary Earth Orbit (GEO) and non-GEO spacecraft through Ground Earth Station (GES) and Internet to the Control and Operations Centres.

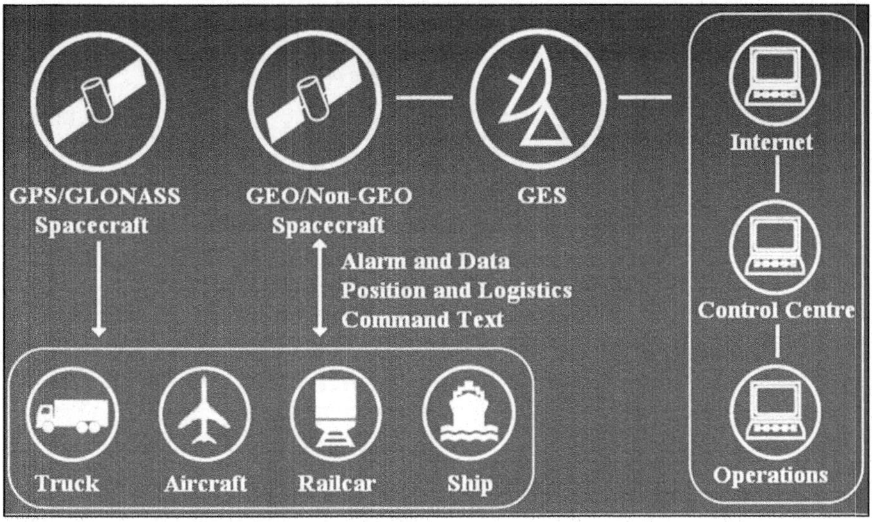

Fig. 9.1 Artifacts empowered by Artificial Intelligence [2]

Because of many incidents in past time with difficulties of searching ships in disaster and for improvement of collision avoidance of ships, the author of this paper is proposing new tracking and determination solutions via Satellite CNS and determination systems known as Global Ship Tracking (GST). Similar to the Long Range Identification and Tracking (LRIT), new and more advanced GST solution contains the shipborne GST information-transmitting equipment, such as integrated GPS or GLONASS Rx and GEO or non-GEO such as Inmarsat or Iridium satellite transceivers, namely, Transmitter and Receiver (Tx/Rx).

Because of many incidents in past time with difficulties of searching ships in disaster and for improvement of collisions avoidance of ships the author of this paper is proposing new tracking and determination solutions via Satellite CNS and determination systems known as Global Ship Tracking (GST). Similar to the Long Range Identification and Tracking (LRIT) new and more advanced GST solution contains the shipborne GST information transmitting equipment, such as integrated GPS or GLONASS Rx and GEO Inmarsat or Non-GEO Iridium satellite transceivers, namely Transmitter and Receiver (Tx/Rx).

The shipborne GST onboard equipment receives GNSS determination signals from GPS/GLONASS spacecraft (1) and sends PTV tracking messages of position (2) via GEO satellite to Ground Earth Station (GES) (3) of Satellite Communication and Application Service Providers (Internet) to the Traffic Conttol Station (TCS) processor (4), whose network configuration is shown in Fig. 9.2.

The current LRIT and newly proposed GST have the same services explained above, and in addition both can provide pulling navigation data of any ship from TCS sites. The difference between them is that LRIT is not determined to provide that some ship can receive navigation data of nearby ships for enhanced collision avoidance. Thus, the red lines highlighted in Fig. 9.2 can be used for pulling service

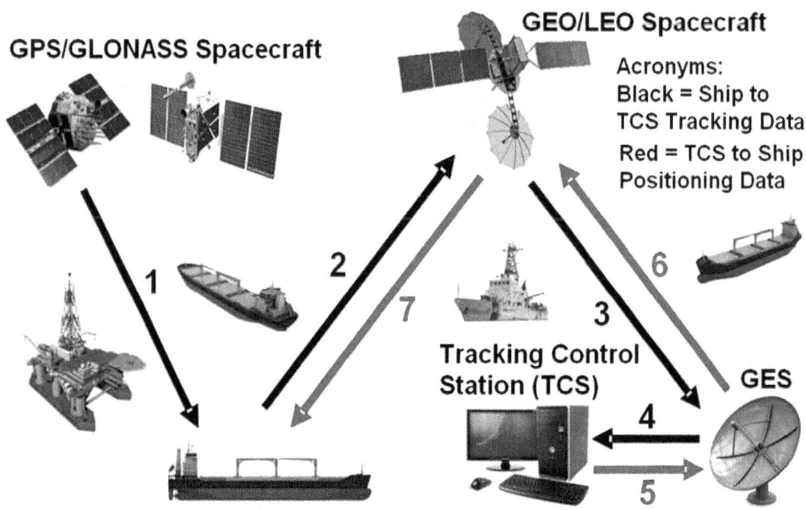

Fig. 9.2 Configuration of GST via GNSS and GEO/non-GEO satellites [2]

and what is more important for sending to any ship on his request the navigation data of adjacent ships in the same sea area.

Onboard oceangoing ships can be installed many of satellite tracking equipment already designated for SAT on board all mobiles such as vehicles, trains, containers, and aircraft, but with simple modification for very harsh weather and severe sea conditions. This equipment has to be suitable for:

- Possibility for installation on board each mobile including ship, and some have to carry 3–5 years batteries, so to work properly even when ship is in emergency situations without any power supply
- Pre-programming for different requirements and to send GPS location and other data on predefined intervals via any GEO or non-GEO satellite systems to shore host application or can be integrated with a mapping application
- Pulling facilities at shore TCS and getting position of any desired ships in vicinity
- Sending positioning messages from TCS to all ships requesting these navigation data of adjacent ships

As stated before, current LRIT system is not projected to do this very important service for collision avoidance of ships. It is important to express that LEO Globalstar and Orbcomm satellite systems are providing both simplex and duplex (two-way) satellite transmission. Duplex satellite system is able to provide sending and receiving of GST data, while simplex system can only enable receiving facilities of navigation data for adjacent ships. However, the third LEO Iridium and GEO Inmarsat mobile satellite systems are providing duplex service only.

At present the following four mobile satellite operators are providing satellite constellations and ground networks for SAT service:

- Inmarsat Geostationary Earth Orbit (GEO) satellite constellation and ground network are providing near-global coverage up to 75° North and South.
- Iridium Big Low Earth Orbit (LEO) satellite constellations and ground network are providing full global coverage because of inter-satellite links.
- Globalstar Big LEO satellite constellations and ground network are providing limited coverage depending on distributed number of available GES or Gateways.
- Orbcomm Little LEO satellite constellations and ground network are providing limited coverage depending on distributed number of GES or Gateways [2–6].

Inmarsat GST Equipment and Data Network

Inmarsat was established as a not-for-profit company in 1979 as the International Maritime Satellite Organization (Inmarsat) initially for development maritime satellite communications. It began trading in 1982 via GEO satellite constellation for almost global coverage, but is not covering both poles. Afterwards Inmarsat started with development service for land (road and rail) and aeronautical applications. Today Inmarsat is transformed in private operating company providing duplex satellite communication at the following radio frequency (RF) bands: at 1.6/1.5 GHz of L-band (Service Link) and at 6.4/3.6 GHz of C-band (Feeder Link).

The former-Inmarsat D+ transceiver with successor IsatData and IsatM2M is developed on basis of Inmarsat-C standard, which is the best solution for GST via satellites including AIS. It is able to transmit and receive GST data anywhere via Inmarsat satellite constellation and is ideal for determination, asset tracking and security in navigation, fleet management, and SCADA applications. In additional, the GST satellite transceiver is low powered by onboard ships and batteries' power supply with possibility to work even if ships are grounded somewhere without main power. In fact, this unit is an integration of GPS Rx and Satellite Transceiver with both antennas.

Features of these units are two-way messaging up to 25-byte message data size from terminal and up to 100-byte message size to terminal. Fast message delivery is in 1 min to terminal and rapid response in 10 s from terminal. This unit can be integrated with GPS Rx providing speed and position data. With external additional sensors, it monitors consumption of fuel, mileage, temperatures, etc.

Recently Inmarsat has developed IsatM2M and IsatData as two-way Short Burst Messaging (SBM) service that enables a wide range of SAT and SCADA (M2M) solutions for tracking and monitoring fixed or mobile assets on a global basis, whether at sea, on the land, or in the air. The new-generation IsatM2M and IsatData Pro satellite telematics is based on Inmarsat D+ standard, offering faster data forwarding rates, quicker responses to polling requests, and shorter time to first transmission. Inmarsat offers two models of unpackaged satellite SAT of SkyWave (today Orbcomm) producer:

1. **Inmarsat-D+ DMR-800L Terminal** – This Inmarsat-D+ satellite transceiver is dedicated for many SAT and GST solutions via both GPS and GLONASS navigation signals, which is depicted in Fig. 9.3 (left). This device provides a flexible, unpackaged assembly of satellite transceiver integrated with GPS or GLONASS receiver, which uses in-unit or separate satellite/GNSS antenna and discrete input and output feeds. In addition, it contains built-in processor/controller board which allows the unit to work as a simple modem and to interface a set of sensors and actuators. This mobile terminal is able to support GST powered by ship power supply or can be an alternative easily packaged with long-life batteries to provide satellite communication service when ship does not have own power supply.
2. **Inmarsat IsatData Pro IDP 600** – This satellite modern IDP 600 series terminal is fully programmable and environmentally sealed that uses two-way GEO Inmarsat IsatData Pro satellite service to provide visibility and communications with people and fixed or mobile equipment even in the world's harshest environments, which is shown in Fig. 9.3 (right). In particular, IsatData Pro is the fastest low-data rate satellite communications equipment especially suitable for vessel tracking and management, such as to enhance maritime safety and to simplify Guide Rescue Operations; to provide Vessel Monitoring System (VMS); to reduce vessel fuel costs and monitor engine performance; and to monitor vessel performance and reduce paperwork. This mobile tracking device, as stated above, is integration of Inmarsat satellite transceiver and GPS or GLONASS (or both) Rx. In fact, if this device is implemented as GST equipment onboard oceangoing ships and connected to special GST network will be able to provide satellite tracking and detection of missed or hijacked ships by pirates, and what is very important as well as to provide enhanced service for collision avoidance [2, 3, 7–9].

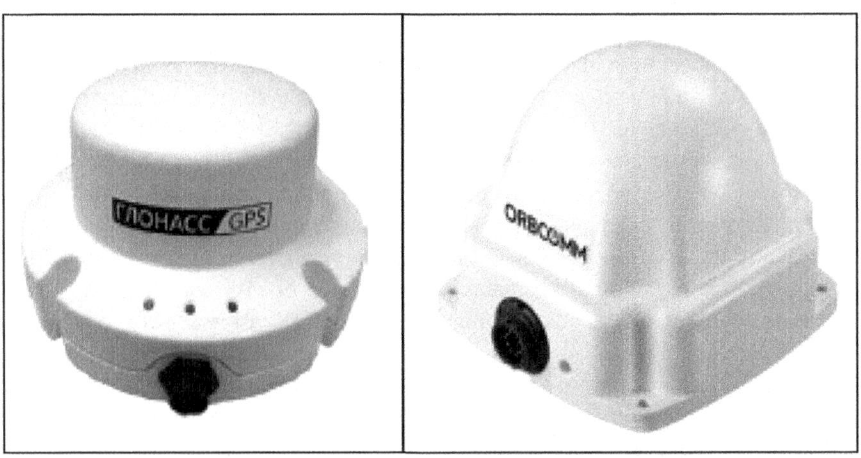

Fig. 9.3 Two generations of Inmarsat-D+ and IsatData Pro Systems [3]

Iridium GST Equipment and Data Network

The Iridium is situated in a near-polar Low Earth Orbit (LEO) at an altitude of 780 km. The Iridium satellites is situated in a near-polar Low Earth Orbit (LEO) at an altitude of 780 km, so they circle the Earth once every 100 min travelling at a rate of about 26,856 km/h. Thus, each satellite is cross-linked (inter-satellite service) to four other satellites, namely, two satellites in the same orbital plane and two in an adjacent plane.

The Iridium Big LEO satellite constellation consists of 66 operational satellites and 14 spares orbiting in a constellation of 6 polar planes, providing real global coverage including both poles and roaming via 48 spot overlapping beams, and the diameter of each spot is about 600 km.

The Iridium as a real global mobile satellite operator provides two-way voice and data communication service including SAT for all mobile applications via RF links at 1621.35–1626.5 MHz, feeder links at 29.129.3 GHz of Ka-band (uplink) and at 19.4-19.6 GHz of K-band (downlink), and satellite cross-link or inter-satellite link at 23.1823.38 GHz of Ka-band. Among the rest, the Iridium satellite network is providing SAT and communications service for all fixed and mobile applications including maritime with the following satellite tracking devices:

1. **Quake Q4000i Terminal** – The Quake Q4000i Terminal is a small enough to fit in hand produced by the US Company Quake. It is a two-way rugged industrial grade modem that can combines dual-mode operability over multiple satellite constellations and GSM terrestrial networks with GPS into a versatile, all-in-one mobile and remote asset tracking solution. The same SAT unit can be optionally supplied for service over Inmarsat, Globalstar and Orbcomm integrated with GPS Rx and with optional GSM cellular service. Technically this is a Short Burst Data (SND) transceiver designed for use as basic unit for many trackers

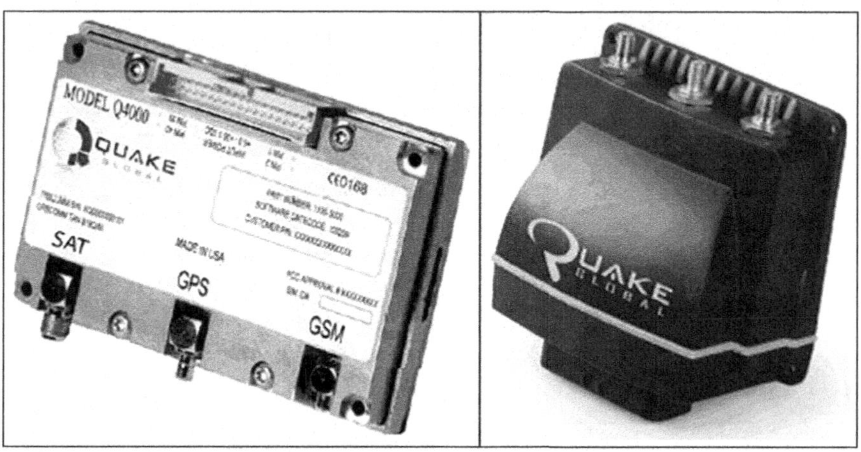

Fig. 9.4 Iridium satellite trackers [4]

using the Iridium Network, which is depicted in Fig. 9.4 (left). This unit can be used for ocean ships and container tracking and for land vehicles and aircraft tracking.

2. **Quake Q-Pro Multipurpose Tracker** – This Iridium transceiver is a small (119.2 × 119.4 × 57.6 mm and 390.6 grams) integrated Iridium, Globalstar, Orbcomm and GSM satellite modem with GPS receiver, which is illustrated in Fig. 9.4 (right). It can be used for GST including for containers, trucks, trains, and aircraft tracking and monitoring [2, 4, 7, 9, 10].

Globalstar GST Equipment and Data Network

The US Loral Space and Communications with Qualcomm Incorporation company developed the concept of Globalstar system at a similar time to Iridium. Globalstar gained an operating license from the USA FCC in November 1996.

The first launch of four Globalstar satellites occurred in May 1998 building space segment of 48 Big LEO spacecraft. The system uses Code (CDMA) and Frequency Division Multiple Access (FDMA) methods with an efficient power control technique and multiple beam active phased array antennas for multiple access and frequency reuse of voice and data transmission.

Globalstar is not providing inter-satellite links and therefore needs a number of GES terminals worldwide. Otherwise, this system started to provide coverage for South Africa in 2015. Globalstar is providing service for users via satellite at 1.610–1.621 GHz (uplink) and at 2.483–2.500 GHz (downlink) and from satellite to GES at 5.091–7.055 GHz (feeder link).

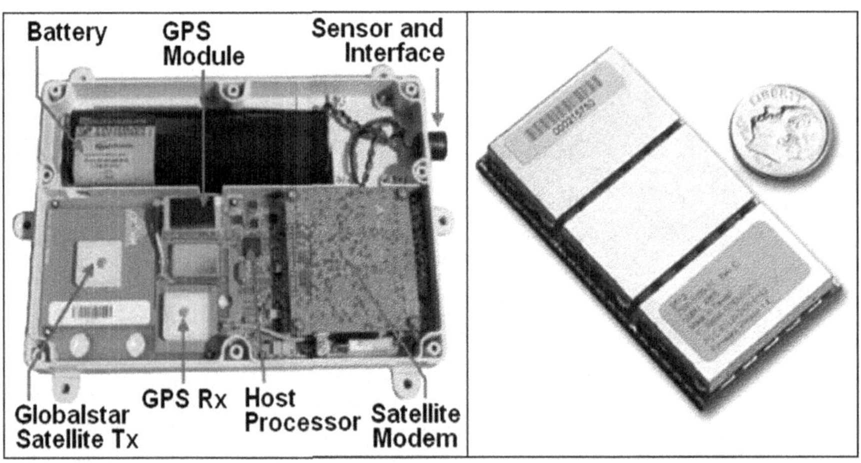

Fig. 9.5 Configuration of Globalstar GST equipment [5]

The Globalstar satellite configuration for data communication, illustrated in Fig. 9.5 (left), is using Big LEO Globalstar satellite network for simple (one-way) data transmissions. The main parts of SAT device are GPS Rx for receiving of GPS tracking data and Satellite Tx for sending PVT data to the TCC via GEO or non-GEO satellites. This device may be powered by onboard power supply or via own long-term batteries. In Fig. 9.5 (right) is depicted another sample SAT of former-Axonn satellite terminal of very small sizes [1, 5, 7, 10].

Orbcomm GAT Equipment and Data Network

The Little LEO Orbcomm satellite communication system is a wide area packet switched and two-way data network providing satellite communication, tracking and monitoring services globally for fixed and mobile assets via 36 LEO satellites. Since 2012 Orbcomm is covering Southern Africa offering messaging services via small GPS/Orbcomm satellite trackers even for GST on VHF-band at 148.0–150.05 MHz (service/feeder uplink) and at 137.0–138.0 MHz (service/feeder downlink).

The Orbcomm operator is also developing system that will provide Satellite Automatic Identification System (S-AIS) for broadcast and ship's identification, position, and other critical data for improving safety and security at sea. The Orbcomm network is providing GST and other container and mobile employing the following units:

1. **Orbcomm OG2-GPS Modem** – This SAT unit delivers connectivity over the LEO Orbcomm VHF satellite network for marine, heavy equipment, transportation, agricultural, and other markets, which is depicted in Fig. 9.6 (left).
2. **Orbcomm GT 1100 Modem** – This small satellite data unit powered by solar rechargeable batteries enables full control of mobile assets and containers, which is shown in Fig. 9.6 (right) [6, 7, 10, 11].

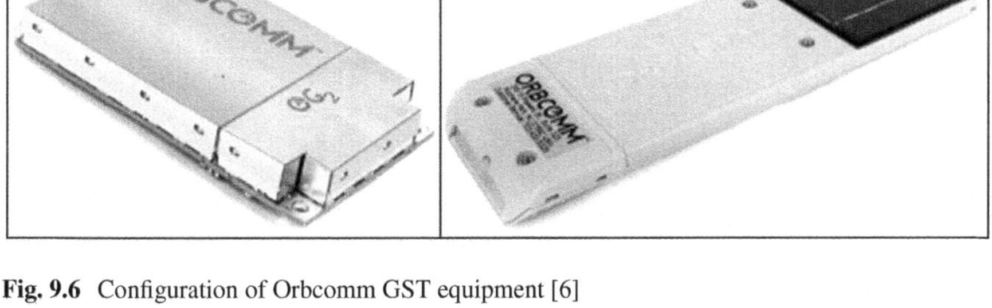

Fig. 9.6 Configuration of Orbcomm GST equipment [6]

9.3 Global Ships Tracking (GST) Network

The LRIT satellite transceiver system is compulsory satellite communication and tracking equipment on board ships established by IMO for vessel tracking worldwide. This system consists of the shipborne data/information transmitting satellite equipment (similar to the above-stated satellite tracking devices); the Communication Service Provider(s); the Application Service Provider(s); the LRIT Data Centre(s), including any related Vessel Monitoring System(s); the LRIT Data Distribution Plan; and the International LRIT Data Exchange.

As stated earlier, disadvantages of LRIT system are that it cannot transmit navigation data of adjacent ships on request of any ship sailing in certain sea area for collision avoidance and that LRIT is not able to provide tracking of missing or hijacked ships.

On the other hand, the proposed GST as integration of satellite communication and GNSS positioning system is able to provide all service as LRIT including to provide tracking of missing and hijacking ships and to determine positions of all ships in vicinity of ship requesting this data for collision avoidance as the best for vessel tracking worldwide.

In Fig. 9.7 is introduced new and simplest concept of GST for worldwide vessels tracking and monitoring. In fact, using satellite links of GEO or non-GEO satellites, any ship is able to send automatically its PVT data, provided by GPS or GLONASS satellites, via Gateway (GES) and Internet to the Tracking Control Station (TCS) and Ships Operations. In opposite direction Control Centre can provide pulling navigation data from any ship, and what LRIT cannot do, Control Centre can send to any ship on his demand position data of all ships in his vicinity for collision avoidance and enhanced safety and security at sea [1, 7, 12, 13].

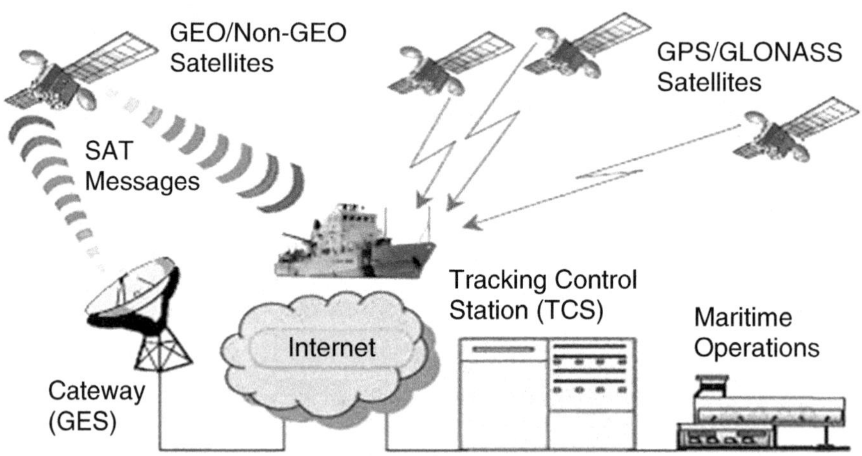

Fig. 9.7 Global Ship Tracking (GST) network [7]

9.4 Satellite Data Link (SDL) Network

The GEO and non-GEO (LEO) satellite constellations can provide Satellite Data Link (SDL) or Satellite AIS (S-AIS) for onboard ships broadcast solution that transmits a ship's identification, position, and other critical data that can be used to assist in ships' navigation and tracking facilities for improvement maritime safety and security. **The SDL system can provide transmission of** Short Burst Messages (SBM) between mobile stations or terminals with GES, Control Centre, and Maritime Operation, whose network is illustrated in Fig. 9.8.

In mobiles, such as ships and surface vehicles in seaport, can be installed special satellite transponders or already stated satellite tracker devices. Mobile transponders can operate autonomously inside the coverage of certain Gateway (GES). The SDL transponder can support the similar services that provide Radio VDL4, but if is using Iridium transponder will be able to provide global coverage including both North and South Poles.

The SDL transponder allows captains on board ships and maritime traffic controllers to receive all vessels traffic data in ocean or coastal navigation, approaching to seaport, and inside of harbors including vehicle movements with the highest possible precision. The receiving SDL units can receive all SDL messages and process them in sophisticated processor. Thus, the receiving SDL messages can be transfer and monitor on special display looks like radar screen. In the same way will be processed and monitored GST data in TCS. The SDL system may drastically improve safety and security at sea and in seaport area as well.

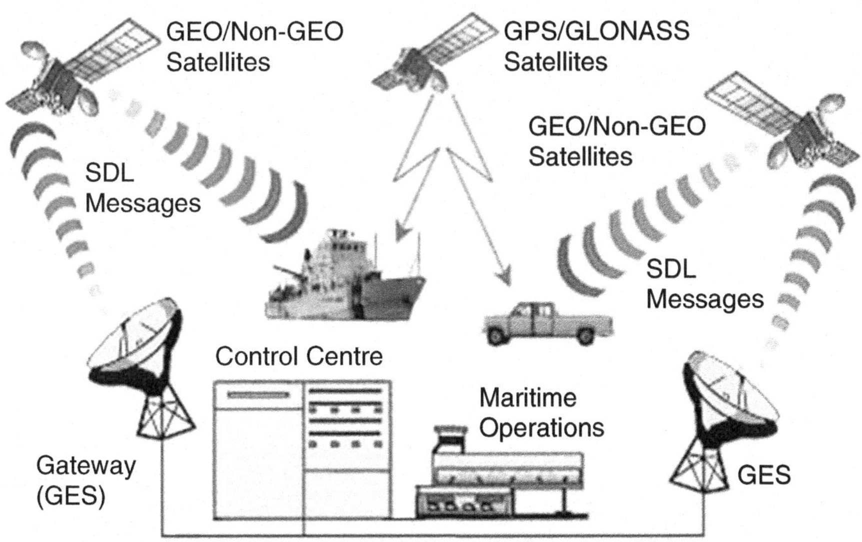

Fig. 9.8 Satellite Data Link (SDL) network [7]

The Gateway (GES) terminals can cover certain satellite systems, such as Inmarsat, Iridium, Globalstar, or Orbcomm networks worldwide and easily interface with other space surveillance systems through the standardized Asterix protocol, which will enable a complete tracking surveillance picture at the seaport derived from several sources. The GES terminal and a ground network will provide increased functionality and capability for wide area coverage of advanced STC Monitoring and Management.

The functionality of the GES terminal is tailored to the specific service applications by its software configuration. Therefore, in Fig. 9.8 is presented that ships and vehicle SDL terminals receive GPS or GLONASS positioning (PVT) signals and automatically send this data via GEO/non-GEO satellites and GES to Control Centre and Ships Operations. In vice versa direction, Control Centre can send instructions to ships on how they can move more safely at sea and seaports and to vehicles in seaports only. Therefore, to get SDL service working, each ship and ground vehicles have to be equipped with SDL transponders or satellite communication devices and in such a way will be able to send and receive SBD or High Speed Data (HSD) for CNS and collision avoidance purposes [1, 7, 13, 14].

9.5 GNSS Augmentation SDL (GASDL) Network

The Regional Satellite Augmentation System (RSAS) of the US GPS or Russian GLONASS, both integration part of GNSS-1 infrastructure, is a combination of ground and space equipment dedicated to provide augmentation of standard GPS or GLONASS signals, whose infrastructure is illustrated in Fig. 9.9.

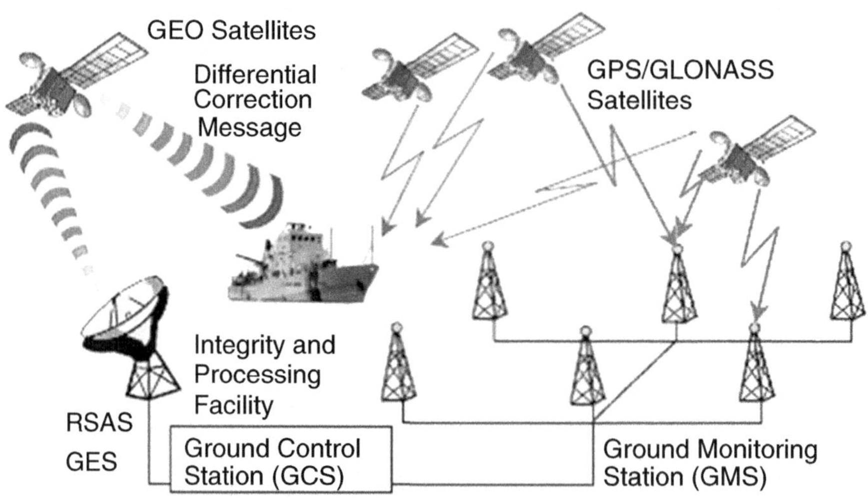

Fig. 9.9 GNSS augmentation SDL [7]

The functions being provided by RSAS are:

- Differential corrections are determined to improve GNSS signal accuracy.
- Integrity monitoring is predisposed to ensure that errors are within tolerable limits with a very high probability and thus ensure safety.
- Ranging is proposed to improve availability.

The numbers of Reference Stations (GMS) are receiving not augmented signals of GPS or GLONASS satellites, processing and forwarding this data to Master Station (GCS). The GCS terminals provide processing of GNSS data to determine the differential corrections and bounds on the residual errors for each monitored satellite and for each area. Therefore, GCS terminal is providing determination of the clock, ephemeris, and ionospheric errors (ionospheric corrections are broadcast for selected area) affected during propagation. The corrections and integrity information from the GCS terminal are then sent to each RSAS GES and uplinked to the GEO satellites. Thus, these separate differential corrections are broadcast by RSAS GES through GEO satellite data link via GNSS transponder at the same frequency used by not augmented GPS receiver.

For instance, augmented GPS Rx is receiving augmented signals of GPS satellite and determining more accurate position of ships. Not augmented GPS Rx can also receive augmented signals if it is provided an adequate software or hardware. The most important stage in this network is to provide technical solution that augmented position of aircraft can be sent automatically via SDL or voice to GES and Control Centre. Finally, these positioning signals can be processed by special processor and displayed on look like radar display, which traffic controller is using for STC and management for enhanced ship traffic control and improved collision avoidance in certain monitoring sea area [7, 12, 14].

9.6 Conclusion

In this paper were described and shown some quite spread, embedded, and sustain systems for maritime satellite tracking applications very important for oceangoing ships', crew's, and passengers' safety and security in all phase of ocean, coastal, and passage navigation.

At this point, every ship operator can deploy any satellite Ship Earth Stations (SES) and ships tracking equipment according to the IMO recommendations and Safety of Lives at Sea (SOLAS) regulations and cost-effective sense. In that manner, the major point of GST network is to find out the best solutions for more reliable global maritime satellite communications, tracking, determination, monitoring, and enhanced collision avoidance system with priorities of ships' safety and security.

Today, Inmarsat GEO satellite operator for mobile satellite communication is only professional system providing near global coverage up to 75° North and South. In any case, with regard to available ocean area coverage, this system and equipment can be used for any types of oceangoing ships in any stage of navigation.

Presently, oceangoing ships are not sailing in Arctic Ocean, but Russian government is proposing these routes, where can be used HF communication systems instead. However, Iridium satellite operator as not professional system as Inmarsat is providing full global coverage, thanks to inter-satellite links; however, Globalstar and Orbcomm LEO have limited coverages and cannot cover polar areas.

Therefore, the future of maritime and other mobile satellite communications has to be combination of GEO, LEO, and other orbits, like Medium Earth Orbit (MEO) and High Elliptical Orbit (HEO) in so-called Hybrid Satellite Orbits (HSO), which can provide a professional service globally and over Arctic Ocean.

References

1. D.S. Ilcev, *Global Mobile Satellite Communications for Maritime, Land and Aeronautical Applications* (Springer, Boston, 2005)
2. D.S. Ilcev, *Global Mobile Communications, Navigation and Surveillance (CNS)* (DUT, Durban, 2012)
3. Inmarsat, *Inmarsat GEO Satellite Equipment and Network*, London, (2018).
4. Iridium, *Iridium Big LEO Satellite Equipment and Network*, McLean, VA, (2018).
5. Globalstar, *Globalstar Big LEO Satellite Equipment and Network*, Covington, LA, (2019).
6. Orbcomm, *Orbcomm Little LEO Satellite Equipment and Network*, Rochelle Park, NJ, (2019).
7. D.S. Ilcev, *Global Mobile Satellite Communications for Maritime, Land and Aeronautical Applications*, vol 1 & 2 (Springer, Boston, 2016)
8. L.Y. Kantor, et al., Sputnikovaya svyaz i problema geostacionarnoy orbiti, Radio i svyaz, Moskva, (1988).
9. D.S. Ilcev, *Global Ship Tracking, Automatic Identification System (AIS) and Determination*, Presentation, Space Centre (SSC) and Postgraduate Studies, DUT, (2011).
10. J. Bendisch, H. Hoffmann, T. Wieneke, H.G. Niemöller, *Aeromission – An Airborne System for Maritime Surveillance* (Aerodata AG, Braunschweig, 2005)
11. D.S. Ilcev, *Satellite CNS for Maritime Transportation Augmentation System (MTAS)*, CriMiCo Conference, IEEE CFP09788, Sevastopol, Russia, (2009).
12. B. Wellenhof, *Global Navigation Satellite Systems (GNSS)* (Springer, Wien, 2008)
13. G. Maral et al., *Satellite Communications Systems* (Wiley, Chichester, 2009)
14. D.E. Kaplan, *Understanding GPS Principles and Applications* (Artech House, Boston, 2006)

Part II
Legal Issues

Chapter 10
A Legal Analysis of the Collision and Strict Liability Framework for the Shipowner of Unmanned and Autonomous Vessels (South Africa)

Ivana Surian

10.1 Introduction

Technology is changing the world incredibly fast, and the digital age is now upon us. The introduction of artificial intelligence and complex algorithms is altering the way people live, work and interact with each other daily. The occurrence of the covid-19 pandemic has further shown people how easy and effective technology can make our lives, thereby pushing the world even closer towards the fourth industrial revolution.

Considering how far technology has come, it brings no surprise that the world is faced with the fact that unmanned and autonomous vessels may be navigating on the high seas in the near future. The reality is that these forms of vessels already exist and, although not yet widely in service, debates surrounding the issues and challenges that they bring with them have been ongoing since as far back as 2012. The most groundbreaking projects arising from these technological developments are the ones currently in the process of developing unmanned and autonomous cargo-carrying vessels.[1]

[1] The most prominent projects aimed at unmanned and autonomous vessel technology developments include Maritime Unmanned Navigation through Intelligence in Networks (MUNIN), the MV Yara Birkeland (by Yara International and Kongsberg) and the Jin Dou Yun 0 Hao (China's first autonomous cargo ship).

I. Surian (✉)
University of KwaZulu-Natal, Durban, South Africa

MACS Maritime Carrier Shipping (Pty) Ltd, Cape Town, South Africa
e-mail: ivana.surian@macship.com

S. Bauk, S. D. Ilčev (eds.), *The 1st International Conference on Maritime Education and Development*, https://doi.org/10.1007/978-3-030-64088-0_10

In order to understand and analyse the legal issues associated with these forms of vessels, it is first important to understand the definitions and distinction between an unmanned and an autonomous vessel.

10.2 Definitions

What Are Unmanned and Autonomous Vessels?

Unmanned vessels have been defined as 'vessels without crew on board, but which are controlled remotely from the shore' [1]. They are vessels that are 'able to navigate from point A to point B, without requiring the support from a crew aboard the ship' [2]. This broad definition indicates that these types of vessels still require some degree of human intervention and control in order to function properly. On the other hand, autonomous vessels have been defined as 'pre-programmed vessels that operate using algorithms' [1]. This indicates that these vessels have no human intervention whatsoever and contain no degree of control over them. Some tests currently being conducted in Norway involve cargo-carrying vessels that have the ability to alter the degree of autonomy with which they operate [3]. Thus, in order to not exclude any particular vessel of this nature, this paper will include both types of vessels and will use these terms interchangeably unless specified otherwise for a particular context or discussion. Only unmanned and autonomous merchant ships will be examined.

Internationally Recognised Definition

The International Maritime Organization (IMO) has acknowledged that the introduction of these forms of vessels is revolutionary for the maritime industry and has thus commenced investigations into the safety, security and environmental feasibility of these new forms of vessels. The IMO refers to the unmanned and autonomous vessel as MASS (Maritime Autonomous Surface Ships) and has defined them as 'ship[s] which, to a varying degree, can operate independently of human interaction' [4, 5].

Will Unmanned and Autonomous Vessels Be Considered as 'Ships' Under South African Law?

Under South African law, the Merchant Shipping Act 57 of 1951 (MSA) defines a 'ship' as: '… any kind of vessel used in navigation by water, however propelled or moved, and includes— (*a*) a barge, lighter or other floating vessel; (*b*) a structure that is able to float or be floated and is able to move or be moved as an entity from one place to another; and (*c*) a dynamically supported craft; and "vessel" has a corresponding meaning'.[2]

The South African Admiralty Jurisdiction Regulation Act 105 of 1983 (AJRA) defines a 'ship' as: '... any vessel used or capable of being used on the sea or internal waters, and includes any hovercraft, power boat, yacht, fishing boat, submarine vessel, barge, crane barge, floating crane, floating dock, oil or other floating rig, floating mooring installation or similar floating installation, whether self-propelled or not'.[3]

These definitions are significantly broad and allow for a large variety of vessel classifications to fall under their application and operation. Thus, it can be concluded that unmanned and autonomous vessels (including remote-controlled vessels) will also be considered as ships under South African law [6].

10.3 Legal Issues Relating to Unmanned and Autonomous Vessel Regulation

The main concern with the unmanned and autonomous vessel is how legal liability will be attributed to it should there be a defect or malfunction with its software which then causes a collision. In other words, since there are no longer a crew and master on board, the question that arises is whether liability can be solely attributed to the shipowner of the unmanned vessel and autonomous vessel.

10.4 Liability for the Negligent Operation of Unmanned and Autonomous Vessels

The law demands that legal liability be placed on a specific actor who bears the responsibility for when things go wrong [7]. Since an unmanned and autonomous vessel has no master or crew on board, this raises the issue of whether the negligence of any other party involved in their operation may then be attributed to the vessel for the purposes of the *in rem* action. There are currently two main actors who are responsible for and will be held liable for the damages caused by the collision of a manned vessel. These two traditional actors are the master and the shipowner. With the introduction of unmanned vessels to the maritime industry, however, we find ourselves faced with new legal actors who may be found liable for the same damages. These new actors are the shore-based control operators, the voyage programmers, the manufacturers and the software providers.

[2] Section 2 of the South African Merchant Shipping Act 57 of 1951 (MSA)

[3] Section 1 of the Admiralty Jurisdiction Regulation Act 105 of 1983 (AJRA)

Shipowner's Liability

The shipowner is generally the party that is held liable for collision damage and is the one who is indirectly brought before the court to defend an action against his/her vessel(s). It has been suggested that the law as it stands regarding a shipowner's liability should be left as it is, as the shipowner will be considered liable no matter how the vessel is operated [7]. The question that arises here, however, is whether the shipowner could still be held liable for collision damage where the unmanned vessel was operated and navigated independently from a SCC and by an independent SBCO or voyage programmer.

Liability for collision damage is currently determined using a fault-based liability system, in which a test of negligence is conducted. In general, maritime laws nationally and internationally have accepted this to be the correct method for determining liability. To determine collision liability on the current fault-based liability system means that there must be an enquiry into the conduct of the shipowner relating to the management of the vessel as well as an enquiry into the conduct of the shipowner's 'servants' (being the master and crew) in the navigation and operation of the vessel. However, where the shipowner chooses to conduct his/her business using an unmanned or an autonomous vessel, the enquiry will no longer involve an investigation into the conduct of the master and crew as they will no longer be present on board the vessel.

Negligence, Strict Liability and Vicarious Liability Under South African law, liability for negligence is based on the principle that the law disapproves of the defendant's conduct in his/her actions towards causing the harm [8]. The enquiry, therefore, involves an evaluation of the defendant's conduct compared against a standard of conduct that is socially acceptable by the public. Where the defendant's conduct falls outside this socially acceptable standard, the defendant will be considered negligent and will be held liable for causing the resultant harm [9]. Thus, the defendant's conduct is tested against an objective standard of the reasonable person (*bonus paterfamilias*) which is based upon the principle that: 'a person is blamed for an attitude or conduct of carelessness, thoughtlessness or imprudence because, by giving insufficient attention to his actions, he failed to adhere to the standard of care legally required of him' [8].

On the other hand, strict liability is a form of liability with no fault. It expresses the viewpoint that a society must hold a person liable for his or her conduct where he or she has chosen to act in a certain way, or where there is a risk associated with his or her action that then causes harm to another. Loubser and Midgely et al. describe the nature of strict liability (liability without fault) as follows: '…society determines that the nature of the conduct, or the risk associated with the conduct is such that the responsible person or entity should compensate anyone who suffers harm as a result of the conduct.' [9]

There are various general characteristics that make up the essence of liability without fault. Neethling and Potgieter et al. set them out as follows: '*(a)* Fault is not required for liability in claims for compensation; *(b) Vis major* (act of God) and fault on the part of the prejudiced person are generally recognised as defences; *(c)* Strict liability is usually imposed – either by legislation or judicial pronouncement – in cases involving activities which as a rule create extraordinary increases in the risk of harm to the community; *(d)* In instances where strict liability has been imposed by legislation, the extent of the liability is usually curtailed by fixing maximum amounts of compensation; and *(e)* Liability without fault is restricted in most cases to damage to life, limb and property (and therefore does not include pure economic loss).

In Continental systems, liability without fault originated primarily from legislation, while in Anglo-American law, case law played the dominant role. In South Africa, both the legislature and the courts have contributed to the development of liability without fault' [8].

There are two theories behind liability without fault. The first is the interest/profit theory. This theory states that 'where a person acts in his *own interest*, and causes harm to another, he bears the burdens and disadvantages which his activities bring about' [8]. Thus, someone who is acting in his/her own interests, and benefits as a result of that activity, should then bear the responsibility for any harm that results from that activity.[4] The second theory is the risk/danger theory. This theory states that 'where a person's activities create a considerable increase in the risk or danger of causing harm, ie, an increased potential for harm, there is sufficient justification for holding him liable for damage even in the absence of fault' [8].

Since the core principle behind delict law in South Africa is that there can be no liability where there is no fault, [8, 9] the question that arises, then, is whether there are any restrictions on imposing strict liability in South Africa. The answer is that there are many areas of South African law that already impose strict liability on individuals who undertake an action with an associated risk that could potentially have a harmful outcome on others. The most obvious example is found under labour law in the relationship between employers and employees.[5] An employer can be held vicariously liable for an employee's actions. Loubser and Midgely et al. state that vicarious liability can be seen as a form of strict liability: 'This is where the employer is held liable without fault for the wrongdoing of an employee, and the liability of the employee, determined according to the normal principles of delict (including fault), is transferred to the employer' [9].

This form of liability is already used under international maritime laws where the shipowner can be held liable for the wrongful conduct of his/her 'servants', being the master and crew on board the shipowner's vessel. The issue that arises here is whether the SBCO of the unmanned vessel (or the voyage programmer of the autonomous vessel) will be considered as 'servants' of the shipowner for the purposes of determining collision liability. This further leads to the question of whether the

[4] This theory has been criticised by Neethling et al. [8].

[5] Other examples of strict liability in South Africa can be found under the Aviation Act 74 of 1962; the Post and Telecommunication-related Matters Act 44 of 1958; and the Genetically Modified Organisms Act 15 of 1997 [8, 9].

current fault-based liability regime will continue to be the sufficient test to use when determining collision liability, or whether a new system should be adopted (such as the strict liability test). These two aspects will be discussed hereunder.

Employees Versus Independent Contractors Employees and independent contractors have been distinguished by South African courts by the type of contract of service that they hold. The main distinguishing feature is that: 'an employee is under the control of the employer in respect of the nature of the work and the manner in which it is to be done, whereas an independent contractor is not subject to the control of the person paying for the services' [9].[6]

An employee's contract of employment usually terminates upon death or when the agreed period has expired, while an independent contractor is contracted to perform certain specified work which may not have a specified deadline or time limit. The courts will, however, take into consideration all the relevant factors under the circumstances in order to determine whether the person is an employee or an independent contractor. Some of the factors taken into consideration are the nature of the work; the manner of performing that work; the manner of payment; the state of the social and economic interdependence of the person; the authority to give instructions; whether any membership of medical or pension fund is involved; whether there is any provision for paid vacation; the number of working hours required; the use of the employer's premises and equipment; and to what extent the person performing the work forms an essential part of the business organisation [9].[7]

The Shipowner's Liability for the SBCO and Voyage Programmer In applying these general principles of fault liability to a Maritime Autonomous Surface Ship (MASS) collision, it is clear that each case will need to be decided on the basis of its own facts. This would entail at least two potentially complex factual enquiries.

Firstly, should it be decided that the SBCO, or voyage programmer, is in fact an employee of the shipowner, the shipowner could then be held vicariously liable for the damage caused by a collision in which the SBCO negligently operated the unmanned vessel, or the voyage programmer negligently input erroneous information into the autonomous vessel's system.

On the other hand, should it be decided, instead, that either one of or both the SBCO and voyage programmer are not employees of the shipowner, but are rather independent contractors, the question arises as to whether a similar conclusion can be reached (but which requires an additional step in the analysis) under South African law as: '…a person is liable for the acts of an independent contractor only in respect of operations where there is a 'non-delegable' duty, in other words, where

[6] *Smit v Workmen's Compensation Commissioner* 1979 (1) SA 51 (A)

[7] See the test used in the cases of *Midway Two Engineering & Construction Services v Transnet Bpk* 1998 (3) SA 17 (SCA); and *Stein v Rising Tide Productions CC* 2002 (5) SA 199 (C).

engaging a contractor does not absolve the employer from a duty not to harm third parties, such as when the operations involve an abnormal level of danger' [9].[8]

South African courts have expressed that this enquiry is not a question of vicarious liability but rather a question of whether the employer had personal liability[9] on the basis that he/she had the duty to take reasonable precautions to ensure that no danger ensued from undertaking the dangerous activity. In the case of *Langley Fox Building Partnership (Pty) Ltd v De Valence*,[10] the court states as follows: '... if work entrusted to an independent contractor is of such a character that, if the contractor does the work and no more, danger will ensue, then liability for damages remains with the employer on the failure of his contractor to take precautions in addition to doing the work. It is the duty of the employer to take such precautions as a reasonable person would take in the circumstances'.[11]

On the other hand, where an employer undertakes the services of a skilled independent contractor, 'where the extent of the danger and the reasonably practicable measures to minimise it can only be determined by such skilled person',[12] the employer can be discharged from liability as he/she will be considered to have taken all reasonable measures to eliminate or minimise the potential dangers associated with that activity. This is an example of where the South African courts have recognised that there may be situations where it would be reasonable to rely solely on the expertise of an independent skilled contractor. The court states this as follows: 'In my opinion, therefore, the duty to take care where the work undertaken is *per se* dangerous could in some cases be discharged by delegating its performance to an expert. In my judgment, the correct approach to the liability of an employer for the negligence of an independent contractor is to apply the fundamental rule of our law that obliges a person to exercise that degree of care which the circumstances demand'.[13]

There are earlier decisions setting out the position that there is a non-delegable duty on an employer who appoints an independent contractor for an inherently dangerous activity to ensure that the proper precautions are taken. However, the SCA has made it clear that this is not an invariable rule as in certain circumstances it may be reasonable to rely solely on the skill of the independent contractor. In the latter circumstance, the employer would not be liable even if the independent contractor had acted negligently.

In any case, South African courts are required to consider the 'extent of the danger, the degree of expertise available to the employer and the independent contractor

[8] *Langley Fox Building Partnership (Pty) Ltd v De Valence* 1991 (1) SA 1 (A). See also *Chartaprops 16 (Pty) Ltd v Silberman* 2009 (1) SA 265 (SCA) at 272A; and *Pienaar v Brown* 2010 (6) SA 365 (SCA).

[9] *Chartaprops* ibid at 278E-F, para 29

[10] *Langley Fox* supra note 8

[11] *Langley Fox* supra note 8 at 11B

[12] *Langley Fox* supra note 8 at 11C. See generally the full discussion at 10A–13C.

[13] *Langley Fox* supra note 8 at 11D–E

respectively, and the reasonably practicable means available to the employer to avert the danger' [9].[14]

10.5 An Argument in Favour of the Strict Liability Approach

Neethling and Potgieter et al. state that the traditional basis in the law of delict is that of the fault theory, which stipulates that there can be no liability without fault. However, due to the industrial and technological revolutions from the eighteenth to the twentieth centuries, a new approach was developed to accommodate appropriately the new challenges that machinery and technology brought with them. This new approach is the liability without fault approach. Thus, it can be seen that liability without fault was a reaction to the technological developments of this new era, which is evidence in itself that liability without fault (also referred to as strict liability) has become the better method in determining liability for technological malfunctions [8].

In applying the general characteristics of strict liability (discussed above) to the unmanned and autonomous vessel scenarios, the following can be seen: Firstly, fault will not be a requirement when determining liability in claims for compensation for unmanned/autonomous collision damages; acts of God (*vis major*) and fault on the part of the prejudiced person will be recognised as defences; strict liability can be imposed through international conventions (governing collisions with unmanned and autonomous vessels) in cases involving activities which as a rule create extraordinary increases in the risk of harm to the community (such as the use and operation of an unmanned or autonomous vessel); in instances where strict liability has been imposed by legislation, the extent of the liability can be curtailed by fixing maximum amounts of compensation;[15] and the liability without fault approach will be restricted in most cases to damage to life, limb and property (all of which the operation and navigation of an unmanned and autonomous vessel may involve).

Secondly, the two theories justifying the use of strict liability will apply to the unmanned and autonomous vessel scenario as a shipowner will be both investing a personal interest and benefiting from the profits thereof in conducting his/her business with autonomous technologies; and by choosing to conduct his/her business with an unmanned or autonomous vessel, the shipowner will be increasing the level of risk and danger associated with navigating a vessel at sea.

In order for this to apply, it needs to be expressly stated in international conventions that the operation of an unmanned and autonomous vessel should be

[14] *Langley Fox* supra note 8 at 13B

[15] This would entail the various national laws imposing a monetary limit on collision liability. Under an international maritime context, however, this would entail a consideration of the issue on whether there should be an amendment to the current convention on limitation of liability for maritime claims or whether a new separate limitation of liability convention dealing with unmanned and autonomous vessels should be developed.

considered as an operation that involves an abnormal/increased level of danger. It has been suggested that such an approach should be taken for these forms of vessels [7]. Zampella indicates that adopting this strict liability approach may be the direction to go towards for collision liability involving unmanned and autonomous technology: 'It might be considered as a solution to avoid the issues deriving from the investigation of fault and negligence regarding new kinds of activities closely dependent on technology. After all, if we consider the employment of an unmanned vessel as a type of activity inherently dangerous, this would justify the adoption of a strict liability regime, for a protection of the other users of the sea from the natural risks deriving from unmanned shipping …' [7].

This would not be the first time that South African law has recognised the implementation of strict liability for an activity that is considered as inherently dangerous. The South African Aviation Act (AA)[16] imposes strict liability on the owner of an aircraft where material damage has been caused by his/her aircraft. Section 11(2) states: 'Where material damage or loss is caused by an aircraft in flight, taking off or landing, or by any person in any such aircraft, or by an article falling from any such aircraft, to any person or property on land or water, damages may be recovered from the owner of the aircraft in respect of such damage or loss, *without proof of negligence or intention or other cause of action* as though such damage or loss had been caused by his wilful act, neglect or default' (emphasis added).

10.6 Conclusion

To devise a new strict liability system under either domestic or international maritime law would mean that liability will be linked directly to the actions of the SBCO or voyage programmer where their actions were the factual and legal cause of the collision. It would, however, not require a determination of their fault (negligence) or the shipowner's fault in relation to such actions. Using this system would also mean that the negligence of parties such as the manufacturer and software provider would be excluded from the enquiry into the liability of the shipowner.[17]

Since the shipowner has chosen to undertake the risk of conducting his/her business using an unmanned or autonomous vessel, where such an activity is considered as inherently dangerous, it can be justified that the law adopts a strict liability approach in holding the shipowner liable for any damage resulting from the use and operation of these new forms of vessels, such as where there is a system malfunction with the autonomous software/autonomous onboard systems, or even where there is negligence on the part of the shipowner's 'servant', 'agent', 'employee' or 'independent contractor'. Such a collision liability regime can appropriately

[16] Act 74 of 1962

[17] A strict liability system should not, however, exclude the possibility of the shipowner seeking indemnification from manufacturers and software providers where technology failure or errors in the programming are showing to have contributed to the collision.

accommodate the use of this new autonomous technology in the maritime industry and can ensure that the general use of such vessels on the high seas is regulated and provides equivalent safety to the operation of conventional vessels.

The strict liability system has not been welcomed by many maritime nations in the past.[18] It would not, however, be the first time that international maritime laws recognise the use of such a system when determining compensation for damages.[19] In any case, should this new method be accepted by the various maritime nations, there is still the issue of uniformity. The determination of collision liability for the unmanned and autonomous vessel may begin to differ from one country to the next, which will promote a divergence from one of the core principles of international maritime law, that of uniformity,[20] and this should be avoided across the international plane. Lastly, it must be borne in mind that introducing a strict liability approach will have additional implications on the shipowner's cost of insurance cover for civil liability. While there is a great need to implement an efficient system to determine collision liability, the stricter the legal regime is, the more it becomes possible to discourage the adoption of these vessels, making them commercially unviable.

References

1. Shipowner's Club, *Unmanned and Autonomous Vessels – The Legal Implications from a P&I Perspective*. Retrieved from https://www.shipownersclub.com/unmanned-autonomous-vessels-legal-implications-pi-perspective/. Last access 30 June 2020.
2. Pol Deketelaere, *The Legal Challenges of Unmanned Vessels*, Unpublished Master dissertation, Universiteit Gent, Belgium, 2016–2017, pp. 2.
3. C. Ellingsen, E. Tøndel, *Maritime Law in the Wake of Unmanned Vessels* (2017). Retrieved from https://svw.no/contentassets/f424f309bd304e99b39f11355e98571f/svw_maritime-law-in-the-wake-of-the-unmanned-vessel.pdf. Last access 30 June 2020.
4. IMO webpage, *IMO takes first steps to address autonomous ships*. Retrieved from http://www.imo.org/en/MediaCentre/PressBriefings/Pages/08-MSC-99-MASS-scoping.aspx. Last access 30 June 2020.

[18] This is evidenced by the reaction of the many maritime nations (under which the large carriers operated) to the United Nations Convention on the Carriage of Goods by Sea, 1978 ('the Hamburg Rules'). These particular maritime nations refused to ratify the convention for fear of the strict liability it imposed on the Carriers in the exercise of their duties and responsibilities relating to the carriage of goods by sea [10].

[19] One example is the International Convention on Civil Liability for Oil Pollution Damage, 1969 (CLC), which places a strict liability on the shipowner to compensate for oil pollution damage caused by his/her vessel [7].

[20] Stated by Justice McReynolds in 1916 in the US case of *Southern Pacific Co. v Jensen* 244 U.S. 205, 215, 216 (1916) as follows: 'The general maritime law as accepted by the federal courts constitutes part of our national law applicable to matters within the admiralty and maritime jurisdiction … no such [state] legislation is valid if it … works material prejudice to the characteristic features of the general maritime law or interferes with the proper harmony and uniformity of that law in its international and interstate relations' [11]

5. Luci Carey, Report on BIMCO Autonomous Ships Seminar (Report 19/01) NUS Centre for Maritime Law, 2019, pp. 6.
6. CMI Maritime Law for Unmanned Ships, Questionnaire (2017) *Completed by the Maritime Law Association of South Africa*. Retrieved from https://comitemaritime.org/work/unmanned-ships/. Last accessed 30 June 2020.
7. P. Zampella, *Maritime and air law facing unmanned vehicle technology*, Unpublished PhD thesis, Università degli Studi di Cagliari, Italy, 2019, pp. 17, 215, 216 – 217, 219.
8. J. Neethling, J.M. Potgieter, et al., *Law of Delict*, 7th edn. (LexisNexis, Durban, 2015) pp. 137, 379, 380–381, 401–402
9. M. Loubser, R. Midgley, et al., *The Law of Delict in South Africa*, 3rd edn. (Oxford University Press, Cape Town, 2018) pp. 154, 458, 469, 471
10. F. Reynolds, The Hague rules, the Hague-Visby rules, and the Hamburg rules. Maritime Law Assoc. Aust. New Zeal. J. **1990**, 30–33 (1990)
11. Harvard Law Review Notes, Uniformity in maritime law. Harv. Law Rev. **37**(8), 1114–1118 (1924)

Chapter 11
Overtaking Vessels: COLREGs vs Practice

Luka Grbić, Mate Barić, Dalibor Ivanišević, and Ana Gundić

11.1 Introduction

Collisions are the most common type of maritime accidents, whereas human factor is their major cause [1, 2]. When there is a risk of ship collision, proper avoidance actions must be taken in accordance with COLREGs [3, 4]. The 1972 COLREGs Convention [5] was designed to update the internationally agreed rules on safe navigation, safe speed, navigation in narrow channels and navigation in separation schemes. It outlines the requirements for navigational lights, shapes and signals and their usage as well. The sole purpose of COLREGs is for them to be obeyed by navigation officers in order to avoid collisions at sea. However, although the rules have existed for almost half a century and have been widely accepted and implemented in the formal education of seafarers, collisions at sea still happen. As indicated in the project "Avoiding Collisions at Sea (ACTs)" carried out by six European maritime universities and academies, "many of the basic principles of COLREGs are improperly understood and applied. It is also common practice to use the VHF Radio, although it is not prescribed by COLREGs. C4FF's recent investigation into COLREGs has revealed that almost 50 percent of seafarers throughout the world disregard/ignore COLREGs at sea when taking actions to avoid collisions" [6, 7]. Mistakes referring to COLREGs are usually made because rules and regulations have been ignored [8].

The Rules describe simple situations between two vessels. However, situations that are more complex frequently occur at sea as well (including three or more vessels). Therefore, it is of utmost importance to understand the Rules and to apply them correctly. However, research [9] has shown that some rules are difficult to

L. Grbić (✉) · M. Barić · D. Ivanišević · A. Gundić
Maritime Department, University of Zadar, Zadar, Croatia
e-mail: lugrbic@unizd.hr; mbaric@unizd.hr; divanisevic@unizd.hr; agundic@unizd.hr

S. Bauk, S. D. Ilčev (eds.), *The 1st International Conference on Maritime Education and Development*, https://doi.org/10.1007/978-3-030-64088-0_11

understand [10]. One of these rules is the Rule 13 (Overtaking), whose understanding, especially among the students, is rather low. Furthermore, there is no specific Model Course covering solely COLREGs [11].

The question then arises as to why it is important to understand this specific Rule. The following simple situation [12] can help us understand the problem better; vessel A overtakes vessel B on the starboard side and the distance between them is 5 miles. When vessel A comes alongside vessel B, she turns her course to port. Is this still an overtaking situation or is it now a crossing situation? More complex situation [13] is when vessel A overtakes vessel B on her port side while vessel B is in crossing situation with vessel C on her starboard side, and the risk of collision exists. In this situation, vessel B is the vessel being overtaken, and not in risk of the collision with vessel C, she maintains her course. Vessel A cannot turn to starboard side because of the vessel B and cannot turn to the port side because of shallow waters. In other words, bad judgment and poor understanding of the Rules lead to the collision between vessel A and vessel C.

COLREGs do not specify the side a vessel should overtake another vessel on. The procedure itself depends on many factors such as fairway width, other vessels, approaching side, planned route, water depth, etc. In practice though, seafarers usually perform overtaking on the port side of a vessel to keep clear of her starboard side. However, this action is not the result of obeying the Rule but an example of a good seamanship. The authors have tried to determine prevalence of this practice among maritime professionals and among students by using the online survey.

Understanding the Rules and their proper application are the main goals of seafarers' learning and training. The important segments in learning and understanding the Rules are learning systems and tools that maritime schools have been using, [14] onboard training and many online sites using various tools such as video animations, navigational simulators, images, vessel small-scaled models, etc. [15]. During the period of studying and practical training on board, knowledge and proper understanding of the Rules are tested. Overtaking situations where delaying a simple action may result in close quarter situation or collision were put a particular emphasis on. Therefore, it is important to teach the students how to apply the Rules properly, keeping in mind, at the same time, that overtaking can be carried out on both sides of a vessel depending on the situation.

11.2 Research Methods

In order to find out whether there is a disproportion in terms of the side of overtaking, seafarers and students' knowledge and understanding of the Rule 13 were queried by using the online survey. In order to attract as many respondents as possible, the survey was distributed online and it was intentionally kept simple. Besides the basic question about seagoing experience and onboard position, there was only one question asked: "According to COLREGs, which side is it allowed to overtake another vessel on?"

During the 4-month period that the survey was online, 181 maritime professionals and 269 maritime students participated in it. Posting an online survey made it possible for respondents from 26 countries worldwide to participate in the research. The most of the respondents, as far as maritime professionals are in question, were from Croatia (89 respondents), India (17 respondents), Turkey (14 respondents) and the UK (13 respondents). As far as students are concerned, majority of them were from Croatia (222 respondents) and Bulgaria (43 respondents).

11.3 The Analysis of the Results

The term maritime professional refers to masters (43% of the respondents) and officers of the navigational watch (OOW) (the remaining 57% of the respondents) (Fig. 11.1). In total, 56% of the respondents had more than 10 years of seagoing experience, 15% had from 5 to 10 years of seagoing experience, 23% had 1 to 5 years of seagoing experience and 6% had less than a 1 year of seagoing experience (Fig. 11.1). This diversity in seagoing experience and the number of student respondents provided a reliable data for the comparison of learning processes, knowledge and application of the Rules in practice.

The analysed answers have shown that 69% of maritime professionals think that a vessel can be overtaken on both sides, 17% think that a vessel can be overtaken only on her starboard side whereas 15% think that a vessel can be overtaken only on her port side. As far as the students' answers are concerned, 32% of them have replied that a vessel can be overtaken on both sides, 42% think that a vessel can be overtaken only on her starboard side and 26% think that a vessel can be overtaken only on her port side (Fig. 11.2).

These results show a disproportion between the opinions of maritime professionals and students. Moreover, the results indicate that the Rule 13 is one of the Rules difficult for the students to understand [5].

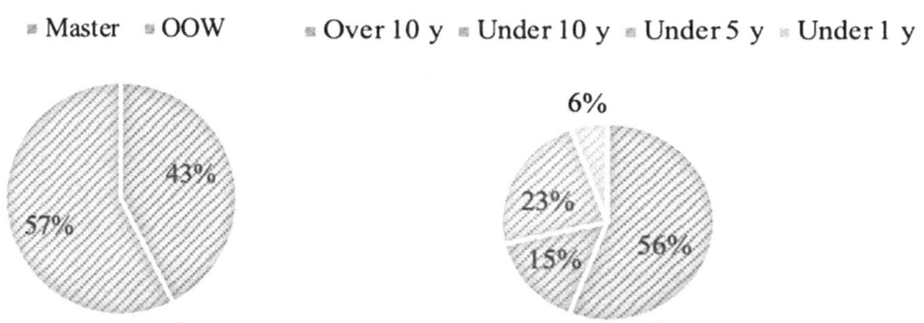

Fig. 11.1 Percentage of maritime professionals and their years of onboard experience

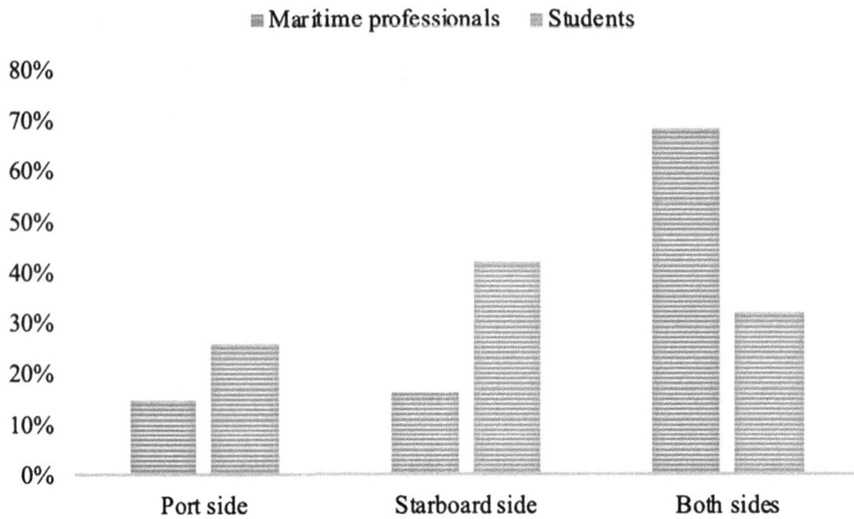

Fig. 11.2 According to COLREGs, which side is it allowed to overtake another vessel on?

As seen in Fig. 11.2, there is a large disproportion between the opinions of students and maritime professionals. That result is somehow expected because no student can match a competence of an experienced seafarer. Most of the seafarers know that they can overtake another vessel on both sides of a vessel, depending on the situation.

Maritime professionals have answered the question correctly in considerably greater percentage than maritime students, 31% of which have chosen the wrong answer. This result indicates that the Rule 13 (among some other Rules) is either poorly written or inadequately lectured. Regardless of onboard experience (including watchkeeping experience), there are still more than 30% of maritime professionals who do not understand the Rule 13.

When comparing these results with the ones referring to seagoing experience, it can be concluded that seafarers with long onboard experience do not understand this Rule completely (Fig. 11.3). Most of them have answered that they would overtake a vessel on her starboard side. This situation could be associated with the Rule 15 – crossing situation.

As far as the students who participated in the survey are concerned, most of them have answered that they would overtake a vessel on her starboard side, and the result is almost the same for all three answers. Twenty per cent of the students had an

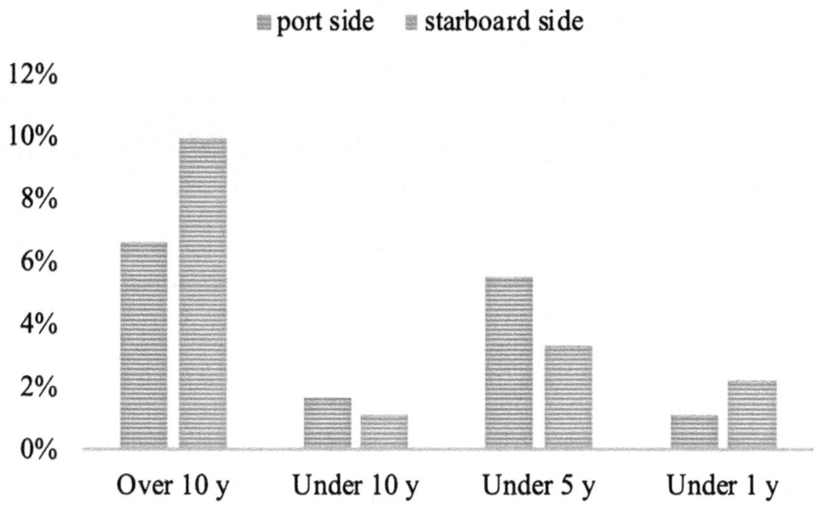

Fig. 11.3 Percentage of seafarers' answers referring to the side of overtaking

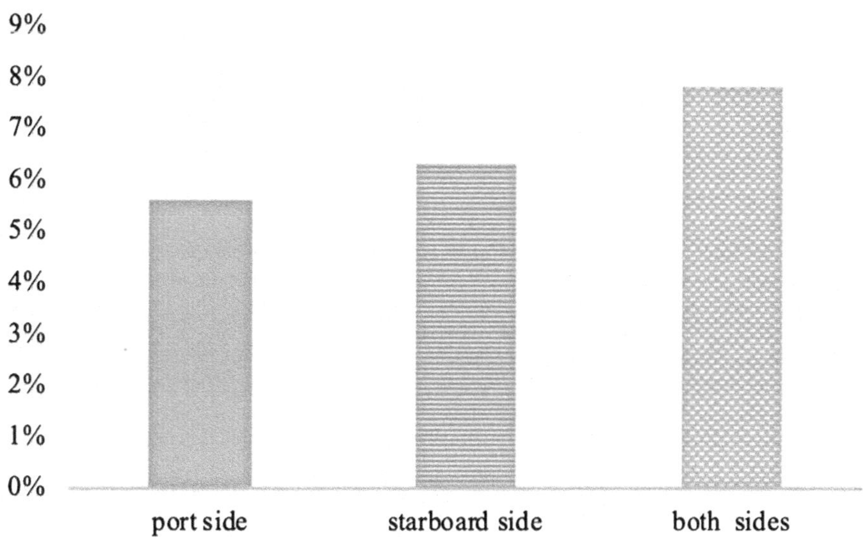

Fig. 11.4 Answers of the students with onboard training experience

onboard training experience (Fig. 11.4), and most of them have shown better knowledge than students without onboard training experience.

However, very similar distribution of answers of different target groups shows that this topic should be studied more during the onboard training.

11.4 Conclusion

The analysis of only one COLREGs rule, referring to overtaking at sea, has clearly shown a discrepancy in knowledge, in understanding and in application of the rule in practice regardless of the level of knowledge and experience. The results have also shown that proper learning process combined with onboard training may improve considerably the understanding of the rule. Since learning processes have been improved by using modern educational tools, it would be expected for misunderstandings of a particular Rule to be eliminated or reduced. Furthermore, the goal must be absolute as well as proper understanding and application of the Rules. However, this is still not the case. The authors are of the opinion that the Rules have to be edited and some of them rewritten in order to avoid misunderstandings and wrong application. Besides the modernisation of the Rules, learning process should be improved by using tools such as e-learning platforms. They could help teachers and students to analyse the Rules by using navigational simulators that create realistic onboard environment. The authors hope that other rules difficult to understand will be scrutinised as well and that action would lead to general review of the Rules. Continuous improvement of learning process would lead to thorough understanding and application of the Rules.

References

1. M. Majid, J. Mehdi, *Marine Accidents Investigation Methods Based on the Role of Human Factors in Accidents* (2018)
2. M.A. Rothblum, *Human Error and Marine Safety, USCG Research and Development Center*
3. Y. He et al., Quantitative analysis of COLREG rules and seamanship for autonomous collision avoidance at open sea. Ocean Eng. **140**, 281–291 (2017)
4. M. Baldauf, et al., *A Perfect Warning to Avoid Collisions at Sea?*. Zeszyty Naukowe Akademii Morskiej w Szczecinie (2017).
5. Convention on the International Regulations for Preventing Collisions at Sea, 1972.
6. ACTS Project, Aims, Rijeka, (2015), http://ecolregs.com/
7. U. Acar, R. Ziarati, M. Ziarati, An investigation into COLREGs and their applications at sea. Safe Ret Port, 40 (2012)
8. R. Ziarati, *Safety at Sea–Applying Pareto Analysis*. Proceedings of World Maritime Technology Conference (WMTC 06), Queen Elizabeth Conference Centre. Vol. 94 (2006).
9. D.J. Mohovic, R. Mohovic, M. Baric, Deficiencies in learning COLREGs and new teaching methodology for nautical engineering students and seafarers in lifelong learning programs. J. Navig. **69**(4), 765–776 (2016)
10. Đ. Mohović, R. Mohović, R., M. Barić, *Identifying Skill Gaps in the Knowledge and Teaching of COLREGs*. 17th International Conference on Transport Science (2015).
11. D. Ivanišević, A. Gundić, Đ. Mohović, COLREGs in STCW convention. Pomorski zbornik **54**(1), 23–35 (2018)

12. A. Zekić, Đ. Mohović, R. Mohović, Analysis of the level of knowledge and understanding of regulations for preventing collisions at sea. Pomorstvo **29**(2), 143–149 (2015)
13. Marine Accident Investigation Reports, *Collision of Chemical Tanker "Sulphur Garland" and Oil Tanker "Wakomaru No. 2"*, Japan Transport Safety Board, (2017, February).
14. D. Ivanišević, A. Gundić, Đ. Mohović, Differences in understanding the COLREGs among the students from different systems of education for seafarers. J. Mar. Res. **14**(2), 3 (2017)
15. Đ. Mohović, M. Barić, R. Mohović, *Innovative Way of Teaching COLREGs*. 7th International Conference on Maritime Transport. (2016).

Part III
Educational and Training Challenges

Chapter 12
Smart Shipping Needs Smart Maritime Education and Training

Anatoli Alop

12.1 Introduction

The rapid development of digital technologies in recent years and the exponential growth of computing power and storage capacity of "wise machines" lead inevitably to the increasingly important role of artificial intelligence (AI) in many fields of activity, including maritime transport [1, 2]. These developments, combined with growing global challenges, give us reason to believe that, if not in the coming years, then in the coming decades, we definitely will face major changes in the world in general and in maritime affairs in particular. They, among other things, will force us to abandon the beliefs and paradigms that have so far seemed to be "waterproof."

Before discussing in a practical way the problem posed in the title of this paper, it is essential to ask some to a certain extent philosophical questions, which inevitably arise when analyzing this topic and which cause a slight dissonance in the author. There may not be (or even cannot be found at all) good answers to these questions, especially within the scope of single paper, but defining these questions and "bringing them into play" seems important to the author of the paper.

The first question is what we mean by talking about something or someone "smart" (smart person, smart behavior, smart economic, smart shipping, or even smart education). It is clear that the answers to the question about the "smartness" of the listed objects and subjects, even if they could be found, would differ significantly from each other. The author of this paper does not attempt and even probably is not able to give an exhaustive definition of "smart" for all these objects and subjects (and possible many others). However, the author would like to clarify what he

A. Alop (✉)
Centre of Maritime Studies, Estonian Maritime Academy of Tallinn University of Technology, Tallinn, Estonia
e-mail: anatoli.alop@taltech.ee

S. Bauk, S. D. Ilčev (eds.), *The 1st International Conference on Maritime Education and Development*, https://doi.org/10.1007/978-3-030-64088-0_12

believes could be the "smartness" of future shipping and maritime education that directly related to it, at least in view of some important aspects.

The title of the paper explicitly suggests that the need (or at least one of needs) to make maritime education "smarter" lies in the fact that shipping is becoming, or at least it is expected to become, "smart," so defining "smart shipping" in the context of this paper is important. Only by clarifying what does it mean will it be possible to understand what education is needed in this context and what the "smartness" of that education should be.

In general, the use of the word pair "smart shipping" refers today primarily to the taking into use of the latest technological solutions, i.e., digitalized and algorithmic shipping, mainly the deployment of so-called intelligent ships at sea. The term "intelligent ship" implies that all or the vast majority of decisions for such type of ship are made not by humans (who are probably not enough intelligent) but by AI. Presumably, the role of seafarers will be taken over by artificial intelligence, either in full or in part. Consequently, if digital shipping is a priori smart, it inadvertently leads to the conclusion that the lack of wisdom or even stupidity is the prerogative of the human, and the lower the share of human participation or intervention, the smarter shipping is.

According to this, there would be nothing wrong with shipping itself as an essential component of the global supply chain. Shipping or, more broadly, maritime transport perfectly serves today the needs of stakeholders and economic system in whole that built on the postulates of infinite growth and making a profit. In other words, it is assumed that making shipping smarter is not a question of changing shipping itself, i.e., its priorities and goal settings, but only a question of improving the form of performance. The more digitalization and the less human intervention, the more wisdom and less stupidity in the system will be. In principle, the author has nothing against such an approach, although this position is quite controversial, but the author would like to approach it from a little different angle.

It is supposed that smart or otherwise digital shipping will serve the system better, enabling shipowners and other beneficiaries to make greater profits. Yes, of course, it will serve better, but the question is, is the current economic system and the way humanity lives on Earth in general smart? If smart solutions serve not a very smartly designed system and faulty goals, can we call those solutions "smart"?

In the opinion of the author of this paper, we cannot. Not only and not so much reducing the costs of beneficiaries and helping them to win in the competition, thanks to digital solutions, should be the sign of wisdom and the main purpose of attainment. Humankind developed so far that all the fields of activity, industries, paradigms, and theories need a radical revision, finding out fundamentally new and nonstandard solutions. Shipping, like virtually all other human activities, must become smarter at a fundamentally new, meaningful higher level, and for achieving of this goal, the education, including maritime education, needs to be revised.

In the author's opinion, now is the right (if not the last) time to review the basic (first of all economic and social) theories and postulates and give ourselves honest answers about what is happening and what is smart and what is not. Perhaps the shipping that, thanks to digital solutions, will be able to transport more and more goods with the same or even lower costs, make more profit, and be more competitive on the market cannot be called "smart" for that reasons only. It seems that the smart is shipping that allows, thanks to digital solutions, to find more optimal ways of reducing the negative impact on the environment, to make the consumer society more sensible, and to reduce ultimately the risk of humankind self-destruction. Such an approach should also significantly change the question of defining smart maritime education.

The second question is who is the target group for "smart" maritime education? The answer is obvious, the reader can say. Those who study in the education institutions are the target group. The answer is correct, but it seems that those who study become predominantly maritime transport employees, in other words more or less mindless performers, not decision-makers. As a rule, those who make a profit and make decisions in maritime do not study something that is not directly necessary for their businesses and other similar activities. It can be said that in maritime affairs (as in many other areas), the principle of stratification may be observed, according to which at least three parallel "worlds" coexist: stakeholders and other beneficiaries or "moneymakers," politicians and other officials or "decision-makers," and employees on different levels or "job makers."

The latter, who are definitely the target group for (smart) maritime education, must serve the first and partly the second group. Thus, they have little opportunity to make a significant impact on the solving of real large-scale problems using advantage of their smart education; the primary achievement for them is to be able to earn a living for themselves and their families. Science that closely intertwined with and based on education is in the same position. Science needs money to function, and more and more money is needed today to do good science. As well known, who pays, that orders music.

The actors of first and second groups could benefit from an eye-opening and making them smarter professional maritime education, and even more universal education; thanks to that, they would not make moneymaking decisions without considering the issues of expediency and sustainability. As rule, they are usually not interested in such kind of education. They probably do not have time to make such a mess because they are busy on moneymaking 24/7. Of course, they may learn something: business administration, financial management, etc., but there is little point in contributing to the solution of problems of humankind.

Thus, the author realizes that no matter how much he would like the maritime education (and education in general) to become smarter and thus contribute to the improvement of the world, it seems to be more a dream than reality. Many educated people may understand what is going on, but if the few ones who can really make decisions and change things are not among them, their understanding can change nothing.

12.2 Development Trends in Education

A Brief Historical Overview

One of the fundamental principles of how the world works is that major developments take place in the form of a spiral [3]. There is a well-known saying that "everything new is a well-forgotten old one." This is true, but since everything is constantly in the processes of changing, the new never is the same as old, but this is old transformed into new conditions, at a new level of development, so the processes may be shaped like a spiral. This also applies to education, both in the most general sense and in terms of different areas of education.

If we look back in time, we see that humanity's capacity has grown incredibly over the last 500 years, both in scale and speed, and in most areas, this growth has been exponential, and it is continuing. Suffice it to say that during last *ca* 500 years, humanity has grown about 14 times, production 240 times, and energy consumption 115 times [4, p. 322].

This ever-accelerating process started with the so-called scientific revolution, which began the era of modern times about 500 years ago. A key feature of the past years is that humanity has invested more and more resources in research and science during this time, through which it has acquired unprecedented new capabilities and gained more and more new resources as a result. This change can be named "revolution," because before that, both governments and wealthy people allocated some money for science and education, but primarily with a view to maintaining existing capabilities, not acquiring new ones [4]. Before scientific revolution, the high-quality education was the privilege of rich and talented individuals, and the ways and means of acquiring it were their own concern or, more precisely, were determined primarily by their status in society.

While before era of modern times, education was the privilege of the individuals, the scientific revolution led to the emergence of educational institutions, i.e., universities (which became also research institutions), but the education received there was not initially intended for the masses. The purpose of that was not only, or even not so much, the acquisition of knowledge for a specific profession, but rather it was the acquisition of universal knowledge. As a rule, what was learned at universities covered a wide spectrum; university graduates were the people who were interested in many fields of science and the connections between them, and some of them were the ones who carried forward science in the broadest sense of the word and made major discoveries and inventions during the centuries.

Although a general division between the humanities and the real sciences was already emerging at the universities, the line between them was initially blurred, and it was more common to acquire a complete picture of the world rather than a narrow discipline. In modern terms, people with such an education could be named as "generalist" (that did not express contempt yet at that time); it would be preferable to call them "universalists" or even more precisely "holists."

However, they should be honored, because thanks to talented individuals' curiosity and thirst for knowledge of some of them, they were finally behind a breakthrough and an ever-accelerating triumph of science.

Taking into use a water and steam power *ca* 260 years ago to mechanize production and introduction of electricity and taking into use fossil fuels *ca* 150 years ago were named accordingly the first and the second industrial revolutions. The third industrial revolution was the vigorous development of electronics and automation, and computer sciences as well, which came to the aid of mass production automation about 50 years ago [5]. Today we are talking about the fourth industrial revolution (Industry 4.0), which is actually a continuation of the third one at a new qualitative level; the second name of it is the digital revolution. From the point of view of education, the key word in all these periods is "mass"; that means first at all mass production, the other side of which is mass consumption. As education began to serve more and more the industry (in the broadest sense of this word) during the industrial revolutions, we can speak of education becoming mass education from the second half of the eighteenth century.

Specialist vs. Generalist

The key word for mass production serving education has been "specialization." Ideally, owner of the production unit wants to see people in his production facilities who can perfectly perform a single operation or set of certain operations. Even towards creative people (e.g., engineering staff, product developers, managers, etc.), the word "specialist" has a positive meaning; a specialist is one who knows everything or almost everything about one specific thing or area and is able to develop them in the most optimal way.

The opposite of the "specialist" is naturally the "generalist", who, on the contrary, acquired a little negative background during the industrial revolutions' times. The generalist may know in total no less than the specialist may; in fact, generalist knows more, but he or she knows as rule only partially about any particular things, and there is nothing he knows all about (otherwise he would no longer be named as a generalist). Much of knowledge of generalist cannot be directly useful in mass production; at the same time, effective visionaries and "idea generators" are often generalists who have a broad horizon and can see the "big picture" or, otherwise, we would name them "universalists."

Today's formal education (universities, vocational schools) has reached the peak of maximum in specialization. International Standard Classification of Occupations (ISCO-88) contains thousands of occupations divided into nine major groups [6]. Each university has hundreds of curricula, many of which can have in turn several specializations; thousands of several specialties are studied in vocational training institutions as well. A graduate in one specialty is usually more or less ignorant in all other specialties. This is also understandable because the high technical and technological level in most areas means that in order to become a specialist of excellence, one has to dedicate nearly half of active life to acquiring this specialty, i.e., learn and acquire knowledge and practical skills for decades. The situation is similar in science: the more and more narrowing specialization of research topics leads to the fact that professional articles in one field are poorly understood or not understood at all by researchers in other fields, not to mention so-called ordinary people.

One may ask, what is wrong with that? Every person learns one thing as clearly as possible and becomes a top specialist in it. Everyone is highly valued in his or her workplace and does not fear to become a miserable loser in the competition. Top specialists do good job; their products and services are of high quality. Top-level researchers with narrow specialization are able to effectively develop their field of research and eventually achieve success and, if they are doing well and lucky, become laureates of Lifetime Achievement Award and even, why not, Nobel Prize laureates. Everything seems to be fine.

However, such narrow professionals, be they in industries or in sciences, have one major drawback: their excessing commitment to one discipline does not leave them much room to see the big picture. Sometimes they are unable (often unwilling) to relate their actions to potential consequences of them, which can have a much wider and more negative impact than they might think. There is one good saying about it: "No seeing the forest behind the trees." The development of human civilization has reached a stage where all the people have to think about what they are doing and where their actions will eventually lead. The visionaries or universalists with a wide horizon and nontrivial thinking are again may be not less, but even more important, than specialists for the survival of humankind. Moving along the spiral, we are back, but on a new level.

12.3 Smart Maritime Education

New "Old" Priorities

In publication [7] the author examines, inter alia, maritime safety issues in the coming era of smart shipping. Ensuring the safety of human life at sea has always been one of the main priorities of shipping; the second priority is the preservation of property, first with regard to ships and goods. The protection of the marine environment may be mentioned as a third priority; unfortunately, until a few decades ago, this priority was clearly below the first two in terms of importance.

These priorities will certainly not become less important even if full-autonomous, semi-autonomous, and remote-controlled vessels become more and more powerful in occupying their rightful niches in shipping. At the same time, the relationship and interaction between these priorities can change significantly, and it is vital that decision-makers and main actors in maritime understand this in a timely manner and reshape their personal and corporate goals and attitudes according to changing circumstances.

Primarily it concerns the safety of human life. If the ship is autonomous or remote-controlled, there shall be no persons on board during the voyage and, in the event of an accident, no one will be endangered immediately, at least until the ship in distress endangers other ships, other facilities, and persons on them. At the same time, for people who come to the rescue of an unmanned ship in distress and the goods on board, there will still be a danger. However, important difference is that at least there is no difficult choice as whether and how much to risk the lives of some people in order to save lives of other people. Rather, it is a choice of whether or not to risk human lives to save property. This second option seems much easier.

Pollution of the marine environment with oil and oil products, plastics, etc. leads initially indirectly and locally but later ultimately directly and globally (through drastic changes in the living environment) to endanger people's lives and health. It is therefore important to be aware of the possible consequences of one or another activity, not only in the short term but also in view of the far-reaching results. It is hard to believe that an AI programmed at any price to protect the property "entrusted" to it will be able to make the right strategic decisions of such kind. These decisions remain the prerogative of the human beings, and the smarter the education they receive, the smarter decisions they make.

One of the important problems of people who have studied narrow specialized professions is that they often are not able to think of themselves as an integral part of big system (society, humanity). They do not realize that everything they do and all the decisions they make not only affect their own life and well-being but in one way or another extend to other people and further to the community, ecological system, and thus all humankind. Many know the famous theory about the "butterfly effect" [8], but few are able to identify themselves with this effect.

The Growing Importance of Diversity

When going to work on any conventional ship, a ship's officer today must know and be able to do mainly the same things; the differences due to the type, size, etc. of the ship are not very large. Hence, the level of standardization and unification of modern legislation governing maritime education, first the STCW 78 Convention, is high [9].

One of the features of the near future is that both in life as a whole and in its various fields, more or less applicable several solutions are possible, the choice between which is not always as unambiguous as one might think. In standardized and unified (maritime) education, it applies usually "the only one right solution" principle, which means that there is always one right course of action in every possible situation, which must be clarified during theoretical and practical studies and even better played through on the simulator; in any such situation, this solution must be implemented without much thought. In general, this principle is the basis of an unmanned ship control algorithms. However, already today, and even more so in the future, there will be the implementation of the right solutions in shipping an increasingly multivariable activity. A choosing between these variables is not an easy task; the author dares to say that in some cases the AI may not be able to perform this task perfectly. It still has to be done by the captain (in the case of an unmanned ship, the shore operator).

As one such example, we can imagine a situation where we choose between a bad choice and an even worse one which must be done: either collide with another ship or steer your ship to the ground to avoid a collision. All solutions are bad, but less bad one have to be chosen, as a rule in the time scarcity conditions. An AI that is programmed to protect anyway the shipowner's interests may choose to steer own ship bow to the center of the other ship's side, which is likely to ensure relatively little damage or at least a floating position for the ramming ship. For another ship, this would mean much more serious consequences, with high probability a shipwreck. Moreover, if this other ship is a fully loaded tanker, severe consequences for the marine environment are likely to be expected. Can for specific response programmed AI consider all this and, more importantly, make a decision, which contradicts his algorithm? Probably not. However, a captain or shore operator must be able to do it, and his or her fundamental knowledge, attitudes, and, most importantly, personal characteristics play a crucial role in this.

If we talk about the increasing diversification of the knowledge and skills needed at sea, there will be more different options in the near future, than there are now. When acquiring maritime education, branching points should already emerge during the studies; from those points, students go to one degree or another in different ways by acquiring their future specialty. Ultimately, those who go on board conventional vessels and those who will operate remote-controlled vessels from shore-based control centers must have not quite the same knowledge and skills.

These two are opposite variants: one is for those whose entire working life is spent, as it is today, on board the ship, and the other is for those who perform their duties in full on shore (yet we must call both those seafarers). However, there may be a considerable number of intermediate variants between them, for example, those who are employed in regiments serving fully autonomous vessels and have to deal promptly with any kind of problems that may arise with such kind of vessels at sea.

The author of this article believes that developments in the ships' digitalization and becoming "more intelligent" will not be sudden and rapid from one marginal case to another; so-called semi-autonomous solutions are more likely to emerge in the coming years, e.g., the 3M and MP variants described in the author's publications [10, 11]. Even this will also take a long time, perhaps decades, to reach the mass application of such solutions, not to mention fully autonomous ones.

Initially, several variants will be made for gradually increasing the role of digital solutions and AI in the management and operation of manned ships. More than one intermediate variant will emerge, which differ in the degree of AI interference in the control and operation of the ship. Computers will take over more and more ship operation and management functions; they will not only reduce the need for human intervention, but also the functions of people on ships will change all the time. This requires increasing of the ability of crewmembers to constantly learn and retrain and, among other things, to work effectively together with AI.

This means that the qualifications and knowledge of future seafarers must also evolve, depending on the degree of automation and digitalization of vessels, on which they are to be employed. The knowledge of these people must be universal on the one hand, because they must be able to cope with new types, including unexpected situations, and on the other hand, it must be specific, i.e., they must be experts in the intelligent systems that the ship probably will be.

Supposedly, operators of remote-controlled vessels are unlikely required (and probably they cannot) to be top-level IT professionals, but they should have IT education on level that gives them certain advantages. First, it would allow them to understand what is happening in the first moments of a cyberattack or other major failures in order to implement the most basic preventive measures in order to prevent or at least slow down as far as possible the subsequent negative developments; the speed of response can be crucial here. Second, they must be able to successfully communicate and collaborate with relevant IT and AI professionals, both in routine and emergency situations.

The qualifications of the developers of "smart shipping" algorithms are also important. It is vital that the algorithms are not only based on the results of statistical processing of Big Data but also take into account the specificities of the field, including possible effects and consequences. The algorithmic developers should either be maritime experts to one degree or another or receive qualified assistance from experts with maritime training and maritime experience as well as IT expertise.

Key Concepts for Future Maritime Education

The focus of maritime education must shift from what is taught to how. In his previous publications [7, 11], the author discussed the essential professional and human qualities of the people working in shipping today and especially in the future and described them in terms of diversity, personal achievements, social values, and preparedness for future challenges. Here, the author would like to add a sense of responsibility and thinking ability. Highly standardized and formalized maritime education must become education that develops more creativity and independent thinking.

In addition to the need for diversity in maritime education, as described in Sect. 12.3.2, the author highlights three key requirements described in his publication [11] that he believes should characterize the changes that will need to take place in the near future in the education of seafarers. The first is to make specialty subjects and the whole learning process more complex. The second is closer and more meaningful cooperation both between teachers and between teachers and students; it is clear that without it, the first goal cannot be achieved at all. Third is flexibility, i.e., quick response to changes in professional life and life in general and bringing education in line with them.

In order not to repeat himself, the author will not explain these key concepts in detail; those interested can be acquainted with them in the publications [7, 11]. However, the author would like to describe briefly some of the changes in the learning process of his home institution that he has undertaken to implement his concepts.

Case study As the author of this paper is the Head of the master's study program "Maritime Studies" in Estonian Maritime Academy and one of the lecturers of this curriculum, he has a good opportunity to launch so to say a pilot project in cooperation with two other lecturers of the program. The aim is to introduce in teaching process a new complex course (hereinafter referred to as CC) based on the three courses of special studies currently being separate in the curriculum, applying the principles described in general terms before. Due to the limited volume of the paper, it is not possible to describe this project in detail here; the author only outlines its essence and main features.

The CC is formed on the basis of the following three courses: Maritime Cluster and Shipping Economies, Port Operation and Basis of Management, and Maritime Law. The volume of each subject is 6.0 ECTS, so CC is in total 18.0 ECTS, and it lasts for one semester (16 weeks). The main form of studies is project learning. It is planned to involve graduates of previous years with whom the academy has traditionally good relations and who are usually in key positions in their companies and institutions. They are intended to play the role of observers and guest lecturers. During the CC, lecturers and students play through situations and try to solve problems, which can take place in the work of modern shipping.

Setting and completing tasks is made more complicated by bringing into play various influencing factors, some of them may be classified as expected (more or less) and some as unexpected. For example, expected ones may include the integrated digitalization of ships and ports, influencing of drastic climate change, and so on. Several global factors affecting the world economy as well as the state of local affairs, such as pandemics (why not?), major power outages due to various natural factors, etc., may be classified as unexpected.

Two additional factors favor the use of this methodology. Firstly, the curriculum "Maritime Studies" has two main specialties: Technical Exploitation of Ships and Navigation and Shipping Management, so that two professional groups can be selected within the study group: seafarers and shore personnel. It mimics the realities of shipping, where these groups are closely linked and cooperating with each other. Secondly, almost all students work on ships or in shore companies during their studies, which provides a good opportunity to link studies closely to real working life and solving the more real problems.

12.4 Conclusion

It can be said that education goes hand in hand with the major changes in the economy and social life; the only question is, who plays a leading role in this trilateral union? At first glance, it seems that education and closely related to it science is the initiator, activator, and guarantor of continuous development. In this paper, the author expressed his position that this may not always be the case today, i.e., some things may be "turned upside down." Not always common sense and an understanding of what leads to what is winning; it is often the "money world" that dictates future solutions, including paradigms and directions in education.

Now the author's third philosophical question: how can some destructive aspects of human nature be overcome? Moreover, could it be at all? There is reason to doubt that. Why have Adam Smith's theory and other mainstream economic theories found such a fertile surface and flourished for centuries, even though it is clear that the economic system built on them is unsustainable and ultimately leads to a general collapse rather than a general prosperity? Is it not because it resonates very well with human qualities such as greed, ignorance, etc.?

Besides, people do not see anything what they do not want to see; they just close their eyes. It will soon be 50 years since publication of the so-called Meadows Group report entitled "The Limits to Growth" [12]. Its content and the predictions of human development modeled in it were shocking. However, many, especially the political and business circles, did not want to and still do not want to hear about it; the global business-as-usual (BaU) scenario is continuing. We are now quite close to the environmental and economic collapse predicted in the report, but for the most part, it does not worry people.

A good example of human ignorance is the lesson of the coronavirus: for decades, scientists have warned that a global pandemic with very serious consequences is quite likely, if not inevitable. Did anyone pay attention to this? The pandemic came and no one was ready for it.

From all this, it can be a disappointing conclusion that (a) in the struggle between common sense and the desire to make more money, common sense does not usually prevail and (b) the most common (if not the only) lesson of all lessons for humanity is that no lessons teach people, society, and humanity as a whole anything.

The author sees the challenges in changing maritime education; it can no longer be taken as just a way of acquiring certain knowledge and skills. The educated person's worldview and attitudes become important, or rather, what he or she decides to do with his or her knowledge and skills, whether he or she is able to see the big picture and assess the effects and consequences of his or her actions. This is not only a challenge, but it is also a problem.

Figuratively speaking, the time of just executors of orders is getting over. In fact, "smart machines" will more and more be taking over the functions of executors. However, humans are programmers of these machines, and humans will decide on how their wisdom may be used. Thus, the role of human wisdom increases drastically, because the price of stupidity and other "nice" human qualities may be too high.

Author believes that the only means to make people smarter are right upbringing and smart education. Maritime education is no exception, and although those coming to study maritime education have already acquired their main worldviews and beliefs in their families and in secondary school, it is never too late. Maritime education institutions have a role to play in increasing the share of wisdom in the world, at least in the maritime affairs.

References

1. Brief Guide to Artificial Intelligence in Shipping., https://thetius.com/brief-guide-to-artificial-intelligence-in-shipping/
2. A. Love, *AI in Shipping: Areas to Watch in 2020*, https://www.ship-technology.com/features/ai-in-shipping/
3. D.E. Beck, T.H. Larsen, S. Solonin, R. Viljonen, T.Q. Johns, *Spiral Dynamics in Action: Humanity's Master Code* (Wiley, Chichester, 2018)
4. Y.N. Harari, Sapiens. A Brief History of Humankind (in Estonian). Äripäev, Tallinn (2016)
5. Timeline of Revolutions., https://www.industrialiotseries.com/newsroom/timeline-of-revolutions/
6. ISCO-88 International Standard Classification of Occupations., https://www.ilo.org/public/english/bureau/stat/isco/isco88/major.htm
7. A. Alop, *On the Way from Unified Standardized MET to Personalized Maritime Education. Modern Challenges in Maritime Education and Training* (Batumi State Maritime Academy, Batumi, 2019), pp. 8–17
8. J.L. Vernon, Understanding the Butterfly Effect. Am. Scient. **105**(3), 130 (2017)
9. International Maritime Organization., http://www.imo.org/en/About/Conventions/ListOfConventions/Pages/International-Convention-on-Standards-of-Training-Certification-and-Watchkeeping-for-Seafarers-(STCW).aspx
10. A. Alop, The main challenges and barriers to the successful "Smart Shipping". TransNav Int. J. Mar. Nav. Saf. Sea Trans. **13**(3), 521–528 (2019)
11. A. Alop, *The Challenges of the Digital Technology Era for Maritime Education and Training. 2019 European Navigation Conference (ENC)* (IEEE Press, Warsaw, 2019), pp. 1–5
12. D.H. Meadows, D.L. Meadows, J. Randers, W.W. Behrens, *The Limits to Growth; A Report for the Club of Rome's Project on the Predicament of Mankind* (Universe Books, New York, 1972)

Chapter 13
Enhancing Maritime Education Through Online Distance Learning in Developing Environments: Case Study of South Africa

Margaret B. Masuku

13.1 Introduction

Maritime cluster is one of the key enablers of growing global industry and trade. It requires skilled and competent people in administration, business and industry, including those who operate modern ships and port equipment for cargo transportation and handling. Maritime cluster should respond effectively to permanently growing demand of international trade and ongoing changes in shipping business and industry, which are mostly based on virtual intelligence and advanced technology. Seafarers are on the first line to implement the conventions and regulations developed by maritime entities in such dynamic setting. Improving seafarers' competences by all means would help in enhancing safety and efficiency of navigation and marine environment protection.

The International Maritime Organization (IMO) reported that the human element and poor competence are among the main causers of accidents at sea,[1] whereas competence of the seafarers can be described as the worthy performance on board. This *steams* the need for more effective Maritime Education Training (MET) system aiming to overcome the problem of human errors and keep pace with rapid changes in maritime.[2] Ships are only good as the officers who operate them correct-

[1] IMO (2014). Maritime Education and Training. World Maritime Day, IMO: London

[2] Li (2017). Implications of Distance Learning Competence-based, maritime education and training

M. B. Masuku (✉)
Durban University of Technology, Maritime Studies, Durban, South Africa
e-mail: margaretm@dut.ac.za

S. Bauk, S. D. Ilčev (eds.), *The 1st International Conference on Maritime Education and Development*, https://doi.org/10.1007/978-3-030-64088-0_13

ly.[3] For this reason, the need for proficient seafarers seems to be a global concern for maritime effectiveness, especially in the light of shifting shipping trends, increasing ships' size and speed, increasing their cargo capacities, as well as the demand for marine transport.

Due to some research studies,[4] the need for 38,500 skilled officers by the end of 2018 has been projected. The IMO endorsed Maritime Education and Training as the theme for World Maritime Day in 2015.[5] The same source reported that maritime education should be focused on, considered, discussed, analysed, and continuously improved upon. Moreover, during the 112th session of the IMO Council meeting (16–19 June 2014), the IMO Secretary General, Mr Koji Sekimizu, has pointed that: "effective standards of training remain the bedrock of a safe and secure shipping industry, which needs to preserve the quality, practical skills and competence of qualified human resources".[6]

The national (South African) legislation has to be modernised in the sphere of higher education in terms of recognition and proper interpretation and implementation of the STCW Convention requirements and in terms of faster deployment of virtual learning as a supplement or substitute to the traditional education and training of the seafarers. Within the context we should not lose sight of the fact that STCW Convention itself calls for a proper education as the foundation of successful training and acquiring competences. In order to confirm this observation, the quotations from the STCW Manila Amendments, Chapter II, Section B-II/1, Paragraph 14 are given: "Scope of knowledge is implicit in the concept of competence. This includes relevant knowledge, theory, principles and cognitive skills which, to varying degrees, underpin all levels of competence. It also encompasses proficiency in what to do, how and when to do it, and why it should be done. Properly applied, this will help to ensure that a candidate can: work competently in different ships and across a range of circumstances; anticipate, prepare for and deal with contingencies; and adapt to new and changing requirements". Additionally, of importance within the context is that the newest STCW Code amendments concern and not only concern but strongly recommend the introduction of modern training methodology including distance learning and web-based learning in seafarers' knowledge acquiring and upgrading.[7]

---

[3] Alop (2004). Education and training or training contra education. In safety at sea through quality assurance in MET institutions. IMLA Conference. St Petersburg: Admiral Makarov State Maritime Academy

[4] Drewry (2014). Manning 2014, Annual report. London: Nigel Gardiner

[5] IMO (2014). Maritime Education and Training selected as World Maritime Day Theme for 2015. Retrieved from: http://www.imo.org/en/MediaCentre/PressBriefings/Pages/21-council112wmdtheme.aspx#.XhnqcX9KiM9 (last access: 11th January 2020)

[6] Ibid

[7] Bauk S., Kopp M., Avramović Z., "A Case Study on Introducing E-learning into Seafarers' Education", JITA - Journal of Information Technology and Applications, Vol. 3, No. 1, June 2013, pp. 34-43

13.2 Motives for Implementing ODL

It should be noted that many ODL and e-Learning initiatives and proposals for South Africa and other developing nations are still presented with numerous core social, political, economic, environmental, cultural, experience and other disadvantages, constraints and challenges, needing to be considered or resolved preconditions for their success. One study on postgraduate black students in South African higher education mentions the difficulties not only in accessing Internet and resources but also in being able to adapt towards full digital literacy and learning participation for those from traditional learning and communities [1].

The need for effective institutional and senior management support is perceived as vital to aid e-Learning's functionality along with sound leadership/commitment from the tutors/lecturers and active responsibility for self-learning among the students. Core identified challenges aside from the digital divide and technology/access to computers are issues of securing data, limited Internet bandwidth, initial starting costs and lack of capacity to enforce student participation. Even when facilities are provided, students may face high queues for printers and computer laboratories. Lecturers also express concern about being subject to peer critical appraisal in placing their materials online and losing potential copyright as the institution asserts intellectual property ownership. Teachers also lack financial incentives to participate, given the time pressures involved and low prestige in valuing teaching capacity far less than focusing on publishing score at most colleges and universities.

South African Maritime Safety Authority (SAMSA) have not specifically proven their commitment to considering ODL for the maritime sector. It remains conspicuously absent in their Annual Performance Plans and reports among their maritime operations, governance, sector development, seafarer development and welfare programme [2].

13.3 Purpose of the Study

South Africa is a developing country. Uprising the level of people's education, knowledge and skills is a key for overcoming transitional period and making South Africa a developed country. Twenty-five years ago, people in South Africa did not have access to higher education. Today circumstances are positively changing, but there is still a room for improvements. People need sound education in their young ages, but also during the whole life, they need to learn and acquire new knowledge and skills to keep up with modern, technology-driven, complex and growing society. Lifelong learning becomes necessity. It can be realised through formal and informal learning channels. Internet as a huge virtual library can help a lot in this regard. However, people should have a certain level of knowledge and education in order to use its potentials purposefully. A good way of using Internet smartly for educational purposes is online distance learning.

> The purpose of this study is to explore the lecturers' and students' awareness regarding the benefits and impediments of online distant learning at a high maritime educational institutions in South Africa, along with their readiness to adopt this form of knowledge transfer.

The research has been conducted among lecturers and students/ex-students-seafarers at Maritime Studies Department (Faculty of Applied Sciences, Durban University of Technology (DUT)). Choosing this specific focus group has been motivated by the fact that there is no online distance learning (ODL) programmes at high maritime education and training (MET) institutions in South Africa. Maritime students have to upgrade their knowledge and skills permanently to be competent at world maritime labour market. Since they have to work as seafarers, upgrade their knowledge and refresh the certificates at the same time, ODL might be a good solution.

The findings of this study aim to assist South African METs to model and evaluate the feasibility of pioneering ODL programmes using innovative educational methods and technology. Once successfully implemented, ODL programme could serve as a model for METs that function in different developing environments.

13.4 Advantages of ODL

Online distance education programme in MET can give a flexible alternative on time and location, whereas the traditional classroom training programme requires the seafarers to fix time and location. Distance education can also relatively decrease the training fees and allow the students to learn without coming to school. Moreover, with the rapid change of maritime technology and legal requirement, many refresher courses can also be delivered through distance education [3].

According to Chen, it is believed that online distance learning can save these costs; again it can not only significantly reduce the cost of classroom infrastructure but also reduce travel expenses and shorten the training time. Additionally, students can study while working through in-service training. This will not affect the work, but will save on many cost types and improve the efficiency [4]. It should be noted that online distance learning has strongly support personalised learning. Furthermore, e-learning can provide more independent learning space for different learning style students. According to their working time, study habits, learning needs and learning capacity, students can independently adjust learning style, and they can learn the knowledge and skills over and over again, thereby, to improve their efficiency of learning. In addition, e-learning system could be combined with the characteristics of adult training, which emphasises students don't have to learn step by step. To make learning become more free and more flexible, they can flexibly choose between autonomous learning and collective learning.

Aside from the previously stated advantages, it is gradually proving to assist in ensuring competency-based training and assessment. Blended learning advantages have been identified as saving time, ensuring more effective use of class time, easier differentiation of individual student needs, more active students and more creativity for students. It is perceived as being great in aiding students to prepare better, teaching twenty-first-century digital skills and involving less paperwork which is more ecologically sustainable and cost-effective (Its Learning Inc 2015). It centralises learning resources in one remote location so they can be accessed anywhere by anyone as needed. It retains them longer than physical copies may exist and can help better inform parents, family and others not directly involved in the education.

Lastly, it can greatly improve the learning effect. E-learning training system can not only through a simple document permeate traditional theoretical knowledge to students but also can provide numerous picture, audio and video files to make training courses become lively and interesting. It also however allows students to feel fun in learning and creates a relatively relaxed learning environment and atmosphere.

13.5 Disadvantages of ODL

It is noteworthy that though e-Learning has advantages, there are also existing disadvantages. Lack of social interaction has been identified as one of the main disadvantages. While the student can have some interaction through email, chat rooms and other online platforms, it is relatively different than traditional classroom education. Moreover, not all courses can be offered online. Some courses with practical skills are hard to deliver by distance education; therefore, they need direct contact. These provide implications for seafarers requiring social skills as part of their duties and standard English.

Furthermore, online distance learning goes against face-to-face communication between teachers and students or students with each other. Online distance learning has been recognised as the technically supportet interaction between teachers and students, such as tele-conferences. However, when compared with the traditional face-to-face interaction, online distance learning still deprives students the opportunity of direct emotional communication and more reflects the ″human-computer interaction″. Increased global digitisation trends and the Fourth Industrial Revolution, creating increasingly automated vessels, digital ports and interconnected maritime logistics supply chains, further increase the pressure for virtual rather than real-time training and need for fewer, more technically sophisticated seafarers/staff. Associated studies have shown that when students are in a team learning environment and atmosphere, their learning effect will be better and learning proficiency is also higher. Based on the findings, it should be noted that more and more organisations have been attentive of this; therefore, they began to accept blended learning theory, which still emphasises traditional face-to-face training methods besides using the e-learning for network training and learning.

An alternative source focuses on the current challenges presented by e-Learning for teachers less familiar with it [5]. Comparatively few teachers and lecturers across many countries are quick and responsive to utilising all the potential of myriad forms of e-Learning. Acquiring information, social and administrative uses deserve more attention than dynamic and interactive communication and learning platforms. Implementation challenges remain as few universities, colleges and high schools which utilise e-Learning actively monitor their lecturers' and student's efficient utilisation of these platforms, tools and technology and subsequently take remedial measures or actions. Nor do they focus on the potential to subsequently improve educational outcomes via upgrading technology and e-Learning tools/techniques. Results indicated that the majority of 76 surveys and 15 interviews showed technology such as interactive white boards, Social Network Service and LMS-VLE Cloud-based services/the Internet were employed to locate information and complete and create online activities/lessons and social communication.

13.6 Durban University of Technology and ODL

Of the few South African maritime-orientated tertiary institutions, Durban University of Technology (DUT) are becoming more alert and progressive at envisioning e-Learning as an integral part of their curriculum. The Nautical Science degree specifically targets navigation, meteorology, shipboard management, construction and stability. This extends to short professional courses with simulators in ECDIS, electronic navigation, maritime education and human elements in thought and leadership. It appears on their website and within their handbook [6]. Students and lecturers employ blended and web-based learning via Blackboard where lecturers are trained and encouraged to upload course materials. It even asks as part of the recruitment process lecturers' ideas and potential approaches towards integrating e-Learning techniques. However, DUT does not currently focus on distance learning… This further motivates the subsequent methodology and survey participants recruited from locations such as DUT.

Other maritime education and training institutions within South Africa have yet to fully embrace e-Learning but include Cape Peninsula University of Technology, University of Stellenbosch, South African Maritime School and Transport College, SAMSA and several TVET colleges.

13.7 Applied Methodology

In order to highlight the most significant issues regarding the importance of ODL, quantitative analysis, by the form of survey questionnaires, was carried out with students and lecturers. The choice of the research methodology techniques empowered the respondents to express their genuine opinions regarding implementation of ODL in MET.

In developing countries, there is very little preliminary research on the adaptation of Cloud resources in education. The model proposed here (Fig. 13.1) has been inspired by a study which has been carried out in sub-Saharan Africa [7]. This model represents the basis for designing a questionnaire, by means of which the readiness of the high education institutions in the developing country (South Africa) to implement this type of education could be analysed. The model is based on triangulation (reconciliation) of two theories of adoption and expansion of ICTs: theory of diffusion of innovations [8] and the theory of a technologically acceptable model [9]. The model, which is proposed here, includes one dependent variable: intention to adopt Cloud into education. The independent variables in the model are organised in several subgroups: innovative, economic, technical, contextual and organisational factors (attributes). The last, but not the least, is the independent variable: actual use of Cloud in high education. In Fig. 13.1, direct and indirect links between dependent and independent variables are presented.

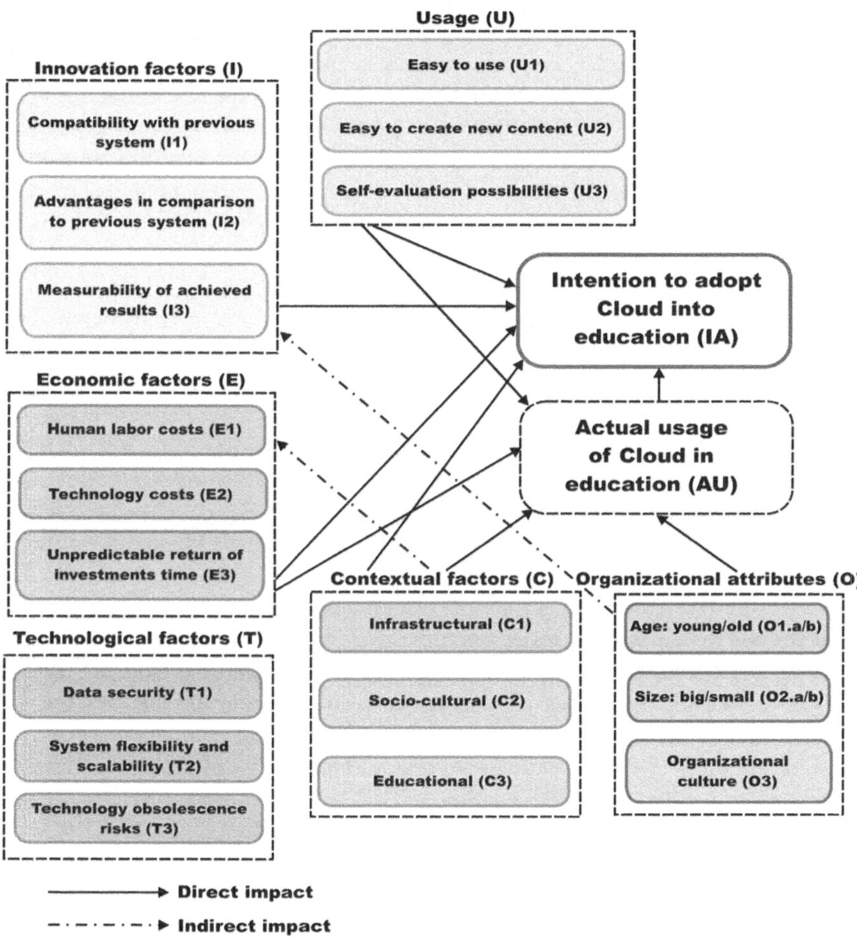

Fig. 13.1 Relations between relevant factors for moving education into Cloud. (Source: [10])

13.8 Data Collection and Analysis

Data will be collected through a questionnaire. A questionnaire is a preformulated written set of questions to which respondents record their answers. Respondents will have to answer the questions in such way to choose one number of Likert interval scale due to the best of their knowledge, experience and/or intuition. The Likert scale is designed to examine how strongly respondents agree or disagree with the statements on a five-point scale with the following anchors, which can be modified depending on the question formulation: strongly disagree (1), disagree (2), neither agree nor disagree (3), agree (4) and strongly agree (5). The Word document with questionnaire will be sent via email to the respondents. After completing it, the respondents will send it back. In the case on need, responders will be kindly reminded via mail to send their response in due time.

The preliminary version of questions contains question for lecturers 37 and 35 for students; after the consultation with the experts, that number is reduced to 25 for teachers and 25 for students. Questionnaires at the end were in accordance with code of conduct at WMU and ethical requirements. The final version of questionnaires for both lecturers and students is given in Appendix (1). After getting approval, questionnaires were sent via email to 40 teachers and 200 students at DUT, CPUT and UMA. Then the prospective despondences were reminded twice to send back their response in due time. Thanks to their kindness after less than a month, all responses were collected. The collected data were stored, and their validity was tested as well. Majority of the responses were numerical (Likert scale), and all responses passed preliminary test. Responders' additional narrative comments were left for the discussion in conclusion.

Since we have applied hypothetico-deductive approach, it commonly corresponds to quantitative analysis of collected data. The multiple regression analysis has been used, since the whole set of independent variables regarding both lecturers and students attitude towards ODL and e-Learning, were considered. Prior to multiple regression analysis, basic statistical descriptors of the data set were tested. Quantitative analysis has been conducted in SPSS and Excel Module for multiple linear regression graphical analysis. Simulation has been done. They were realised in a very short time frame over a couple of minutes.

Proposed hypothesis was tested upon the strength of correlation between dependent and independent variables when it comes to lecturers' and students' sets of responses. Statistical relevance of observed correlations has been tested via ANOVA in SPSS, and it has been proven over the set of analysed data. Upon the proven hypothesis, we offered a theoretical proposal for high MET stakeholders in terms of faster and more effective implementation of ODL and e-Learning. It is to be emphasised once again that emergency situation such as Covid-19 urges the whole process. All teaching and learning activities at DUT currently are moved to virtual platforms as Moodley and MS Teams, while all communications between lecturers and students are conducted via Zoom, WhatsApp and social media (Facebook, Twitter and Instagram).

13.9 Results and Discussion

The research and discussion are drawn from the responses of the lecturers and students interviewed through questionnaires sent to them via email. In total 35 responses from the lecturers (from DUT, CPUT and UMA MET higher institutions) are received in due course and 99 responses from students (from DUT, CPUT and UMA MET higher institutions).

Due to the extensive literature review including textbooks, journals, theses, conference proceedings, reports, newspapers, personal experiences and informal and formal discussions with colleagues, two sets of questions for both lecturers and students at examined METs were conceived. The questions are organised logically and neatly in appropriate sections along with the instructions on how to complete them. This helps the respondents to answer the questions without difficulty. All questionnaires are personally administrated. Through such approach doubts can be clarified, respondents can be motivated easily, almost 100% response rate is ensured and anonymity of respondents was high. The questionnaires were sent to the respondents via mail. Consequently, a wide geographical area in South Africa was reached. In addition, respondents can take more time to respond at convenience.

As a measurement tool, Likert scale has been used as a commonly exploited method of measuring opinions and attitudes. It measures the extent to which participants agree or disagree with given statement and typically ranges from 1 (strongly disagree) to 5 (strongly agree) with a neutral point in the middle (i.e. neither agree nor disagree). It is in the same time semantic differential scale and the numerical scale.

The applied approach was quantitative. After collecting data, in the next step, the data were analysed in order to answer the research questions. Before statistical analysis, data accuracy, completeness and suitability for further analysis were ensured.

On the basis of key statistics, in particular mean values of examined variables, upon the sets of 35 lecturers' response and 99 students' response, the following has been concluded:

(i) Lecturers: The interviewed lecturers have assessed real needs for introducing ODL in MET in SA as relatively high one (3.6 at 1–5 Likert scale). They have emphasised the need for free Internet access to lecturers and students (4.5), as well as the need for permanent institutional technical support in realising ODL (4.4). On another side, they express scepticism towards hypothesis that ODL can enable access to higher maritime education to the students living in rural areas and to those who are somehow socially marginalised. Also, lecturers are sceptical towards the assumption that ODL can upraise lecturers' and students' digital skills. This opens room for further investigation through in-depth interviews with lecturers.

(ii) Students: The interviewed students have assessed the real need for introducing and adopting ODL at METs in South Africa as a high one (4.03 at 1–5 Likert scale). They believe that ODL can support them to reach higher digital skills (4.29) and to upraise their thinking skills (4.14). But, they are also highly aware that the number of South African maritime higher education institutions, which can provide lecturers and students Internet access and computer labs, is constrained (4.01). The investigation reveals that insufficient percentage of students has personal tablet, laptop or smartphone (2.28). They are not well informed about similarities and differences in meaning of blended learning, e-Learning, computer-based learning, web-based learning and Cloud learning (2.43). And students are bit sceptic in terms that ODL is a good way of knowledge transfer for seafarers' knowledge refreshment and lifelong learning (2.91). All above noticed should be further interrogated through in-depth interviews with the students.

Due to the cross-correlation analysis among each pair of analysed variables in the model, the following conclusions can be drawn:

(i) Lecturers: Uprising students' thinking skills is in positive linear correlation with integration of formal and informal learning styles; level of ODL adoption in South African METs is in positive correlation with availability of Internet access and computer labs; lecturers' readiness to adapt curricula is positively correlated with availability of online assessments, etc. On another side, easier learning through ODL is in negative correlation with allowing ODL access to socially marginalised groups, which can be explained through the assumption that such way of learning might not be the most convenient one for those learners. Also, reducing costs due to introducing ODL is in negative correlation to making learning easier to socially marginalised students. It might be explained by the assumption that they are anyway prevented to participate due to the absence of Internet access, gadgets, etc. By analogy, the other pairs of constructs which are in strong correlation with certain significance can be argued.

(ii) Students: The level of students' preparedness for ODL is in strong positive linear correlation with their ability to manage their time well, their self-discipline and belief that ODL is more interesting than classical face-to-face learning. Also, upraising students' digital skills through ODL adoption is in strong positive correlation with significance with fostering their curiosity, innovativeness and virtual engagement in global e-classroom. On another side, students' belief that ODL can increase their digital skills is in strong negative linear correlation with the low level of their familiarity with e-Learning terminology and basic principles, including the lack of their personal laptops. In a similar manner, the rest of the positively and negatively correlated pairs of constructs in the model can be discussed.

On the basis of multiple regression analysis over the set of available data, the following has been found:

(i) Mean absolute per cent errors in the analysed cases of lecturers and students are, respectively, 15.08% and 21.80%.

(ii) Calculated $\overline{Y_s}$ value by multiple linear regression can vary based on standard error of regression estimate (SE) for the values ±0.709 in the case of lecturers' response and ±0.859 in the case of students' response.

(iii) Correlation coefficient values (r) are both above 0.56 that indicates strong linear correlation among considered dependent and independent variables in both analysed cases.
(iv) Coefficient of determination (r^2) indicates that $\overline{Y_s}$ is determined in 31% of samples in the first analysed case, suggesting a satisfying linear dependence, while in the second case, it is 40%, suggesting also satisfying linear correlation.

13.10 Conclusion

In conclusion, it should be noted that scattering in research topics and results when it comes to ODL and e-Learning in South Africa is more or less similar in other developing environments. Nevertheless, there is a need for upgrading high and continuous education through employing ODL and e-Learning mechanisms, since it is imperative of digital age in which we live and the necessity of digital literacy and team work in virtual environment advocate for ODL and e-Learning. Furthermore, everything goes globally and classrooms are on the road to become global to at least certain extent, especially when it comes to MET, which should be globally recognised and accredited to ensure safety, efficiency and effectiveness requirements. In addition, developing and implementing agreeable and effective policies in South Africa context still lags behind the countries of the first world, slowing down MET development and its harmonisation with the highest standards in the field.

It is noteworthy to mention that Covid-19 pandemic has proved the necessity of e-Learning in developing environments and the result of the study has shown the positive interest of online distance learning from the respondents.

Recommendations for Further Research

Further investigations should be done among the larger cohort of lecturers and students at METs in South Africa. Also, relevant stakeholders should be involved as respondents from DHET, SAMSA and EMSA, researchers in the field of high education, etc. In some cases, in-depth interviews should be conducted instead of questionnaires. This can give us better insight into the analysed preferences, needs and constraints when it comes to introducing and adopting e-Learning at METs in South Africa. Besides quantitative analysis, in-depth interviews will be better background for additional qualitative analysis. Also, instructional design component shall be involved in further investigation in the field. Benchmarking with METs in some other African countries should be done, as well as benchmarking with some METs from developed countries in Europe, for instance. This will give us broader picture of the issue. In addition, opportunities of collaborative online international learning with other developing and/or developed countries should be explored in some more detail. The issues of lecturers' intellectual properties should be examined as well, since it is an important dimension of the problem. In actual conditions, the impacts of Covid-19 on METs should be explored, since this can reveal some important facts about (un)successful practices in implementing e-Learning in emerging situations. Besides pragmatics of raising competences and achieving positive economic effects, exploring e-Learning adoption in high METs in South Africa should include social dimension, along with lecturers' and students' emotions and feelings if they work separately and in a kind of isolation. This important constructivist component should be among others a part of further, more rigor investigation.

References

1. T. Takalani, *Barriers to ELearning Amongst Postgraduate Black Students in Higher*, (2008)
2. South African Maritime Safety Agency (SAMSA), *SAMSA Annual Performance Plan 2018/2019*, SAMSA Report, Durban, (2019).
3. Y. Jiang, Q. Li, The implication of distance learning in competence based maritime education and training. Int. J. Learn. Teach. Edu. Res. **16**(5), 31–41 (2017)
4. F. Chang, Intelligent assessment of distance learning. Inf. Sci. **140**, 105–125 (2001)
5. O. Sadeck, J. Cronje, A continuum of teacher's eLearning practises. Electron. J. eLearn. **15**(5), 396–409 (2017) Viewed 3 Feb 2020, https://www.ejel.org
6. Durban University of Technology, Faculty of Applied Sciences Maritime Studies Handbook, Durban University of Technology, Durban, (2019).
7. M.S. Humphrey et al., Conceptualizing a model for adoption of cloud computing in education. Int. J. Inf. Manag. **36**, 183–191 (2016)
8. E.M. Rogers, *Diffusion of Innovations*, 5th edn. (The Free Press, New York, 2003)
9. F.D. Davis, Perceived usefulness, perceived easy of use, and user acceptance of computer technology. MIS Quart **13**(3), 319–339 (1989)
10. S. Bauk, M. Kopp, Z. Avramović, A case study on Iintroducing e-learning into seafarers' education. J. Inf. Technol. Appl. **3**(1), 34–43 (2013, June)

Chapter 14
Seafarer Training in the Age of Autonomy

John Abercrombie

14.1 Introduction

Seafarers and most people involved in the maritime industry ashore have over the past several years heard sometimes controversial points in the debate over the future of the maritime sector, concerning the autonomous operation of ships. Much of the discussion focusses on the possibility of ships operating autonomously without crew and with little or no human intervention and all the problems associated with that scenario.

The concept of automation is not new; as early as 1967 the Maritime Safety Committee (MSC) of the International Maritime Organization (IMO) distributed Circular 37 titled "Automation in ships" followed by Circular 55 in 1968, Circular 66 in 1969 and Circular 78 in 1970, thus setting a baseline for the introduction of autonomous ships. In the early 1980s, Japan lead the way in the discussion of intelligent unmanned ships where they envisaged unmanned ships supported by shore-based staff linked via satellite communication [1].

In 2006 the IMO initiated a harmonised E-navigation system which integrates electronic equipment and allows an exchange and analysis of information and data. The advances made in the E-navigation field provided a boost in the interest and perceived feasibility of autonomous ships.

J. Abercrombie (✉)
Maritime Studies Department, Faculty of Applied Sciences, Durban University of Technology, Durban, South Africa
e-mail: johna@dut.ac.za

S. Bauk, S. D. Ilčev (eds.), *The 1st International Conference on Maritime Education and Development*, https://doi.org/10.1007/978-3-030-64088-0_14

14.2 From Perception to Reality

In 2017, industry pressured IMO to play a regulatory role with respect to autonomous ships, and in the same year, the term Maritime Autonomous Surface Ships (MASS) was adopted and referred to future ships whether unmanned or fully autonomous. The degrees of autonomy identified as:

- Degree one: Ship with automated processes and decision support – Seafarers are on board to operate and control shipboard systems and functions. Some operations may be automated and at times be unsupervised but with seafarers on board ready to take control.
- Degree two: Remotely controlled ship with seafarers on board – The ship is controlled and operated from another location. Seafarers are available on board to take control and to operate the shipboard systems and functions.
- Degree three: Remotely controlled ship without seafarers on board – The ship is controlled and operated from another location. There are no seafarers on board.
- Degree four: Fully autonomous ship – The operating system of the ship is able to make decisions and determine actions by itself.

It soon became clear that the current framework of maritime law, including national legislation and international treaties, regulations and codes including an extensive range of issues, such as the human element, safety, security, liability and compensation for damage, interactions with ports, pilotage, responses to incidents and protection of the marine environment, had to be investigated and possibly amended before the operation of autonomous ships could begin to approach legality.

The list of instruments to be covered in the MSC's scoping exercise for MASS includes those covering safety (the SOLAS Convention); the International regulations for the prevention of collisions at sea (COLREGS); loading cargo and ship stability (International Convention on Load Lines); training of seafarers and fishers (the STCW Convention and Code and the STCW-F Convention); search and rescue (SAR); tonnage measurement (Tonnage Convention); International Convention for Safe Containers (CSC); and special trade passenger ship instruments (SPACE STP, STP). The process of assessing IMO instruments to see how they may apply to ships with varying degrees of autonomy continued during the Maritime Safety Committee (MSC) 100th session [2].

The first step has been completed. The next step is to analyse and determine the most appropriate way of addressing MASS operations, taking into account, inter alia, human element, technology and operational factors. The analysis will identify the need for:

- Equivalences as provided for by the instruments or developing interpretations
- Amending existing instruments
- Developing new instruments
- None of the above as a result of the analysis

The aim is to complete the regulatory scoping exercise in 2020.

The Maritime Safety Committee (MSC), 101 session, in June 2019 approved Interim guidelines for Maritime Autonomous Surface Ships (MASS) trials in MSC.1-Circ. 1604.

The Principles and main objectives state that "Trials should be conducted in a manner that provides at least the same degree of safety, security and protection of the environment as provided by the relevant instruments. The following objectives are provided to guide relevant authorities and stakeholders when planning, authorizing and conducting trials of MASS-related systems and infrastructure".

Onboard or remote operators of MASS should be appropriately qualified for operating MASS subject to the trial. Any personnel involved in MASS trials, whether remote or onboard, should be appropriately qualified and experienced to safely conduct MASS trials. Appropriate steps should be taken to ensure sufficient cyber risk management of the systems and infrastructure used when conducting MASS trials [3].

The increasing awareness and evidence of MASS implies there is an increased chance of shared internet digital system being compromised as a result of malicious cyber-attack [1]. The MSC in its 98th session adopted resolution 428 on Maritime Cyber Risk Management in Safety Management Systems to support future shipping's security and operational resilience to cyber risks.

It is asserted that in a shared internet digital system, the human element and careless habits of employees are the weak link in the maintaining the cyber security of the system [4] and that MASS systems should allow the interaction between security risks such as terrorism, organised crime, piracy, cyber/radio frequency/satellite attacks, drone strikes, holding to ransom, diversion of sensitive cargo to hostile actors on the one hand and disaster risk management such as collisions, pollution, loss of business continuity, search and rescue of crew and disaster impacts on the other [5].

14.3 Are Seafarers Dispensable?

During the 100th session of the Maritime Safety Committee held in 2018, a video showing the trial of a fully autonomous ferry operating in Finland between the ports of Parainen and Nauvo in fully autonomous mode and under remote control operation was shown, and the remotely controlled operation of an offshore vessel in August 2017 was also showcased. In this case the vessel was controlled remotely from San Diego by standard bandwidth over a distance of over 8000 km.

Mr. Branko Berlan, Accredited Representative of the International Transport Workers Federation (ITF) to IMO, posed the question: "Are seafarers indispensable?", and his message was that seafarers are still key to safe and secure ship operation. He pointed out that "The accident/incident rate for international merchant ships is less than 5% of all ships per year", and "Seafarers are prepared for new technologies and automation". He added, "It is happening: it is not revolution, it will not come tomorrow or next week; it is evolution" [2].

Some of the most dynamic, traditionally human-judgment intensive aspects of vessel operations are the SOLAS convention provisions relating to "recovery of persons from the water"; carriage requirements for navigational equipment; load line information; emergency towing arrangements; and navigation bridge visibility to mention but a few. As shipping continues its voyage into the next age of technological evolution, as it did from oar to sail, from sail to steam and from steam to diesel, the legal and regulatory infrastructure will continue to adapt [6].

In order to answer the question "Are seafarers dispensable?", it is necessary that the roles of persons involved in the operation of MASS be identified. As the vessels become more autonomous, going from "degree 1" through to "degree 4", the roles of persons will change from a purely operational requirement to a more managerial requirement, and there will still be the need of seafarers at partial automation or digitalized systems because their "cognitive skills and tacit knowledge are essential for safe operation of ships" [7].

In the past 120 years, there have been four so-called industrial revolutions. The first was the introduction of steam to industry marking large-scale mechanisation which radically changed the labour market and the education system. The next revolution was the introduction of electricity to power mass production. The "information age" based on computers and using information technology was the next to cause widespread change. The latest revolution is based on the technological advances made in cyber communication, the Internet and artificial intelligence which has made the reality of autonomous ships possible.

During the previous industrial "revolutions", the introduction of new technology replaced some jobs but at the same time created a new set of jobs requiring different skills and competencies. For example, the introduction of mechanisation in production lines decreased the need for manual labour but created a labour demand for operation and maintenance of the machines. Recently, the introduction of GMDSS made the job of the radio officers on board ships redundant, but the recent technological advances in communication and integrated navigation equipment required the employment of electrotechnical officers to maintain the electronic equipment.

The present revolution is not easy to pigeonhole as it is now happening with the expectation of more innovations in the near future through utilisation of digital power and a combination of technologies. This era is known for its paradigm shift from traditional central production into decentralised, smart, autonomous systems.

These technologies along with intelligent equipment and control systems redefine the way the work is done in the maritime industry which means new job definitions that require a different set of skills and competencies [1].

What would be the possible skills and competencies of the operators of the unmanned and autonomous ships be? It is evident that this must be known before a proposal of what the best possible course of actions would be for the maritime training institutions to be prepared for development of competencies that seafarers may need to man the future maritime industry.

However, in returning to the original question, "No. Seafarers are indispensable. They are the force that continually adapts to the changing environment and challenging technological advances and they will do it once again in this, the age of MASS".

14.4 MASS Era Roles, Skills and Competencies

There are many aspects of autonomous shipping including design and development, providing control systems, logistics services, marine insurance and surveying and vessel operations such as safe navigation in close proximity of other stationary or moving vessels and objects that will not change except for the degree of technology available. The implementation and utilisation of technology does not only depend on the feasibility of technology but also on other factors such as business advances, shipping economics, the regulatory environment and politics.

In shipping, change, such as the introduction of new technology requires the revision of the legal and regulatory regime, the administrative requirements and the development of maritime education and training to provide seafarers with the new skills and competencies required.

In the autonomous era, the challenge is to anticipate the skills and competencies Control Centre Operators will need and to additionally provide training for these roles. While there is no clear indication on the time frame for its full implementation, it is critical that all stakeholders foresee the trend for new roles, skills and competencies in the future development of the maritime workforce, which will involve the use of cognitive skills and tacit knowledge for safe ship operation. The predetermined fact is that future shipping will require a well-trained and highly educated workforce.

The shift in job description and the type of competency required by the maritime workforce including seafarers is staggering. The shipping industry has already initiated the shift through embracing integration of digitalisation into onboard and shore-based shipping operations. It is clear that many of the technical competencies regulated in the STCW Convention and Code still play a significant role in the maritime industry in the short, medium and long terms. On the other hand, technological awareness, computing and information technology skills, environmental awareness and sustainability concerns will be essential competencies for the future seafarers faced with emerging challenges of MASS [8].

The Maritime Unmanned Navigation through Intelligence in Networks (MUNIN) Project was established to develop a technical concept for the operation of an unmanned merchant vessel and assesses its technical, economic and legal feasibility. The project was launched in 2012 and envisaged autonomous operation of an unmanned vessel during deep-sea voyages. A crew on board was to execute the operation in congested waters or during the approach to port.

The MUNIN proposed a concept, where the ship is autonomously operated by onboard systems but the monitoring and controlling functionalities are executed by an operator in a Control Centre ashore. The following systems and entities were implemented as prototypes:

- An Advanced Sensor Module takes care of the lookout duties on board the vessels.

- An Autonomous Navigation System follows a predefined voyage plan, but with a certain degree of freedom to adjust the route autonomously, e.g. due to an arising collision situation or significant weather changes.
- An Autonomous Engine and Monitoring Control system enriches ship engine automation systems with certain failure-pre-detection functionalities while maintaining optimal efficiency.
- A Shore Control Centre manned by skilled nautical officers and engineers continuously monitors and controls the autonomously operated vessel.

The complexity of technical solutions needs to be kept as low as possible in order to minimize the need for human maintenance and interventions. The "Shore Control Centre" (SCC) is continuously manned and thus able to take over remote control in difficult situations. Without an SCC, the ship would require extremely complex software to handle all eventualities. With an SCC, the ship will still be able to handle the majority of all situations autonomously, but can fall back automatically on shore control for the remaining extremely difficult to handle situations.

The Shore Control Centre (SCC), while primarily focussed on monitoring the operational status of the vessel, also provides the capability to take over direct remote control in extraordinary situations. This requires the operator to be aware at all times of the situation on board the vessel and creates a challenge to keep situation awareness of the operator appropriately high despite the physical distance between the vessel and the human.

The MUNIN Project furthermore proposed a system which would allow an operator to monitor six vessels, and assuming a technologically mature state, it was proposed that the SCC be able to monitor 100 ships, 6 ships per operator. In addition, the system allows for operator assistance needed when the vessel is released to, or retrieved from, autonomous operation, as well as in communication with other ships and shore-based services. Provisions for advanced assistance by a SCC "captain" and "engineer" for problem-solving will also be available [9].

Skills and Competencies Required in the MASS Era

From the information presented above, it is clear that many of the technical competencies regulated in the STCW Convention and Code will still play a significant role in the maritime industry in the MASS era. However, technological awareness and computing and information technology skills coupled to environmental awareness and sustainability concerns will need to be considered essential competencies for seafarers faced with emerging challenges of MASS.

The STCW Convention of 1978 was introduced by the IMO in an effort to impose minimum standards of Maritime Education and Training on seafarers. As technology advances were introduced into the shipping industry, the focus of STCW shifted to developing seafarer's technical knowledge and, more recently, soft skills such as teamwork, leadership, decision-making and resource management, bolstered by improved communication skills.

Knowledge of ship operation and safe navigation remain fundamental competencies in the new era although all operations will be impacted by ever-advancing technology. The increasing need for soft skills in the form of resource management, team work, leadership, communication and professional language needs to be expanded to include critical thinking, quantitative reasoning, technical literacy and awareness, computing and interpersonal and social skills, adaptability, flexibility and self-development.

14.5 Seafarer Training in the Age of Autonomy

Internationally, seafarer training has changed fundamentally in the last decade. The now outdated model of shipping companies sponsoring cadets at sea and meeting their training expenses has been replaced by Maritime Training Institutions offering degree programmes where students do not need sea service in order to qualify from the place of learning. The curriculum is still monitored and approved by the local Maritime Authority, and the content of programmes meets the minimum requirements laid down in the STCW Code.

A major advantage of this system as opposed to an insular "Maritime Academy" approach is that the institutions usually have a diverse number of qualified lecturers in diverse fields. This allows these institutions to provide maritime learners with the skills needed in the autonomous era from Faculties such as Electronic or Electrical Engineering, Information Technology, Management and others, while the Maritime Studies Department focusses on navigation, naval architecture and ship construction, ship stability, seamanship and the traditional aspects of shipboard skills and acts as the lead provider for seafarer training.

The professional merchant mariners who operate ships today are the crucial on-scene decision-makers, repairmen and physical security providers who make commercial shipping secure, efficient and inexpensive [10]. For the MASS era, these are the aptitudes needed within remote control centres, coupled with advanced computing skills, technological awareness and an acute sense of situational awareness which only comes with experience.

Industry will be the driver of change in Maritime Education and Training causing institutions to adapt to increased technological demands of the MASS era and the IMO to adapt the STCW Convention and Code to meet these changes. Although the emergence of MASS is not going to be sudden, it is certain, and Maritime Education and Training Institutions and the IMO need to adapt to the new approach.

Recent events have caused MET Institutions to modify their delivery of education and training, and these institutions need to ensure that they remain adaptable to possible changes such as technology manufacturers providing a decentralised education system direct to their customers along with continued professional development and lifelong learning. The MET Institutions need to integrate research into the teaching discipline and provide relevant educational technology and encourage a culture of continued professional development for academic staff.

MET Institutions urgently need to update their curriculum, develop appropriate human resource strategies, enhance cooperation with industry and develop appropriate infrastructure. A critical component of future maritime education and training is the guaranteed provision of suitably qualified lecturers able to teach the competencies required for tomorrows' workplace.

Maritime Authorities need to address the changing environment and proactively manage the regulatory changes that will promote the development of autonomous surface ships. These regulatory and developmental changes cannot occur in isolation but must be undertaken in a consultative environment supported by Industry [1].

14.6 Conclusion

It can be concluded: "Seafarer training has evolved over time as technology has introduced ever more sophisticated navigation systems and equipment and the time to modify our approach to training for autonomous ships has arrived" [7].

This paper has accompanied the reader through the developmental stages of the MASS and the problems associated with the MASS era and proposed the roles and competencies required by seafarers in a MASS environment. Proposed changes to Maritime Education and Training and the Institutions that deliver MET have been made, and these are supported by proposed changes to Maritime Authorities and the International Maritime Organization.

References

1. 21th Marine Industries Conference (MIC2019) 1-2 January 2020 – Qeshm Island, (2020)
2. IMO MSC 100, in MSC 100, (London, 2018)
3. IMO, MSC.1-Circ.1604, in *MSC Circular*, (London, 2019)
4. IMO, MSC 98, in *IMO*, (London, 2019)
5. Maritime Autonomous Surface Ships (MASS) and Framework Development Challenges, (2019). [Online]. Available: https://www.gsdm.global/category/emerging-issues/ [Accessed 10 June 2020].
6. C. Hannan, The Modern MASS-T Head: The Rise and Reality of Maritime Autonomous Surface Ships, (2019)
7. Kitada et al., Command of Vessels in the Era of Digitalization, in *International Conference on Applied Human Factors and Ergonomics*, (2018)
8. K. Cicek, E. Akyuz, M. Celik, Future Skills Requirements Analysis in Maritime Industry, (2018). [Online]. Available: www.elsevier.com/locate/procadia [Accessed 10 June 2020].
9. Maritime Unmanned Navigation through Intelligence Networks, 31 August 2015. [Online]. Available: https://cordis.europa.eu/project/id/314286/reporting [Accessed 10 June 2020].
10. D. Dubay, Why we Will Never See Fully Autonomous Commercial Ships., 25 June 2019. [Online]. Available: https://www.maritime-executive.com/author/commander-david-dubay-uscg [Accessed 3 June 2020].
11. N. Institute, *Automation and autonomous systems* (Nautical Institute, London, 2019)

Chapter 15
Maritime Education and Training (MET) Curriculum Challenges in the Twenty-First Century

Nomzamo C. Phewa

15.1 Introduction

Maritime education and training (MET) has become the key contributor in the development of the maritime industry given that maritime operations play a crucial role in improving the economy of many countries. Over the years trade has increased, competition intensified, cargo volume grew, and ships became larger and more specialized [9]. Due to these changes, seafarers had to realign their responsibilities and adapt to the changes.

Nowadays, education is influenced by technological advancement and globalization. The shift from traditional methods to modern methods of education and training is upon us. The fourth industrial revolution is changing the roles of seafarers and the way maritime industry as a whole operates. Alop [1] highlights that the distinctions between the fourth industrial revolution and the previous industrial revolutions are extent, systemic impact and speed. Currently, the world is witnessing an increase in levels of autonomation on vessels. Smart Ports such as the Port of Singapore (Singapore) and Port of Rotterdam (Netherlands) are already in operation [3]. This is an indication that some parts of the world are already preparing for the fourth industrial and innovation wave.

Education has to take a lead in alleviating the challenges of the fourth industrial revolution, and in this case, maritime education and training is no exception [1]. This disruption requires maritime education and training institutions to relook at their skill set and curriculum development and design. Moreover, with the rise of digitalization, investment in education is becoming more important. According to Lunenburg [8], "Curriculum development is the process of planning, implementing, and evaluating curriculum that ultimately results in a curriculum plan". When

N. C. Phewa (✉)
World Maritime University, Malmo, Sweden

S. Bauk, S. D. Ilčev (eds.), *The 1st International Conference on Maritime Education and Development*, https://doi.org/10.1007/978-3-030-64088-0_15

developing a curriculum, there are factors to be considered of which without these, the product would be incomplete and inconsistent.

15.2 Problem Statement

According to HRDC [6], South Africa is faced with skills crisis, and there is a need for skills development in the maritime sector. Today, about 98% of South Africa's total trade is carried by sea which shows the significance of shipping. It is estimated that South Africa's maritime industry may contribute about R129 billion and R177 billion to GDP by 2033, therefore creating between 800,000 and 1,000,000 jobs.

More challenges were discovered within the maritime industry in South Africa, and they include disconnection between higher education and training institutions and the maritime industry which results in outputs that do not match the demands of the industry [6]. The disconnection between the two mentioned parties led to curriculum developers getting confused on what is required in the maritime sector. There has also been an issue of poor governance which leads to poor quality for education and training, and it can be assumed that poor education stems from not properly developed curriculum.

The study intends to highlight challenges of maritime education and training curriculum and emphasize the importance of curriculum presage.

15.3 Aims and Objectives

The aim of the study is to examine underlying challenges of maritime education and training (MET) curriculum in the twenty-first century. The specific objectives of this research are as follows:

- To identify possible and existing challenges of maritime education and training curriculum
- To assess curriculum development and design and its implementation
- To assess roles of government in responding to technological advancement

15.4 Research Question

To conduct this study and address the aim and objectives of the research, the following questions were used to drive this study and find the specific areas for analysis:

- What are possible and existing challenges of maritime education and training curriculum?
- What strategy is used for curriculum development and its implementation?

- What are the roles of the government in responding to technological advancements?

15.5 Applied Methodology

Research Design The study is based on the curriculum challenges for maritime education and training in South Africa with reference to Norway. In obtaining data the researcher will use descriptive and exploratory research designs. According to Shukla [18], exploratory research design deals with the collection of either primary or secondary data using informal procedures to interpret them. Moreover, exploratory research design includes focus groups, in-depth interview and projective techniques. The researcher will employ this research design with the aim of investigating sources of challenges related to maritime education and training curriculum. "The goal of descriptive research is to describe a phenomenon and its characteristics" ([12] p.129). In this study the researcher will use descriptive research design to gain opinions of participants on the challenges of maritime education and training curriculum.

Research Method The study will employ qualitative research approach as the aim is to examine existing and potential challenges of MET curriculum in the twenty-first century. Monsen and Van Horn [11] state that qualitative research approach generates narrative data that can be explained in spoken and written words rather than in figures. Furthermore, Nassaji [12] explains that one of the main aims of qualitative research is to obtain initial insights into decision issue and opportunities.

Data Collection Data collection is a process by which scholars, researchers and other professionals collect data to check their hypotheses and arguments in order to answer research questions. Gathering of data can be done through interviews, visual observations, surveys, interpretations and experiments. Due to Covid-19, data will be collected through reading of reports, articles, journals, interviews (telephonic) and questionnaires (mostly through Google Forms). Participants will include officials from the South African Maritime Authority (SAMSA) and Department of Higher Education and Training (DHET) and university officials from the Durban University of Technology (DUT), Nelson Mandela Metropolitan University (NMMU) and University of South-Eastern Norway (USN).

Data Analysis Method The researcher will use Google Forms to collect and analyse data as this will be a simple way to reach participants from Norway and other provinces of South Africa. Narrative analysis will be the main analysis method to analyse unspoken and spoken words obtained from reports, journals, interviews and questionnaires related to the study. Narrative analysis is defined as a strategy that recognizes the extent to which the stories people share give insights to their experiences [4]. In the later stages of narrative analysis, researchers turn into storytellers because of the interpretation of gained data [16].

15.6 Curriculum Presage and Foundation of Curriculum Development

Curriculum presage is defined as "Those activities and forces which influence curriculum developers in their curriculum decision-making tasks. These activities and forces are brought with the developers when they come to the task of constructing a curriculum. As such they consist of the curriculum backgrounds, curriculum conceptions, curriculum representations, curriculum foundations of the various curriculum developers and the curriculum context in which they work. This combination of past activities and current forces will have a profound effect upon the final curriculum through the nature of the input from the individuals involved" ([14], pp. 25–6).

Philosophical Forces According to Paul Hirst, philosophy involves interpretation of ideas and theories in which one's experiences and undertakings are intelligible (as cited in Print [14]). Philosophy is essential for selecting appropriate learning objectives. Furthermore, curriculum developers have to understand their own beliefs of the learning process, knowledge and the world in order to make informed decisions regarding curriculum. Therefore, it is vital to realize the importance of philosophy in education. There are three questions to be considered in understanding the influence of philosophy: (1) what is real? (ontology), (2) what is good? (axiology), and (3) what is true? (epistemology). Differing answers to the mentioned questions have an effect on how curriculum developers reckon of the curriculum.

Sociological and Cultural Influences Print [14] states that culture and society have a huge effect on the development and design of the curriculum. One cannot separate education from society and culture as the purpose is to equip societies with necessary skills and knowledge. Curriculum developers have a responsibility of considering these influences in the formation of the curriculum, meaning their task entails transforming traditional knowledge, attitudes, values and assumptions into curriculum learning activities, content, objectives and evaluation. The important influence of sociological factors is on the content of the curriculum because all the objectives are connected to activities taking place in societies. Curriculum formed must complement multicultural societies.

Psychological Considerations Sharma [17] defines psychology as the science of human behaviour and experience. There are psychological factors to be considered when planning for curriculum development [2]. Psychological factors assist developers in determining on types of approaches for learning, objectives and evaluation procedures to be employed at the end of the lesson [15]. According to Print [14], curriculum developers should consider the following psychological forces before forming the curriculum:

- *Educational objectives* – The understanding of the principle of psychology assists curriculum developers in setting suitable objectives for learning.
- *Learning processes* – Psychology helps in improving an understanding of how people learn and behave in a learning setting.
- *Student characteristics* – It is crucial to understand learners' personality and background so as to know what objectives to set.
- *Teaching methods* – By understanding learner's background and personalities, the decision of choosing suitable teaching methods becomes easier.
- *Evaluation procedure* – Psychology also contributes to the evaluation procedures for both learners and teacher performance.

15.7 Curriculum Development Model in the Twenty-First Century (Phases)

Organization The model recommends that the starting point for the curriculum development is formalizing measures of curriculum presage, meaning a closer look at those individuals who will undertake the role of developing the curriculum should be made – this entails checking their background and factors that form the way they think [14]. Maritime education and training institution would involve instructors, curriculum committee and officials from the Ministry of Maritime Affairs. Factors such as philosophical, social and psychological play a vast role on how curriculum developers perceive knowledge.

Development In this phase the group entrusted with the responsibility of forming the curriculum begin to plan for curriculum development document and the required material. In order to achieve the aim of this phase, curriculum developers follow the trail in the middle of the model being situational analysis, aims, goals and objectives, content and instructional evaluation. Print [14] asserts that, through performing situational analysis, educators get the opportunity to be aware of the needs of the students and resources available to meet those needs. Moreover, situational analysis also assists them in devising useful and proper aims, objectives, goals and appropriate content for meeting the philosophical, social and cultural and psychological needs of students. In terms of instructional evaluation, curriculum developers have the responsibility of planning effective assessment processes, and this is done to determine the level at which students have met the intended learning objectives. The focus of instructional evaluation is on the product; in other words it evaluates how well learners have understood during the learning process.

Application In this phase the actual application of the curriculum commences, and it includes implementation, monitoring of and feedback from the curriculum and provision of feedback data to the presage team. For the curriculum project, change is likely to take place, and if it does happen, developers should ensure that they avoid confusion and disturbance by developing a plan for implementing the change.

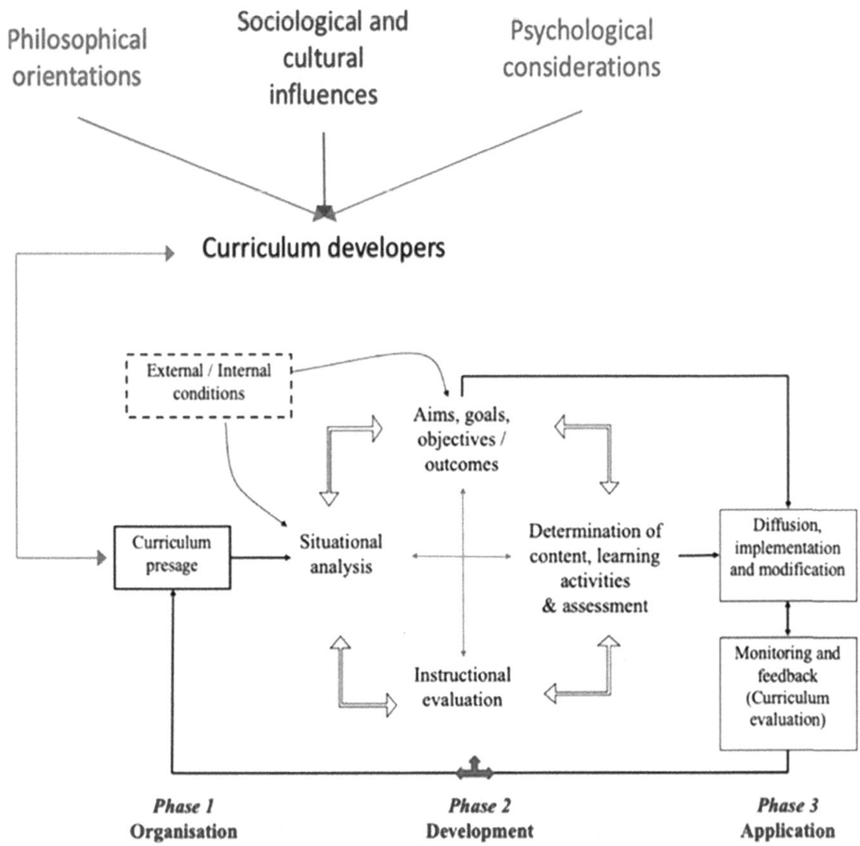

Fig. 15.1 Curriculum development model. (Adapted from Manuel [10] and Print [14])

However, the successful implementation lies in the ability and willingness of curriculum developers to accommodate adjustments to the curriculum [14]. Monitoring stage is very important because it measures the success of the curriculum undertaking; following that, evaluation of the effectiveness of the curriculum takes place. In addition, feedback from evaluation process will be obtained, and it normally takes a number of years. Figure 15.1 below depicts the curriculum development model; each and every phase is important as they are interrelated, and curriculum developers cannot skip the first phase and begin with application.

15.8 Challenges of Curriculum Development

Diversity. Gosper and Ifenthaler [5] highlight that in the twenty-first century, student profile in higher education is more diverse than that of the previous decades. Furthermore, higher education institutions have become more internationalized than in the previous years. Gosper and Ifenthaler [5] further state that diversity is challenging because it comes with different backgrounds, professional experiences and beliefs about teaching and learning. MET is a complicated and dynamic discipline that requires professionalism when it comes to design and development of the curriculum especially now that education is internationalized.

Conflicting Ideas Curriculum design and development involves a number of stakeholders with different perceptions about educating and training. They also have different views in terms of where to best begin with the development process [14]. Different views tend to cause conflict in the process, and in return the quality of the curriculum is compromised.

Resistance to Change Change can cause despair because people fear that their roles will change and some fear to move from their comfort zone. To some, change brings hope and strength [7]. With regard to maritime education and training, resistance may come from lecturers/instructors and seafarers because they are familiar with traditional methods of operating. Manuel [10] asserts that if the need for change is clearly defined and explained to affected individuals, chances of resistance to change are minimized.

Lack of Resources As much as the rise of digitalization presents positive impacts in terms of creativity and innovation, some maritime education and training institutions still lack adequate resources. According to Phewa [13], it is difficult for countries with inadequate resources to train globally competitive maritime personnel in the technological world. Furthermore, maritime education and training institutions particularly in South Africa are still in need for effective funding models for seafarers and other maritime personnel. The study "Maritime Education and Funding Models in Different Jurisdictions: Challenges and Opportunities in South Africa" showed that out of 44 participants, 13 were funded by their employers and another 13 were funded by their parents/guardians through loans. This is an indication that there still needs to be an investment in resources in order to grow and uplift maritime education and training institutions.

15.9 Government Role

Government has a critical role to play in order to ensure that South Africa's maritime education and training is of good quality and that maritime personnel are internationally recognized. The issue of restructuring the maritime education and training curriculum should make it to national government agenda in order for it to get attention, and this is where the legislative institutions from different spheres of government, administration and other public officials come in.

In addition to the above, government has to ensure that maritime education and training personnel are well equipped with relevant skills in order to adjust to the digital world so that technology does not become a burden to them; in this way resistance to change is unlikely to occur. HRDC [6] points out that South Africa's government should ensure that student enrolment standards are aligned with the industry requirements and that there is a closer cooperation between industry and TVET colleges in respect of funding, curriculum development and performance management.

15.10 Conclusion

Selected individuals should be fully aware of the impact they have on the final product and the future of students. This will minimize challenges of curriculum development and conflict during the process. Moreover, communication between involved individuals should be clear and transparent; this will avoid barriers to progress.

Roles and responsibilities of curriculum developers should be well stated in order to reduce confusion; this will help everyone to understand and work towards the common goal. Additionally, there must be a direct communication between maritime education and training institution and the shipping industry in order to ensure that curriculum meets the needs of the industry.

Due to changes in our social settings, education, economy and technology curriculum requires some modification. This is to ensure that aims, objectives and content are in line with the trending issues in the world; thus this also helps learners in adapting to change, and it is even helpful for the growth of their careers. The process of curriculum change is very critical; that is why individuals involved in designing and developing should not skip any of the phases. Additionally, readiness of governments to embrace digitalization or rather technological advancements in the maritime sector determines the success of the curriculum. Issue such as resistance may hinder the success of the curriculum.

References

1. A. Alop, Challenges of the digital technology era for maritime education and training. European Navigation Conference (ENC), 1–5 (2019)
2. A.E. Carl, *Teacher empowerment through curriculum development: theory into practice* (Juta & Company Ltd, Cape Town, 2009)
3. Y. De Clerk, *Scenario planning for an autonomous future: a comparative analysis of national preparedness relating to maritime policy/legislative frameworks, societal readiness and HR development for autonomous vessel operations* (World Maritime University, Malmo, 2019)
4. N. Gilbert, *Researching social life* (Sage, London, 2008)
5. M. Gosper, D. Ifenthaler, *Curriculum models for the 21st century: Using learning technologies in higher education* (Springer, New York, 2014)
6. HRDC, *Maritime Sector Skills Technical Task Team Report: Advancing skills development in the Maritime Sector* (HRDC, South Africa, 2014)
7. G.a. Lachiver, *Fostering and management curriculum change and innovation* (IEEE, 2002)
8. F. Lunenburg, Curriculum development: Inductive method. Schooling **2**, 1–8 (2011)
9. S. Ma, *Maritime economics: Lecture handout* (World Maritime University, Malmo, 2018)

10. M. Manuel, *Curriculum development and design: Lecture handouts* (World Maritime University, Malmo, 2019)
11. E.R. Monsen, L. Van Horn, *Research: Successful approaches* (Diana Faulhaber Publisher, USA, 2008)
12. H. Nassaji, *Qualitative and descriptive research: Data type versus data analysis* (2015), pp. 129–132
13. N. Phewa, *Maritime education and training (MET) funding models in different jurisdictions: Challenges and opportunities in South Africa* (World Maritime University, Malmo, 2019)
14. M. Print, *Curriculum Development and Development* (Allen & Unwin Pty Ltd, Australia, 1993)
15. S.N. Rao, *Educational psychology* (New Age International, LTD Publishers, New Delhi, 1995)
16. K. Rudestam, *Surviving your dissertation: A comprehensive guide to content and process* (Sage, USA, 2015)
17. R. Sharma, *Advanced educational psychology* (Atlantic Publisher and Distribution, New Delhi, 2006)
18. P. Shukla. Essentials of marketing research, (2014).

Chapter 16
European Tempus TATU Project as the Basis for the IT/OT Maritime Engineers Education and Retraining

Vladlen Shapo

16.1 Introduction

The last decade took place true jump in approaches to development, control, and exploitation of different complex technical systems. In maritime field data exchange technologies, based on Industry 4.0, IoT, IIoT, Shipping 4.0, Port 4.0, Life Long Learning (LLL), and other concepts, are being implemented very actively. Thanks to the vast progress in wireless and satellite data transfer technologies and mass appearance of high-performance embedded computer systems and information technologies, it has become possible to maintain inextricable links to the classical approaches and allow to perform absolutely new tasks like truly intelligent remote control and to create fully unmanned ships and complex technical systems. Thereby Unmanned Cargo Ship Development Alliance is founded; Advanced Autonomous Waterborne Applications Initiative autonomous ship research project and Maritime Autonomous Surface Ships direction are created; Distributed Intelligent Vessel Components software, providing new protocol for devices connecting and data transferring, is developed; Digital, Internet, Materials & Engineering Co-Creation technical ecosystem and One Sea Ecosystem Alliance are created; and Safer Vessel with Autonomous Navigation project is realized. As a result of these efforts on maritime branch, new digital transformation plans appeared, which envisage creation of various unmanned, autonomous, and remote-controlled ships by 2025–2035 and even earlier.

V. Shapo (✉)
National University "Odessa Maritime Academy", Odessa, Ukraine

S. Bauk, S. D. Ilčev (eds.), *The 1st International Conference on Maritime Education and Development*, https://doi.org/10.1007/978-3-030-64088-0_16

16.2 Purpose

Thus, it's necessary to improve and deepen learning of information technologies in maritime education and training institutions in the following directions: IoT, IIoT, industrial data transfer technologies, networks, and protocols; wireless data transfer technologies; hardware of wide class computer control systems; satellite data transfer systems, technologies, and protocols; cyber security aspects, technologies, and protocols; big data, artificial intellect, and virtual and augmented reality technologies; and remote control technologies and protocols. For realization of these points, it's necessary to use specific hardware facilities and e-learning technologies.

16.3 Approach

The Theory of Automatic Control and Computing Machinery (TACCM) department of National University "Odessa Maritime Academy" (NUOMA) participated in EduNet program in 2011 and participated in European Tempus TATU (Trainings in Automation Technologies for Ukraine) project in 2013–2017. Goals and obtained results of TATU project are described in paper [1]. During TATU project realization, learning books [2, 3] are created by efforts of Ukrainian and European TATU project participants. Some results, obtained at working at the same time with MOODLE-based e-learning system, are described in papers [4, 5]. Existing EduNet and TATU equipment allows to study the following software and technologies:

1. PC Worx and CODESYS are complex software integrated development environments (IDEs) for development of automation and control systems based on programmable logic controllers (PLCs).
2. Profibus, PROFINET, EtherCAT, and some other technologies/protocols.
3. Wireless data transfer based on IEEE 802.11 b/g/n and GSM standards.

During TATU project NUOMA TACCM department obtained four sets of TATU Smart Lab (TSL) equipment. This is a flexibly configurable mobile set of devices for teaching modern automation technologies. It contains devices from Phoenix Contact, Siemens, and Berghof companies and was developed within the framework of Industry 4.0 concept. There are three independent portable hardware modules (HM) in each TSL set: Profibus, PROFINET, and EtherCAT/Process Modeling (PM) HM. Appearance and internal structures of Profibus HM, PROFINET HM, and EtherCAT and PM HM are presented in Figs. 16.1, 16.2, 16.3, 16.4, 16.5 and 16.6 accordingly.

Profibus technology was created in 1989. Through time it became obsolete for strategic goals and big projects but will be used in future to realize compatibility between different generations of industrial devices. That's why it's necessary to study this technology mainly for maintenance of installed systems. More prospectively to shift attention to modern protocols/technologies where addressing is based

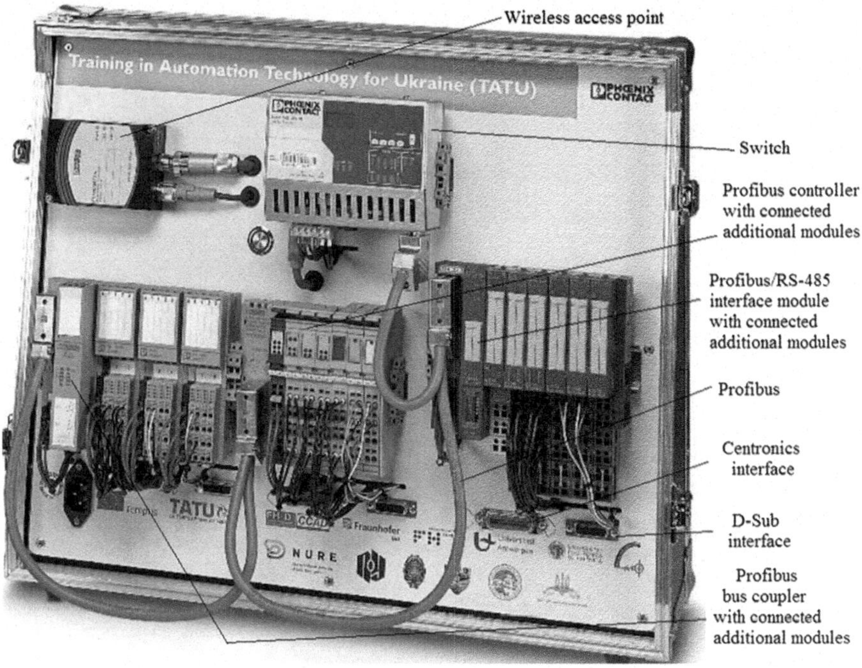

Fig. 16.1 Appearance of TSL Profibus hardware module

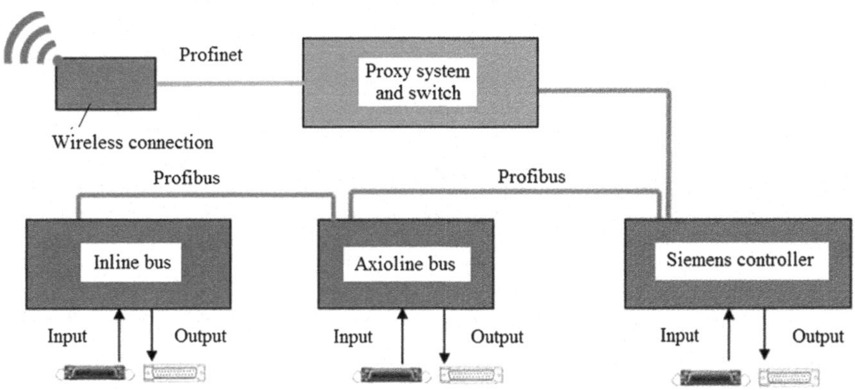

Fig. 16.2 Internal structure and possible external connections in TSL Profibus HM

on IP addresses and devices may be reached using IP network directly without additional proxy system (like PROFINET and EtherCAT).

The third HM (Fig. 16.5) is intended for modeling and visualization of technological processes. Models should be developed in CODESYS of 3.x versions and loaded into the EtherCAT EC2250 PLC. The graphical interface can be seen in the browser.

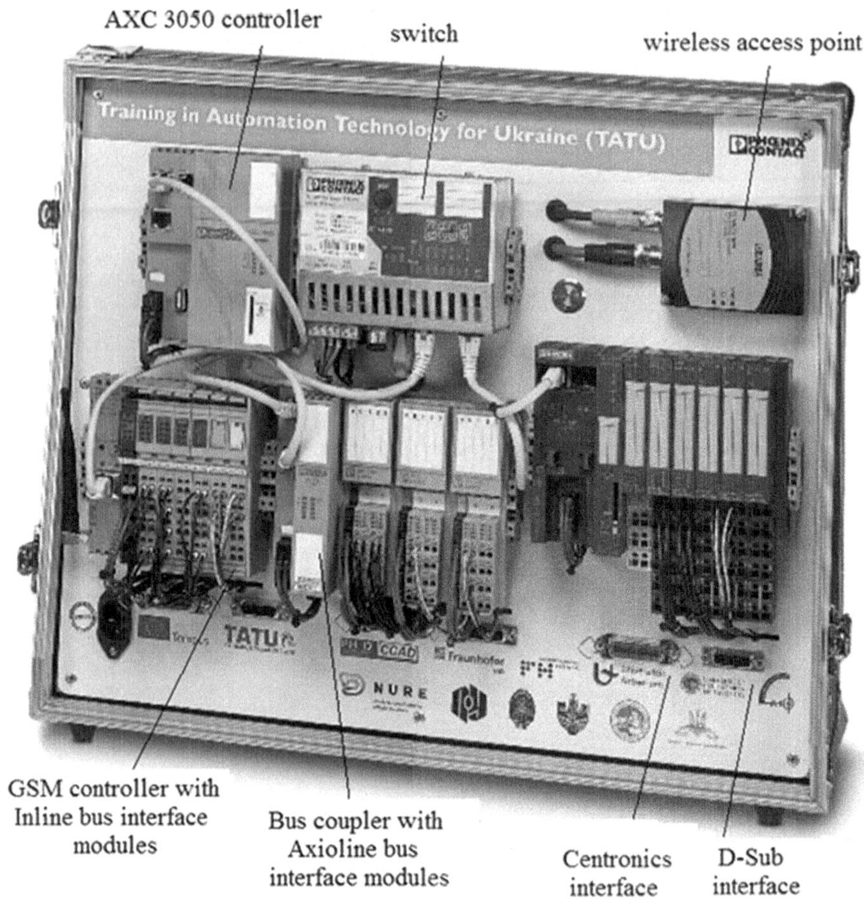

Fig. 16.3 Appearance of TSL PROFINET hardware module

This module also allows to study the programming of PLCs in CODESYS IDE. This HM has various analog and digital inputs/outputs and buttons for simulating a complex control system. Each button has a built-in LED, connected to the digital output. To provide physical connections between HM, 24-pin Centronics IEEE 488 or 15-pin D-sub connectors are used. The analog/digital conversion is performed at 12 bits resolution. Only this HM can be connected to other HMs by standard cables (Figure 16.7). These cables may also be used for connecting other external devices that do not have IP addresses. Devices with IP addresses are connected to RJ-45 ports of Ethernet switches using a standard Cat 5 twisted pair cable.

TATU TSL is equipped with two wireless access points (WAPs), integrated in TSL Profibus and TSL PROFINET HMs. WAP FL WLAN EPA is an interface between the Ethernet/PROFINET cable and wireless networks (Figure 16.8). It allows integration with the networks based on industrial Ethernet (PROFINET,

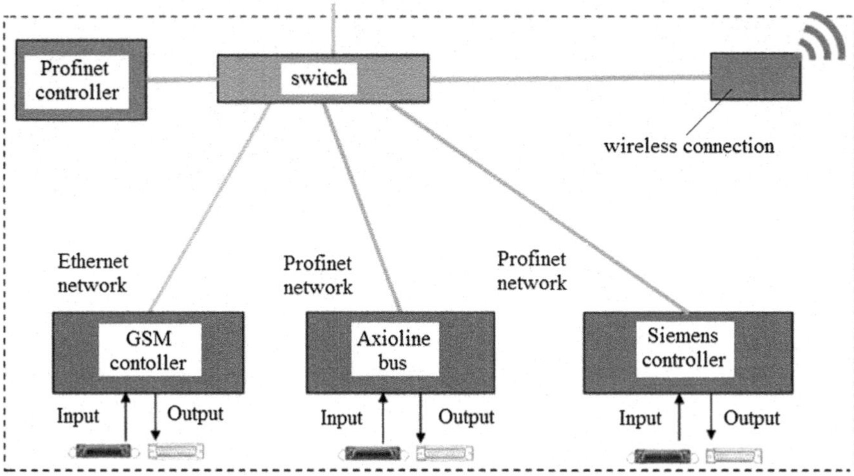

Fig. 16.4 Internal structure and possible external connections in TSL PROFINET HM

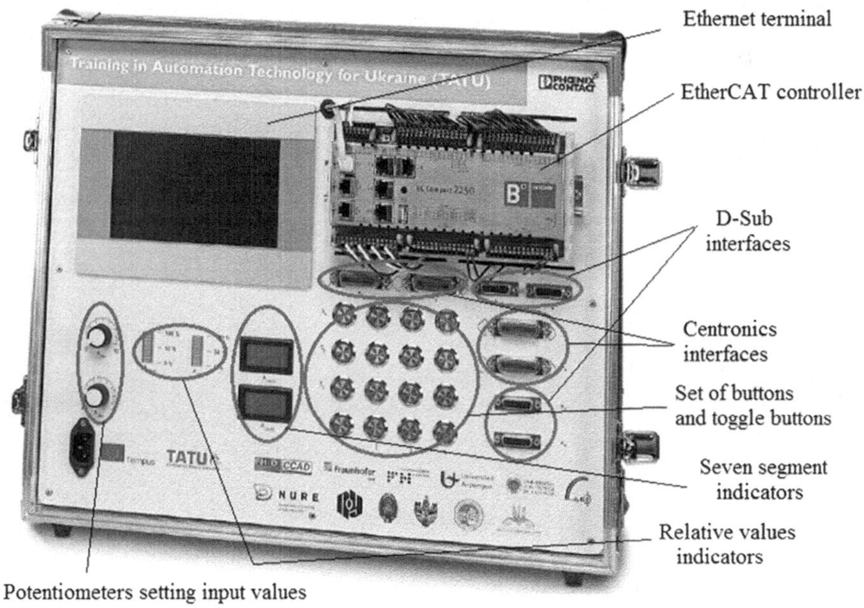

Fig. 16.5 Appearance of TSL EtherCAT and PM hardware module

Modbus/TCP). It has a PROFIsafe profile to avoid data transmission failures and is compatible with IEEE 802.11 b/g/n standard operating at 2.4 GHz and supported by bar code scanners, RFID tag readers, and weighing systems. Mentioned WAP is optimized for process automation. Industrial M12 connector is used for connecting WAP to the wired network with up to 100 Mbps data transfer rates.

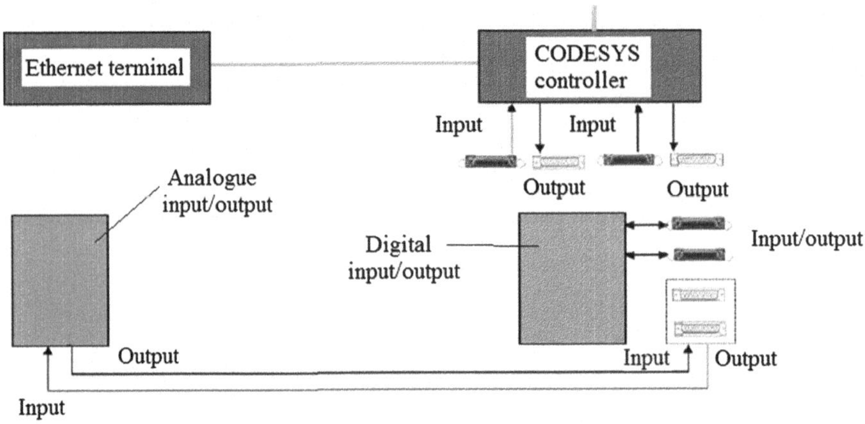

Fig. 16.6 Internal structure and possible external connections in TSL EtherCAT and PM HM

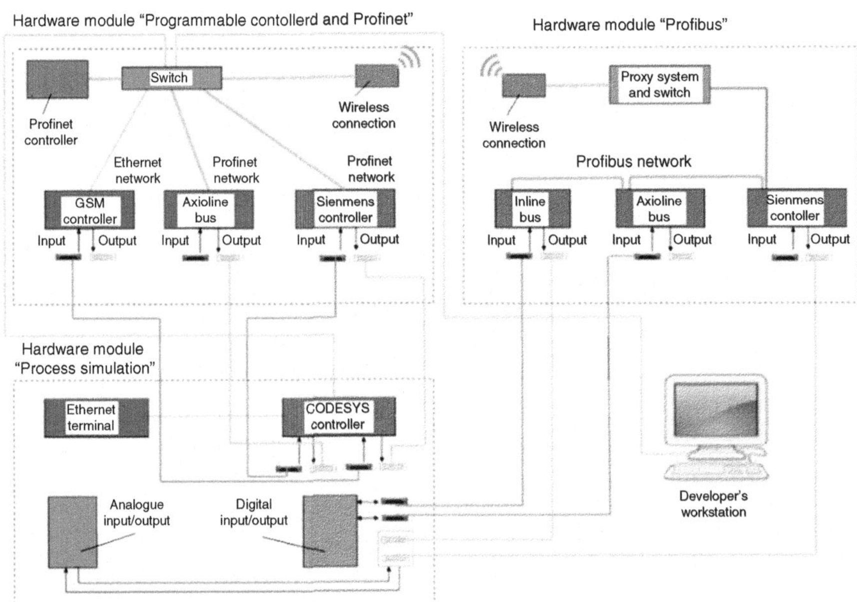

Fig. 16.7 The structure of TSL hardware modules and their connection to computer

The PLC can be configured and programmed in PC Worx IDE using Ethernet in five programming languages in accordance with IEC 61131-3. TCP/IP and UDP/IP are used to get access to the PLC. Using corresponding function blocks in PC Worx IDE, it is possible to organize data exchange (i.e., values of variables corresponding to the measured process parameters and physical quantities) between PLCs. This approach allows implementing distributed and configurable automation solutions.

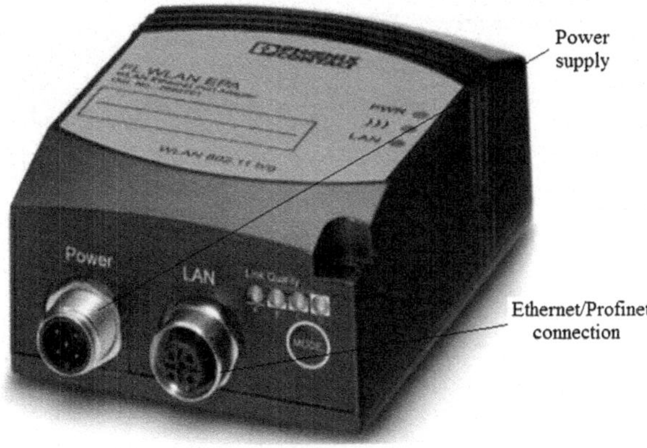

Fig. 16.8 Wireless access point FL WLAN EPA

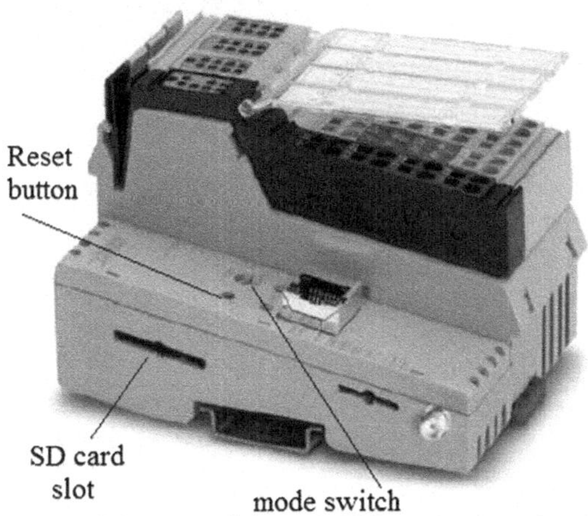

Fig. 16.9 ILC 151 GSM/GPRS PLC

When using the AX OPC server, the PLC is accessible over Ethernet network and can be used in various visualization packages.

The ILC 151 GSM/GPRS PLC (Figure 16.9) may be used for studying wireless data transfer in global networks. It has integrated ports for connecting Ethernet and Interbus networks and an integrated quad-band modem. The PLC can act as a Modbus client and works with external SD format flash memory. Integrated GSM modem allows to perform sending and receiving of SMS and control remote PLC via GPRS (General Packet Radio Service) or CSD (Circuit Switched Data) technologies. The TCP/IP connection in the user project must use the appropriate PC Worx IDE function blocks to implement communication by sending SMS over the GSM network or to set up a GPRS connection. Mentioned wireless possibilities for truly remote control, integrated in ILC 151 GSM/GPRS PLC, allow to simulate the getting controlled object away from developer or supervisor. This is very significant for seafarers and ships which work in such seas which are full of islands like Baltic Sea, Norwegian Sea, Aegean Sea, etc., narrow straits, gulfs, and bays.

EtherCAT EC2250 (Figure 16.10), integrated to third HM, is complex multifunctional PLC which allows to study a lot of data transfer protocols and technologies.

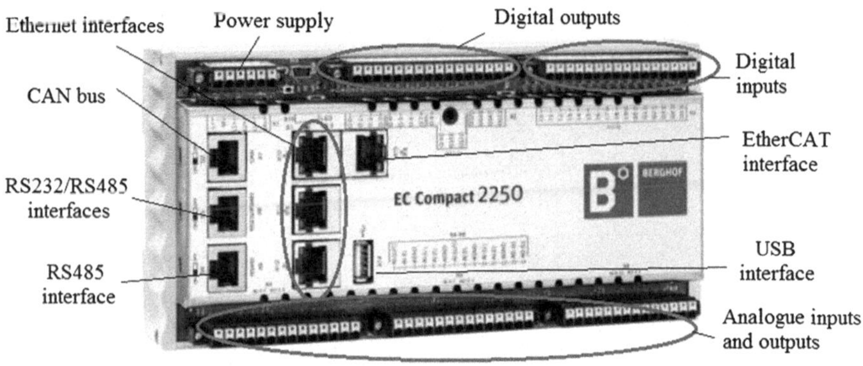

Fig. 16.10 Compact PLC EC2250 EtherCAT

It is designed for hard real-time systems and uses a high-performance Cortex 800 MHz ARM CPU. A lot of digital and analog inputs and outputs are located in the PLC. It has built-in programming languages for working with CODESYS IDE and can be operated together with the CODESYS SoftMotion package, complex applications for controlling multi-axis actuators. The device integrates Ethernet, EtherCAT, CAN, RS 232/485, PROFINET, BACnet, and Modbus communication interfaces and protocols. The EC2250 PLC has 32 digital inputs/outputs and 18 analog inputs/outputs and supports programming, visualization, communication, motion control, master modes in EtherCAT and CANopen technologies, and serial data transfer interfaces.

The AXC 3050 PLC (Figure 16.11), integrated to TSL PROFINET HM, works with Ethernet family networks and the Axioline F local bus, which can be used for the sequential installation of various modules one close to the other (up to 63 devices). It allows to configure the PLC using TCP/IP or UDP (User Datagram Protocol). It has three integrated Ethernet ports X1, X2, and X3 and two USB interfaces. Using PC Worx IDE, it is possible to organize data exchange between the PLCs. This approach allows implementing distributed and configurable automation solutions. By using the OPC server, the PLC is accessible over Ethernet network. The PROFINET technology is always available when connected via the RJ45 connector of the X3 interface. The PROFINET functionality can be activated on the Ethernet interfaces X1, X2, and X3. This PLC can act as a Modbus client and can be configured as a Modbus TCP server. Due to the Web-based management interface integrated into the PLC, the user can visualize the status and diagnostic information from the PLC in the browser. The PLC has 4 Mbytes of internal memory for program storage and 8 Mbytes of memory for data storage; 128 kbytes is used for storing data after power off. The PLC works with SD format flash memory or USB drive. The number of control tasks performed simultaneously is 16. The AXC 3050 PLC may be used as a bridge, gateway, or proxy system for data transmission between different network segments.

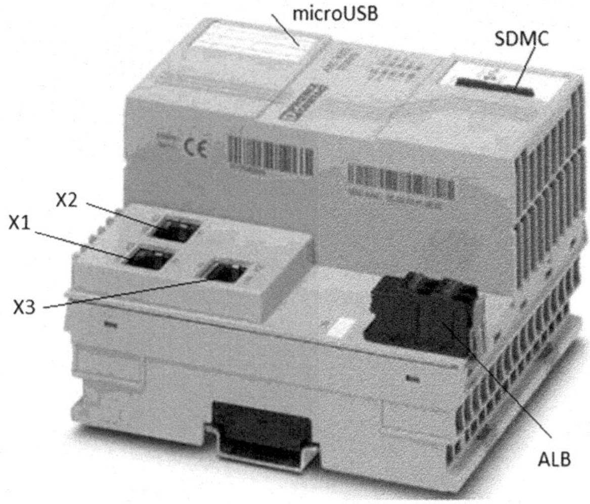

Fig. 16.11 AXC 3050 PLC, X1, X2, and X3 interfaces for connecting to Ethernet family (8-pin RJ45 connectors); SDMC (Secure Digital Memory Card), a slot for connecting flash cards such as Secure Digital); MicroUSB, a connector for connecting USB devices, hidden under a paper tag for information labels; ALB, Axioline bus

Structure of entire laboratory created for maritime engineers studying and retraining is shown in Fig. 16.12.

16.4 Actual or Anticipated Outcomes

Equipment obtained at realization of European Tempus TATU project allows to learn PC Worx and CODESYS IDEs for PLC programming and to study the following network devices: switches, WAPs, different types of PLCs, and some other equipment. An actively working laboratory "Means of industrial automation and network technologies" is created. In general it allows to learn studying disciplines, connected with abovementioned areas. E-learning system based on MOODLE platform is actively used for students' remote access to corresponding teaching materials. Learning books in Russian, Ukrainian, and English languages are created and successfully integrated in the education process. At least 200 students of "Ships' power plants operation," "Ships' power and refrigerating plants operation," "Automated Control of Ships' Power Plants," and "Operation and Maintenance of Ship's Automated Systems" specialties pass corresponding lessons every year.

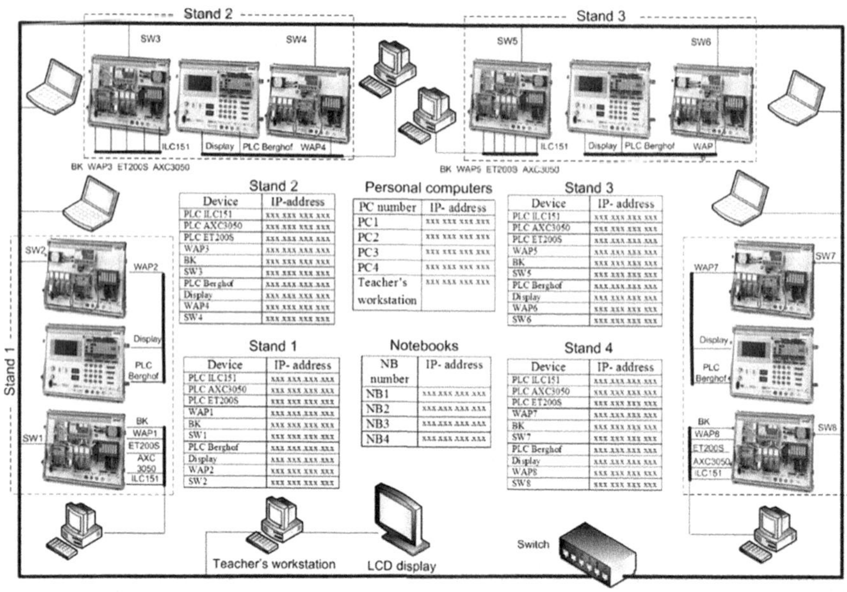

Fig. 16.12 Structure of "Means of industrial automation and network technologies" laboratory

16.5 Conclusions

Ways of modern concepts and technologies implementing in maritime branch are analyzed. Concrete directions of deep studying are shown. Possibilities and technologies, successfully realized in education process, are described. Possibilities of existing equipment may be expanded by additional modules installation.

References

1. R. Langmann, Workshop: The TATU Lab & smart education [Text], in *2016 13th International Conference on Remote Engineering and Virtual Instrumentation (REV).–Madrid*, ed. by R. Langmann, Y. Makarova, L. Rojas-Peña, P. Galkin, I. Klyuchnik, V. Voropaeva, V. Pozepaev, L. Zinyuk, R. Skrypyuk, E. Shaporina, V. Shaporin, V. Shapo, S. Gorb, (2016), pp. 400–402
2. S.I. Gorb, V.V. Nikolskyi, V.F. Shapo, S.H. Khniunin, *Programming Controllers in the Integrated Development Environment: Training Manual. Practice* (National University "Odessa Maritime Academy", Odessa, 2017), p. 164
3. C. Madritsch, W. Werth (eds.), *TATU Study Book. Trainings in Automation Technology for Ukraine* (Carinthia University of Applied Sciences, Villach, 2017)., 211 p
4. V. Shapo, *Distance Learning System Application for Maritime Specialists Preparing and Corresponding Challenges Analyzing* (2018). https://doi.org/10.1007/978-3-319-64352-6_97
5. V. Shapo, *Algorithm and Software Tool for Multiple LMS Users Registering* (2020). https://doi.org/10.1007/978-3-030-11932-4_10

Chapter 17
Digital Transformation of the Education Sector in the Western Balkans

Mihajlo Travar, Igor Dugonjić, Zoran Ž. Avramović, Gordan Bajić, and Saša Ristić

17.1 Introduction

After the steam engine that started the first industrial revolution, the conveyor belt that started mass production in the second industrial revolution, and the discovery of transistors and the Internet that connected the world in the third industrial or digital revolution, the Fourth Industrial Revolution began with digital connection of products, machines, tools, and people in "smart" production. It all started in the eighteenth century with the introduction of technology and the transition from manual to machine work, i.e., industrial mode. The first industrial revolution was started by a steam engine, invented by Thomas Newcomen in 1712, and it pumped water out of mines. However, the beginning of the first industrial revolution dates back to 1769, when James Watt perfected Thomas' steam engine, which since has been used to start locomotives and ships that traveled faster and easier and the machine in production that predominantly made agrarian and rural society become industrial and urban. Other important inventions that characterized the first industrial revolution were the telegraph, the lightning rod, and artificial fertilizer. The second industrial revolution began in 1870, and was based on the discovery of electricity and oil sources, and lasted until 1914. It is characterized by advances in technology and the

M. Travar (✉)
Regulatory Commission for Energy of Republic of Srpska, Trebinje, Bosnia and Herzegovina

I. Dugonjić · Z. Ž. Avramović · G. Bajić
Pan-European University APEIRON Banja Luka, Banja Luka, Bosnia and Herzegovina
e-mail: igor.dugonjic@apeiron-edu.eu; zoran.z.avramovic@apeiron-edu.eu; gordan.z.bajic@apeiron-edu.eu

S. Ristić
Deloitte d.o.o. Banja Luka, Banja Luka, Bosnia and Herzegovina
e-mail: sristic@deloittece.com

S. Bauk, S. D. Ilčev (eds.), *The 1st International Conference on Maritime Education and Development*, https://doi.org/10.1007/978-3-030-64088-0_17

adoption of new knowledge in all industries, especially in the automotive industry where Henry Ford started mass production. Serbian Nikola Tesla, who is the greatest researcher and scientist from the Western Balkans, and one of the world's greatest scientists, has made an immeasurable contribution to technological progress and the use of electricity as the inventor of the rotating magnetic field, induction motor, multiphase alternating current, generator, and complete production and distribution system of electricity. Countries that had good foundations from the first industrial revolution, such as Germany, England, and the United States, achieved the fastest industrial development. Considering that the first two industrial revolutions took place in a relatively short period of time, the beginning of the third revolution was also expected quickly. Yet World Wars I and II halted these processes and left great consequences on the economy and society as a whole. In the years after World War II, the world recovered from its consequences, and the beginning of the third industrial revolution came in the 1960s, and we call it the digital revolution, because it is based on the transformation of mechanical and analog electronic technology into digital electronics, and represents the beginning of the information age, in which digital computers began to be used. It is characterized by the invention of transistors; the advancement of communications; the development of microprocessors, computer networks, and the Internet; etc. The third industrial revolution has no end, given that technologies are being built on top of each other, and we can say that it is still active today. In 2011, Germany launched a process of digital transformation of production, and this project attracted attention in Europe. Developed European countries accepted the project and adopted its concepts and thus started thinking about the Fourth Industrial Revolution. All the systems of the third industrial revolution became the systems of the fourth revolution called Industry 4.0. In recent decades, the development of information technology has seen unprecedented growth and continues to progress. The essence of technological development consists of a complex set of information and communication technologies. The future development of information technology lies in the integration of systems, standardization of equipment, expansion of the Internet, and higher speeds, but it is conditioned by new inventions, changing the world even more than expected [1].

17.2 Industry 4.0: Development Trends and the Most Important Aspects of Action

Industry 4.0 is no longer the future; it happened yesterday, and it is happening today. Today's world is rapidly changing and evolving technologically, and anyone who wants to keep up with the developed world must acquire and apply new technological knowledge every day, because something changes every day. In this paper, we will present the most important aspects of the Fourth Industrial Revolution and point out the importance of implementing new technologies in the education system, i.e., digital educational transformations, and point out how employees in the education system can be carriers of new industrial development. The Industry 4.0 concept was defined in 2011 at the Hannover trade fair, when industry leaders identified 42 elements of a new technological development, the Industrial Revolution 4.0. Some of these elements are artificial intelligence, Internet of things (IoT), cloud computing, cyber-physical systems (CPS), digitalized smart systems, visualization, real-time location system (RTLS), big data real-time analysis, smart factories, augmented (AR) and virtual reality (VR), machine-to-machine communication (M2M), joint work of humans and robots, bionics (biologically inspired engineering elements of biological systems found in the nature are applied in research and design of engineering systems), and others. All these elements enable the production and creation of products according to the expectations of each individual customer. Digital transformation means connecting all factors of production, data collection, and analysis, in order to share information and improve business processes, optimize production and costs, have a better market positioning in relation to the competition, and most importantly meet consumer expectations. The goal is that each machine has a management system that has direct communication with other machines, workers, programs, and products in the network and based on that communication and collected information (everyone listens and takes what they need) performs processes that ensure product differentiation within one production line. What we will emphasize in this paper is that the changes brought by Industry 4.0 are not related exclusively to the production process and monitoring of the life of products with the help of digital technologies, but changes that directly affect the way of working, education and qualifications of workers, social life in general, and, most importantly, the education system.

The fact is that Industry 4.0 will change the way production processes take place, many processes will be fully automated or disappear, but it will not completely displace workers from factories, so it is necessary to prepare and train workers to work with smart machines, and the precondition for that is the modernization of the system of education of students in accordance with the needs of industry. In the higher education sector, technological advances and new technologies can empower universities to invest more in research and innovation, to improve student education and the skills they acquire through non-traditional learning, such as online courses or distance learning. Universities and the Industrial Revolution 4.0 have a positive impact on each other, so the Industrial Revolution 4.0 shapes the way of research, and thus the universities themselves, while they make a significant contribution to the Industrial Revolution 4.0 through their work on technological innovation and artificial intelligence. Big data and new software systems simplify the analysis of complex data, enabling university researchers to significantly reduce manual labor and errors. For these reasons, the University of Cambridge established the Leverhulme Center for the Future of Intelligence (LCFI) and involved mathematicians, engineers, philosophers, computer scientists, and political scientists in its work to examine the impact of the artificial intelligence (AI) in society. Regardless of the fact that digital transformation and AI through automation will lead to elimination of some jobs, the integration of AI into all social activities should not be presented as a threat to highly skilled jobs, but as an innovation that will help social development and create even more highly skilled jobs. We can see this in the example of the introduction of autonomous vehicles (vehicles driven without drivers), where on the one hand a large number of people-drivers will lose their jobs and on the other hand new jobs will be created for the infrastructure and development of these vehicles, which will require highly educated staff. Following the changes brought by the Industrial Revolution 4.0, universities must adapt their lectures and be able to provide needed education for students, to start a variety of highly qualified careers [2].

17.3 Digital Transformation of the Education Sector in the Western Balkans

Today we live in an information world in which the production, processing, and storage of knowledge are a very important factor of complete social progress. Education as one of the basic pedagogical categories envisions professional training and development of life skills by acquiring knowledge [3]. As the end of the twentieth century in the countries of the Western Balkans was followed by a difficult economic situation, social division, and war, education systems are far behind European ones. However, the most important characteristic of all countries is that there are commitment and preconditions for the digitalization of education. Each of the countries is at a different stage of development, and in most of them, all the necessary legal and technical preconditions are provided in order to make a step forward in terms of technology.

The Republic of Serbia, which is a leader in the Western Balkans, is for the most part still focused on creating the preconditions for a successful digital transformation of its education system. This primarily implies continuous equipping of primary and secondary schools with a hardware component. The application software component relies on e-Dnevnik, whose pilot project began in 2017 and later was extended to schools throughout the Republic of Serbia [4]. However, according to available information, the digital transformation of this education system is not yet fully completed, and is burdened by the lack of adequate infrastructure to support the software component.

The Republic of Croatia, which not so long ago had the same education system as other countries of the former Yugoslavia, today the largest part of the Western Balkans, has a much better potential for digital transformation of the education sector. This is primarily the result of Croatia's membership in the European Union, which allows access to various funds and good practice of developed countries of the European Union. Croatia has adopted a number of regulations and created quality preconditions for the digitalization of the education system. One of the most important documents was adopted in 2017 entitled as "Strategy for digital maturation of schools and the school system of the Republic of Croatia" [5]. The mentioned document elaborates in detail the ways of introducing ICT into the education system, with the emphasis that teachers and heads of educational institutions are in fact the bearers of the digital transformation of education. Software application support was developed through a system called e-School.

When it comes to other countries (Albania, North Macedonia, and Montenegro), the digital transformation in the education sector has yet to follow and bring about concrete changes.

Unlike the mentioned countries, the Republic of Srpska (entity in Bosnia and Herzegovina) has gone a step further, considering that for several years there has been a complex information system in the education sector that digitally transforms the entire education sector. More than 10 years ago, the Ministry of Education and Culture of the Republic of Srpska, in cooperation with the largest IT company in BiH, started the implementation of the EduIS project, and today the third version of EduIS is active in primary and secondary schools [6]. EduIS is an integrated information system based on cloud technology intended for educational institutions at the local and national level. The goal of using this system is to collect and systematically organize information, so that educational institutions can efficiently organize their activities through technology. All users are provided with a high level of security and identity protection [7]. Also, segmentation of the application solution according to different types of users is enabled. In the continuation of the paper, we will describe the model that was used in order to digitize education in primary and secondary schools through a practical example in the Republic of Srpska.

17.4 Characteristics of Digital Transformation of the Education Sector of the Republic of Srpska

In order to carry out the process of digital transformation in education in an appropriate manner, it was necessary to respond to several challenges, i.e., to create an information system that will primarily be secure and reliable. Since this is a cloud solution that is available online to all users, in order to protect the identity of users, multiple authentication is provided. All data generated by the system is part of a centralized database that provides a high degree of consistency and prevents data redundancy. The data is stored in a data center that corresponds to the level of the Tier 4 standard. The next important characteristic, that the information system for education should have, is scalability. This, in real time, enables the participation of a large number of participants in the education system and the organizations and conduct of online classes. In addition to the technical characteristics that ensured that the information system functions efficiently, through several versions of the system, a number of different functionalities have been implemented, with special attention paid to ease of use and interface. The functionalities of the system are grouped according to similarity, and the entire system is divided into different segments, modules, while the digital transformation of different participants in the education sector is provided [8]. The education system of the Republic of Srpska is organized in such a way that all records of employees are kept at the level of a special educational institution and the integrated solution had to provide full support to educational institutions and enable the formation of organizational structure, records of employees, work experience, training, etc.

The first segment required by the ministry was curriculum management. This meant the transition from traditional (manual) to electronic records of classes, working weeks of employees, and monitoring the fulfillment of curricula. In order to meet this requirement and for the system to function properly, it was necessary to construct a system based on relational databases, i.e., to enable the integration of all modules into one whole. The most important functionalities that were implemented through the curriculum module enabled each educational institution to have the opportunity to create one or more curricula whose fulfillment can be monitored during the school year.

The second segment relates to teaching staff, thus meeting two basic requirements for digital transformation. The first refers to teachers in educational institutions who are enabled to prepare for class in digital form, with integrated information about each student, such as health condition or difficulties in teaching that the student has. The second requirement relates to the integration of different databases so that the competent ministry has at all times available information on any educational institution in the territory of the Republic of Srpska. In order to meet these requirements, it was necessary to implement a large number of functionalities that could be divided into several groups.

The third module of the platform refers to students, and its task is to enable interactive participation in teaching and the opportunity to create and share literature through online services. The aim of this module was to enable educational institutions to provide their students with access to teaching materials if students did not attend classes, or the materials are part of extracurricular activities. Also, the part of the platform intended for students includes the possibility of mutual communication between students within the same school. In response to the set requirements, the platform implementer has created a dashboard for each student, upon which students, in addition to basic information, at all times have an overview of activities such as student grades or notes entered by teachers.

The fourth module of the platform, which is intended for parents, enables monitoring of the educational process by years, through various media for informing about implemented and newly created activities. The aim of the module is to enable parents to communicate through the platform with teachers and school management, to follow the curriculum according to which classes are conducted, to monitor the child's development, and to have insight into its grades and absences. In this way, the educational institution communicates with the parents by sending various surveys or instructions, if there is a need for that.

In order to complete digital transformation of the education sector, it was necessary to digitize all other functions of educational institutions and create a module that will connect professional associates (e.g., psychologists or pedagogues) with teachers, students, or parents. The special importance of this module refers to students who have developmental difficulties or are recognized as the most talented.

Through the platform, professional associates can specifically target and monitor these groups of students. In addition to all of the above, the digital transformation of the education sector also includes an platform based on Microsoft Office 365 services:

- Online application – Teachers and students have the opportunity to use Web applications such as Word, Excel, PowerPoint, and OneNote, with which it is possible to organize work in real time and form joint documents.
- OneNote Class Notebook – Students and teachers have the opportunity to create their own notes and do control tasks and tests; with the help of OneNote, the teacher can quickly adjust the workspace to each student.
- OneDrive – Teachers can share work materials with students. Each user has the option of using 1 TB of storage space.
- Skype for Business – This allows you to communicate with customers and organize online meetings. Using this tool will facilitate communication of all participants in the education sector.
- Exchange Online – This implies standard functionalities of e-mail and all other auxiliary functionalities.
- Yammer – Standard functionality is offered by the Yammer application.

Finally, we have the administration module, which is a unified database or archive in which the data that is generated and the basis for the operation of all other modules are stored in electronic form, since it is used to configure the entire system.

17.5 Conclusion

We decided on this topic, because Industry 4.0 is characterized, above all, by removal of barriers between people and machines and the education sector takes a special place here and is extremely important for the full implementation of digitalization. Also, the reason is that in Industry 4.0, the education and development of students and employees is more important than the development of the systems themselves. As we mentioned in the paper, the countries of the Western Balkans are in various stages of digital transformation, and it is certain that the entire area will need a lot of time to fully digitize their education systems.

However, it is encouraging that a geographically small and economically underdeveloped area like the Republic of Srpska has managed to create state-of-the-art software solutions and implement the digital transformation of the education sector. The fact that the EduIS platform and the model of digitalization of education in the Republic of Srpska are recognized on the African continent in the Federal Republic of Nigeria shows how great a step forward the Republic of Srpska has made in education. At the moment, the online EduIS platform is in the pilot phase of implementation in dozens of educational institutions in Nigeria.

In the end, we can only state that a new interdisciplinary science has emerged as a synthesis of informatics, mechanical engineering, and electrical engineering –

Industry 4.0, and as its name suggests, it should be implemented as soon as possible, because those who do not do so globally will disappear. Some of the technologies listed in the paper are still in the early stages of development, and some have been brought to life in practice at full capacity and are beginning to complement each other and influence all events in our environment. The aim of the paper was to get acquainted with these technologies, modern trends, and challenges that Industry 4.0 brings with it, as well as with its achievements in the field of education and what has been achieved so far in our region.

References

1. Ž. Pekić, S. Jovanovski, N. Pekić, The impact of Felder's learning styles index on motivation and adoption of information through e-learning. Case Study (2016). https://doi.org/10.7251/JIT1602093PJITA
2. S. Tomić, D. Drljača, *Quality Standard for Electronic Education Programs in Higher Education of Bosnia and Herzegovina* (Pan-European University APEIRON, Banja Luka, Republic of Srpska, Bosnia and Herzegovina JITA, 2019)
3. F. Klein, M. Bansal, J. Wohlers, The megatrends od tomorrow's word. Deloitte (2017)
4. Schmidt, E., Cohen, E., Murrey, J.: The New Digital Age (2013)
5. D.J. Ivanović, M. Antonijević, *E-učenje – Stanje i perspektive u Republici Srbiji* (Institut ekonomskih nauka Beograd, 2020)
6. Vlada Republike Hrvatske,: Strategija digitalnog sazrijevanja škola i školskog sustava Republike Hrvatske (2018)
7. F. Schwade, P. Schubert, *The ERP Challenge: An Integrated E-Learning Platform for the Teaching of Practical ERP Skills in Universities* (University of Koblenz-Landau, Germany, n.d.)
8. D. Rogers, *The Digital Transformation Playbook* (Columbia Business School, 2016)

Chapter 18
Artificial Neural Networks in Creating Intelligent Distance Learning Systems

Dragan Vasiljević, Julijana Vasiljević, and Boris Ribarić

18.1 Introduction

Predicting future events is one of the basic areas within artificial application neural networks [1]. Hereby, these future events are necessary abilities as well as students' achievements. Unlike classic methods, based on models, artificial neural networks belong to the class of self-adaptive methods based on data with only a few model assumptions for problems being studied [2]. The research problem is in correlation with activities in the teaching process that can be promoted. Artificial neural networks throughout data testing and training should provide a projection of students' abilities as well as competence [3]. The subject of research in this paper relates to the development and application of algorithms toward the prediction of neural networks use in distance learning systems. Most research work concerning to the use of artificial neural networks in distance learning systems is one-dimensional just because it focuses on methods and techniques applied to either the electronic learning system only or especially to students and teachers [4–6], while fewer papers and research deal with the problem of learning based on past events [7, 8]. So far, this area has not been sufficiently explored, especially the field of e-learning as indicated by the great number of scientific work published in 2019 which refers to the robotics, economy, as well as natural phenomena [9–13]. To create intelligent distance learning systems, this paperwork will display the method of analyzing data using artificial neural networks.

D. Vasiljević (✉) · J. Vasiljević · B. Ribarić
Pan-European University "Apeiron", Banja Luka, Bosnia and Herzegovina

S. Bauk, S. D. Ilčev (eds.), *The 1st International Conference on Maritime Education and Development*, https://doi.org/10.1007/978-3-030-64088-0_18

18.2 Method

Artificial neural networks are an integral part of artificial intelligence primarily used for numerical prediction [14, 15], classification [16, 17], and pattern recognition [18–20]. The data should be divided into three samples: the one for coaching, cross-validation, and testing [20]. The artificial neural network should be trained based on prepared data to identify the relationship between the data and to be able to predict the output values based on the input values. The learning phase is a process of adjusting network weights that take place in multiple iterations or passing through the network. In the cross-validation phase, the network tends to observe the length training, the number of hidden neurons, as well as parameters. Network testing is the third phase of neural network operation, and it is crucial for network evaluation. The difference relating to the learning and the testing phase is that in this phase the network is no longer in the learning process. Network evaluation is done by calculating the error in a way to compare the network output with the actual outputs [21].

18.3 Results

Lectures, assignments, tests, grading, competitions, and directed leisure activities are input variables important for in-depth data analyses as well as model creation neural networks [22]. The output variable is a satisfactory level of achievement. Defined variables data refer to 102 elementary school students (eighth graders) analyzed during the school year 2018–2019. The data has been updated into the Neural Designer studio database for this paperwork due to further processing and the creation of an artificial neural network. The statistics regarding defined input variables are important information for designing models since they can alert to the presence of false data. Table 18.1 displays the minimum, maximum, middle, and standard deviation of all variables in the data set.

Table 18.2 displays the value of the correlations of all input variables. The minimum correlation is −0.440712 between the lecture and competition variables. The maximum correlation is 0.9555376 between tasks and tests.

Table 18.1 Measures of centeredness

	Minimum	Maximum	Mean	Deviation
Lectures	1.00	5.00	4.04	1.09
Assignments	1.00	5.00	3.93	1.23
Tests	1.00	5.00	4.06	1.19
Grading	1.00	5.00	2.96	0.61
Competitions	1.00	5.00	2.60	0.94
Directed leisure activities	1.00	5.00	2.25	0.88
Level of achievement – satisfactory	0.00	1.00	0.51	0.50

Table 18.2 Cross-correlations

	Lectures	Assignments	Tests	Grading	Competitions	Directed leisure activities
Lectures	1	0.83	0.75	0.24	–0.44	–0.29
Assignments		1	0.96	0.56	–0.19	–0.048
Tests			1	0.6	–0.0052	0.19
Grading				1	0.65	0.33
Competitions					1	0.67
Directed leisure activities						1

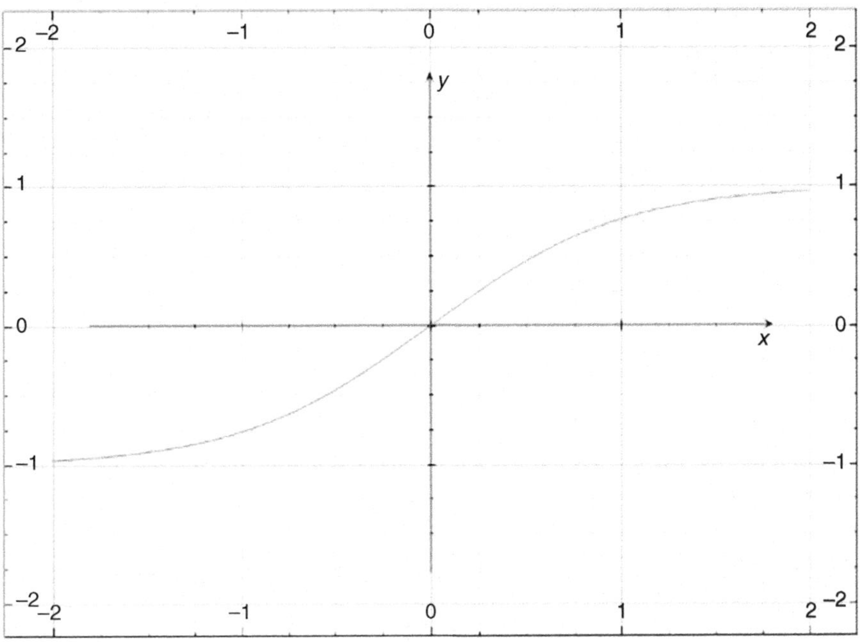

Fig. 18.1 Graphic representation of hyperbolic tangent function

An artificial neural network is defined as a single layer with forwarding propagation, while the activation function is a hyperbolic tangent function which can be mathematically expressed as follows:

$$tghX = \left(\frac{e^x - e^{-x}}{e^x + e^{-x}} \right) \ldots \tag{18.1}$$

where X represents an independent variable size. Graphic representation of the hyperbolic tangent function is given in Fig. 18.1.

The total number of variables for a neural network is seven, of which one variable is a target variable referring to a satisfactory level of achievement. The training strategy over the data set is defined by the optimization algorithm using the quasi-newton method. The optimization process is approximately the same as the ordinary Newton method with Hessian matrix step modification. The following is a brief numerical example of one type of quasi-Newton method that uses the original Hessian inverse matrix for each iteration:

Target function:

$$\min f(x) = 2x_1^2 + 3x_2^2 \ldots \tag{18.2}$$

Starting point selection:

$$\begin{bmatrix} x_1 \\ x_2 \end{bmatrix} = \begin{bmatrix} 1 \\ 1 \end{bmatrix} \ldots \tag{18.3}$$

Hessian matrix inverse calculation:

$$\nabla f(\mathrm{x}) = \begin{bmatrix} 2x_1 \\ 3x_2 \end{bmatrix} \ldots \tag{18.4}$$

$$\nabla^2 f(\mathrm{x}) = \begin{bmatrix} 2 & 0 \\ 0 & 3 \end{bmatrix} \ldots \tag{18.5}$$

$$\mathrm{H}^{-1} = \begin{bmatrix} 0,5 & 0 \\ 0 & 0,3 \end{bmatrix} \ldots \tag{18.6}$$

Finding a new value for variable x:

$$x^{k+1} = \begin{bmatrix} 1 \\ 1 \end{bmatrix} - \begin{bmatrix} 0,5 & 0 \\ 0 & 0,3 \end{bmatrix} \ldots \tag{18.7}$$

Specifying a new value for variable x:

$$x^{k+1} = \begin{bmatrix} 0 \\ 0 \end{bmatrix} \ldots \tag{18.8}$$

Determining the value in case the function converges:

$$\nabla f(\mathrm{x}) = 0 \ldots \tag{18.9}$$

The training of the data set has been performed within a thousand interactions during 1 h of testing time. Upon the performed training, Fig. 18.2 displays the train-

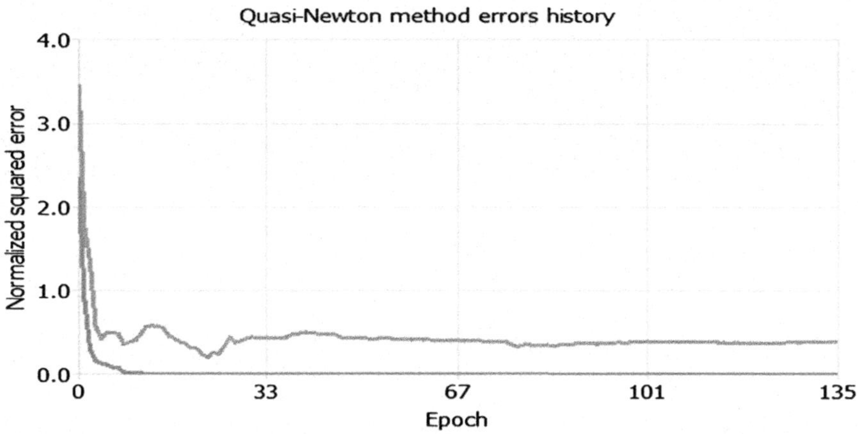

Fig. 18.2 Graphic representation of error history using the quasi-Newton method

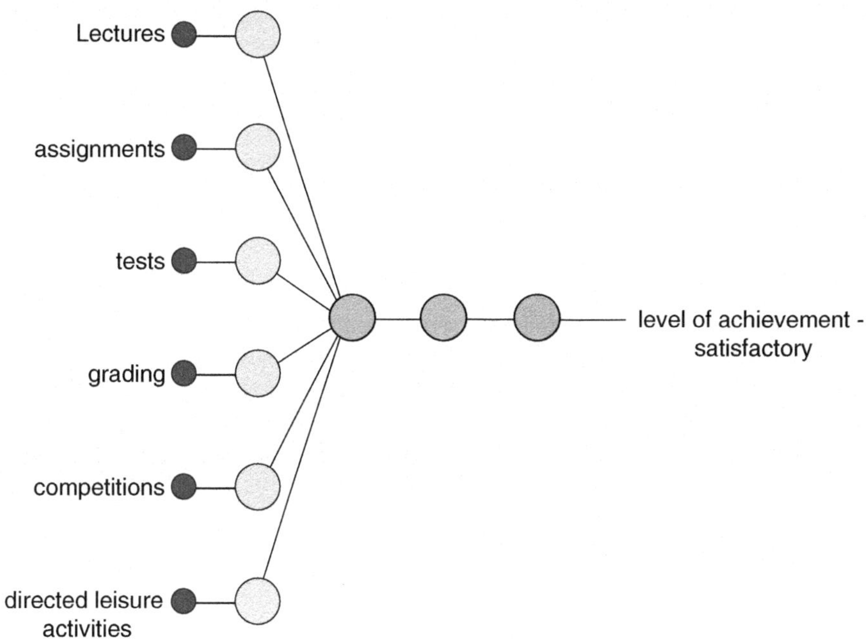

Fig. 18.3 Graphic representation of artificial neural network architecture

ing as well as selection error in each iteration. Blue line represents a training error, and the orange one represents a selection error. The home value of the training error is 2.33562, and the final value after 135 epochs is $9.72136e^{-6}$. The initial value of the selection error is 3.444445, and the final value after 135 epochs is 0,382761.

Figure 18.3 provides a graphic representation of the resulting deep artificial architecture neural networks. It contains a scaling layer, a neural network, as well as a non-scaling layer. The yellow circles represent the scaled neurons, the blue ones represent perceptron neurons, and the red one represents neurons. The number of inputs is six and the number of outputs is one. The number of hidden neurons is one.

The testing has been performed on the data relating to the lecture variable with dedicated value 1. Table 18.3 displays the value for the target variable with value 0.0911 whose correlation with original values has been proved.

Figure 18.4 displays a graph illustrating the dependence of the target variable with input variables.

In order to test the model, we have assigned random variables to the input variables by assigning the value 1 to three variables. The value of the target variable is 0.459, which is closer to 0 than 1, which proves that the model of artificial neural networks provides expected results. Table 18.4 displays the allocated input values as well as the target variable value.

Within creating intelligent distance learning systems, based on the results obtained by the data processed within an artificial neural network of the observed input variables: lectures, assignments, tests, grading, competitions as well as directed free activities represent necessary facilities in achieving a satisfactory level of accomplishment.

18.4 Discussion

Defined input variables and the data processing results, using artificial neural networks, are the basis of creating a distance learning system that needs to provide users with unhindered access to all available resources on the portal distance learning systems [23]. Content organization on the distance learning platform should provide conditions for that clear, transparent, and logical organization of teaching content through lectures, which, in form and content, should be tailored to the target audience. Upon the lectures execution on the platform, tasks related to the completed after-lectures are assigned. Further follow-up of classes is enabled only after tasks have successfully been solved. Tests represent a separate unit in the distance learn-

Table 18.3 Target variable value after corrections

	Value
Lectures	1
Assignments	3.93137
Tests	4.05882
Grading	2.96078
Competitions	2.59804
Directed leisure activities	2.2451
Level of achievement – satisfactory	0.0911282828

	Value
Lectures	1
assignments	3.93137
tests	4.05882
grading	2.96078
competitions	2.59804
directed leisure activities	2.2451
level of achievement - satisfactory	0.0911282828

Fig. 18.4 Graph view illustrating the dependency of the target variable with the input variables

Table 18.4 Inputs vs. target value

	Value
Lectures	1
Assignments	1
Tests	3
Grading	2
Competitions	1
Directed leisure activities	4
Level of achievement – satisfactory	0.459540169

ing system, periodically organized. Due to their form, they can be classified into self-evaluative and formal ones. Assessment is a constant activity whereby the lecturer monitors all the work as well as students' engagement. Contest-related content and likewise directed leisure activities are innovations within distance learning systems that provide trainees with competency development and likewise greater learning motivation. Due to the analyses including 102 elementary school pupils (eighth graders) who have used the distance learning system as an adjunct to traditional teaching, within the school year 2018–2019, as well as 85 elementary school pupils within the school year 2017–2018, taught in a traditional way, the greater success of the pupils using the distance learning system is accomplished and evident.

18.5 Conclusion

The research presented in this paperwork indicates that by application of the neural network in creating intelligent distance learning systems, the achievement and competences of primary school pupils can be significantly improved. This paperwork displays a realized model of artificial neural networks in the function of the development toward organizing teaching content and likewise activities on a distance learning platform. A comparative analysis regarding two-generation eighth graders at their first year of high school education indicates the implementation efficiency of the abovementioned contents of distance learning.

The developed distance learning model has been applied at the elementary school level. However, it can be successfully applied both at a high school or at an academic level. Further development directions for distance learning relate to the application, besides the methods of neural networks and likewise other methods of in-depth data analyses. The aforementioned research results should form the basis for further development of the distance learning systems.

References

1. R. Sharda, Neural networks for the MS/OR analyst: An application bibliography. Interfaces **24**(2), 116–130 (1994)
2. H. White, Learning in artificial neural networks: A statistical perspective. Neural Comput. **1**, 425–464 (1989)
3. K. Hornik, M. Stinchcombe, H. White, Multilayer feedforward networks are universal approximators. Neural Netw. **2**, 359–366 (1989)
4. P. McClure, N. Kriegeskorte, Representational distance learning for deep neural networks. Front. Comput. Neurosci. **10**, 131 (2016)
5. T. Saito, Y. Watanobe, Learning path recommendation system for programming education based on neural networks. Int. J. Distance Educ. Technol. (IJDET) **18**(1), 36–64 (2020)
6. M.R. Syed, Methods and applications for advancing distance education technologies: International issues and solutions. Simulation, 348 (2020)
7. A.A. Heidari, H. Faris, S. Mirjalili, I. Aljarah, M. Mafarja, Ant lion optimizer: Theory, literature review, and application in multi-layer perceptron neural networks, in *Nature-Inspired Optimizers*, (Springer, Cham, 2020), pp. 23–46
8. S. Milinković, M. Maksimović, Case study: Using decision tree classifier for analyzing students' activities. J. Inform. Technol. Appl., Aprerion, Banja Luka, Republic of Srpska, BiH **3**(2), 87–95 (2013)
9. O. Ghorbanzadeh, T. Blaschke, K. Gholamnia, S.R. Meena, D. Tiede, J. Aryal, Evaluation of different machine learning methods and deep-learning convolutional neural networks for landslide detection. Remote Sens. **11**(2), 196 (2019)
10. C.R. Gil, H. Calvo, H. Sossa, Learning an efficient gait cycle of a biped robot based on reinforcement learning and artificial neural networks. Appl. Sci. **9**(3), 502 (2019)
11. H. Nguyen, X.N. Bui, Predicting blast-induced air overpressure: A robust artificial intelligence system based on artificial neural networks and random forest. Nat. Resour. Res. **28**(3), 893–907 (2019)
12. S. Liu, C.W. Oosterlee, S.M. Bohte, Pricing options and computing implied volatilities using neural networks. Risks **7**(1), 16 (2019)
13. B.G. Weinstein, S. Marconi, S. Bohlman, A. Zare, E. White, Individual tree-crown detection in RGB imagery using semi-supervised deep learning neural networks. Remote Sens. **11**(11), 1309 (2019)
14. I.H. Witten, E. Frank, *Data Mining: Practical Machine Learning Tools and Techniques* (Morgan Kaufman, Burlington, 2005)
15. C. Mair, G. Kadoda, M. Le, K. Phalp, C. Scho, M. Shepperd, S. Webster, An investigation of machine learning based prediction systems. J. Syst. Softw. **53**, 23–29 (2000)
16. S.B. Kotsiantis, Supervised machine learning: A review of classification techniques. Informatica **31**, 249–268 (2007)

17. G.A. Carpenter, S. Grossberg, J.H. Reynolds, ARTMAP: Supervised real-time learning and classification of nonstationary data by a self-organizing neural network. Neural Netw. **4**, 565–588 (1991)
18. C.M. Bishop, *Pattern recognition and machine learning*, vol 4 (Springer, Berlin, 2006)
19. S.J. Raudys, A.K. Jain, Small sample size effects in statistical pattern recognition: Recommendations for practitioners. IEEE Trans. Pattern Anal. Mach. Intell. **13**, 252–264 (1991)
20. Марковић,Љ., 2009. Примена backpropagation алгоритма за обучавање неуронских мрежа, Симпозијум о операционим истраживањима, SYM-OP-IS 2009 Београд, (pp. 1–4), ISBN 978-86-80593-43-2
21. М. Зекић-Сушац, А. Фрајман-Јакшић, Н. иДрвенкар, Неуронске мреже и стабла одлучивања за предвиђање успјешности студирања. Економски вјесник **XXII**, 314–327 (2009)
22. I. Isaković, CRM performances accented with the implementation of data warehousing and data mining technologies. J Inform Technol Appl, Aprerion, Banja Luka, Republic of Srpska, BiH **3**(2), 107–112 (2013)
23. S. Tomić, D. Drljača, DDLM - quality standard for electronic education programs in higher education of Bosnia and Herzegovina. JITA – J. Inform. Technol. Appl. **9**(2), 67–79 (2019)

Chapter 19
Ambient Intelligence in the Function of E-Learning Improvement

Julijana Vasiljević, Dragan Vasiljević, and Boris Ribarić

19.1 Introduction

The subject of the research represents analyses of how ambient factors influence the success of students who use e-learning platforms in the teaching process. Ambient intelligence should provide students with optimal conditions for monitoring classes. It should also allow teachers to keep track of how long and successfully involved students are in a particular activity. Recommendations toward enhancing activities can be electronically transmitted to convenient platforms or even to students' wristwatches [1]. The significance of ambient intelligence has been recognized as well as time applied by leading world-class companies such as Siemens, which has invested heavily in smart development buildings and autonomies in the production process in factory halls [2]. Nokia company has also invested significant resources in communication development which do not necessarily relate to smart homes only but have a much wider application for reasons of developed mobile applications [3]. Ambient intelligence [4, 5] is being developed within multidisciplinary fields, thus allowing the benefit of research to be used for a variety of purposes. This paper deals with ambient intelligence in the function of enhancing e-learning. The basic idea behind the concept of ambient intelligence is the adaptation of the environment with the help of information and communication technology for e-learning toward students' needs to achieve better results as well as greater achievement. E-learning system should be built so that it adjusts teaching content to students' needs based on information in real time. Contemporary teaching, especially electronically shaped educational learning processes, and teaching, is getting closer and closer to the concept of students' creation within the pedagogical framework activities. This increasingly signifies that the concept of constructionism is expanding. In the cognitive process

---

J. Vasiljević (✉) · D. Vasiljević · B. Ribarić
Pan-European University "Apeiron", Banja Luka, Bosnia and Herzegovina

S. Bauk, S. D. Ilčev (eds.), *The 1st International Conference on Maritime Education and Development*, https://doi.org/10.1007/978-3-030-64088-0_19

necessary, as well as an instructive and constructivist approach in all teaching situations - from the traditional teaching system to the virtual classroom, as well as e-learning [6–9]. Learning with modern strategies that use the concept of e-learning compared to traditional teaching is characterized by completely different relationships between students and teachers, as well as the possibility of modifying teaching [10]. A student qualified for self-education represents one of the key goals in the process of education. Those methods of work that contribute to a more active attitude of students toward teaching content should be given preference. Training students to use different sources of knowledge as well as becoming independent in the learning process is also very significant [11].

Previous research has not completely provided comprehensive answers to questions regarding the impact of the environment on e-learning, identification, and students' security within e-space. To properly understand the impact of the environment on e-learning as well as predictions of student success, this paper presents an analysis of factors relevant to the success of the teaching process using neural network algorithms.

19.2 Method

While exploring the impact of ambient intelligence on the e-learning process, this paper has defined the largest factor, most relevant to students' achievements in the context of an e-learning environment, with the use of analytical hierarchical methods.

Analysis of the E-Learning Environment

Defined factors are further processed as input variables within an artificial neural network. Results obtained have been used to develop ambient intelligence in the e-learning enhancement function. In order to define the factors of greatest importance toward students' achievements, the authors of this paper have used extended AHP fusion method of triangular numbers that is performed in four steps [12] as follows:

Let X = $\{X_1, X_2, X_3, \ldots\ldots, X_n\}$ be a set of objects and G = $\{G_1, G_2, G_3 \ldots\ldots, G_m\}$ be a set of goals.

Each object is analyzed for each objective, respectively. Accordingly, m values for each object can be expressed as follows:

$$M_{gi}^{1}, M_{gi}^{2}, \cdots, M_{mgi}^{m}, i = 1, 2, \cdots n.$$

where all the M_{gi}^{j}, (j = 1, 2, ..., m) represent triangular fuzzy numbers.

The AHP fusion steps are:

Step 1

The fusion value of the synthetic respective domains with respect to the i-th object is defined as:

$$S_i = \sum_{j=1}^{m} M_{gi}^{j} * \left[\sum_{i=1}^{n} \sum_{j=1}^{m} M_{gi}^{j} \right]^{-1} \tag{19.1}$$

To obtain $\sum_{j=1}^{m} M_{gi}^{j}$, we perform the fusion operation by adding m values for a certain matrix according to the following statement:

$$\sum_{j=1}^{m} M_{gi}^{j} = \left(\sum_{j=1}^{m} l_j, \sum_{j=1}^{m} m_j, \sum_{j=1}^{m} u_j \right) \tag{19.2}$$

To obtain values,

$$\left[\sum_{i=1}^{n} \sum_{j=1}^{m} M_{gi}^{j} \right]^{-1} \tag{19.3}$$

perform the fusion addition operation,

$$M_{gi}^{j} \left(j = 1,2,\ldots,m \right) \tag{19.4}$$

thus obtaining values,

$$\sum_{i=1}^{n} \sum_{j=1}^{m} M_{gi}^{j} = \left(\sum_{j=1}^{n} l_i, \sum_{j=1}^{n} m_i, \sum_{j=1}^{n} u_i \right) \tag{19.5}$$

The inverse vector for a given statement is calculated as follows:

$$\left[\sum_{i=1}^{n} \sum_{j=1}^{m} M_{gi}^{j} \right]^{-1} = \left(\frac{1}{\sum_{i=1}^{n} u_i}, \frac{1}{\sum_{i=1}^{n} m_i}, \frac{1}{\sum_{i=1}^{n} l_i} \right) \tag{19.6}$$

Step 2

If $M_1 = (l_1, m_1, u_1)$ and $M_2 = (l_2, m_2, u_2)$ stand for two fuzzy triangular numbers, the degree of possibilities $M_2 = (l_2, m_2, u_2) \geq M_1 = (l_1, m_1, u_1)$ are defined as:

$$V\left(M_2 \geq M_1\right) = \sup_{y \geq x}\left[\min\left(\mu_{M_1}(x),\right)\mu_{M_2}(y),\right] \tag{19.7}$$

An equivalent statement for (Eq. 19.7) is given by the following:

$$\begin{aligned} V\left(M_2 \geq M_1\right) &= \operatorname{hgt}\left(M_1 \cap M_2\right) = \mu_{M_2} \\ &= 1\, if m_2 \geq m_1, 0 \text{ if } l_1 \geq u_2, \\ &\text{otherwise} \frac{l_1 - u_2}{\left(m_2 - u_2\right) - \left(m_1 - l_1\right)} V\left(S_b \geq S_a\right) \\ &= 1 \text{if} m_b \geq m_a, 0\, if\, l_a \geq u_b, \\ &\text{otherwise} \frac{l_a - u_b}{\left(m_b - u_b\right) - \left(m_a - l_a\right)} \end{aligned} \tag{19.8}$$

Step 3
Possibility degree of a convex fusion number to be greater than k, a convex fusion number, can be defined as follows:

$$\begin{aligned} &VV\left(M \geq M_1, M_2, \ldots M_k\right) \\ &= V\left[\left(M \geq M_1\right) \text{and} \left(M \geq M_2\right) \text{and} \ldots \text{and} \left(M \geq M_k\right)\right] \\ &= \min\left(M \geq M_i\right),\ i = 1, 2, 3, \ldots, k \end{aligned} \tag{19.9}$$

Supposing
d(A_i) = min $V(S_i \geq S_k)$, for the k=1, 2, …, n : $k \neq i$,the weight vector is given as follows:

$$W' = \left(d'\left(A_1\right), d'\left(A_2\right), \ldots, d'\left(A_n\right)\right)^T \tag{19.10}$$

where $A_i(i = 1, 2, \ldots, n)$ represents the number of n elements.

Step 4
In normalization, the normalized weight vectors would be:

$$W = \left(\mathrm{d}\left(A_1\right), \mathrm{d}\left(A_2\right), \ldots, \mathrm{d}\left(A_n\right)\right)^T \tag{19.11}$$

where W does not represent a fuzzy number.

This paper considers attributes as criteria for analyzing the influence of factors on the development of ambient intelligence toward enhancing e-learning as shown in Table 19.1.

Table 19.2 displays, as an alternative, the teaching environment

Upon the implementation of the fusion AHP method as Chang's analytical method [13], we come to results as shown in Table 19.3. It allows us to come to the

Table 19.1 Criteria for developing e-learning ambient intelligence

Criteria	
C1	Audio-video communication possibility
C2	Possibility for continuous teachers' guidance through teaching content
C3	Environment likewise e-learning system provide conditions for dynamic realization of teaching contents
C4	Multimedia presentation of teaching contents
C5	Teaching content design

Table 19.2 Alternatives for developing e-learning ambient intelligence

Alternatives	
A1	Traditional teaching
A2	Hybrid teaching (Traditional and E-learning)
A3	Guided e-learning supported by simulations
A4	Pure e-teaching
A5	Traditional teaching with the use of multimedia content

Table 19.3 Criterion weights and alternatives for developing e-learning ambient intelligence

		Alternatives				
Criterion	Weighted values	A1	A2	A3	A4	A5
C1	**0.14368**	0.237334	0.202872	0.225072	0	0
C2	**0**	0.109059	0.151158	0.130807	0.021366	0.11039
C3	**0.26754**	0.34287	0.303967	0.319555	0.38745	0.36427
C4	**0.22053**	0.154462	0.201578	0.148429	0.38745	0.36427
C5	**0.36825**	0.156274	0.140425	0.176136	0.203734	0.16105
Weights obtained		**0.217443**	**0.206638**	**0.215428**	**0.264128**	**0.23710**

conclusion that criterion C3 "Environment likewise e-learning system provides conditions for the dynamic realization of teaching contents" is weighted by the highest value, while criterion C2 "Possibility for continuous teacher's guidance through teaching content" is weighted by the lowest value. In order to determine the value of alternatives, a convergence consensus model has been applied in this paperwork.

The consensus convergence model [14] has been developed for the sake of decision-making on the effectiveness of ambient factors regarding students' achievements within the e-environment. This model is based on determining the differences in the "weights" of decision-makers on the basis of values assigned by each decision-maker to relevant elements (criteria, sub-criteria, and/or alternatives) [15]. Table 19.4 displays the calculated values of alternatives.

Table 19.4 Criterion weights and alternatives for developing e-learning model

Alternatives	Consensus weight vectors	Rank
A1	0.060	5
A2	0.069	2
A3	0.077	1
A4	0.061	4
A5	0.062	3

Based on the calculation, we can conclude that the alternative A3 "Guided e-learning supported by simulations" is ranked first regarding the importance of the effective development of ambient intelligence in order to enhance e-learning.

Identification and Security Context Analyses

Identification implies observational procedures relating to the notification of students' reactions to the created teaching content likewise the teaching process. Security context implies defining procedures that minimize the possibility of teaching content as well as users' personal information misuse. Stated contexts in this paper have been analyzed by the same method as the context of the environment for e-learning with the maximum assurance of the prescribed quality of e-learning standards [16]. An alternative to the emerging value identification context is "Guided e-learning supported by simulations." An alternative with the highest possible value for context security is "Traditional teaching using multimedia." Criteria considered for all three contexts are C1, Audio-video communication possibility; C2, Possibility for continuous teacher's guidance through teaching content; C3, Environment likewise e-learning system provides conditions for the dynamic realization of teaching contents; C4, Multimedia presentation of teaching contents; and C5, Teaching content design. Comparative analyses have found that "Guided e-learning supported by simulations" represents the optimal environment for e-learning. Considered criteria are the basics for the development of ambient intelligence. Required and sufficient conditions for the C1 criterion "Audio-video communication possibility" can be the existence of compatible audio-video equipment for teachers as well as for students. Required and sufficient conditions for criterion C2 "Possibility for continuous teacher's guidance through teaching content" can be installed interactive platform which is able to track students' work and, of course, when needed, informs the teacher regarding the difficulty in mastering the teaching content. Required and sufficient conditions for C3 criterion "Environment likewise e-learning system provides conditions for the dynamic realization of teaching contents" can be the existence of e-learning platform as well as appropriate microclimate conditions toward the realization of dynamic content and sensors installed for monitoring basic parameters of general conditions like students' pulse, temperature, blood pressure, etc. Required and sufficient conditions for the C4 criterion "Multimedia presentation of teaching contents" can be the existence of e-platform which allows you to create, set up, or use the multimedia presentation of teaching content. Required and sufficient conditions for C5 criterion "Teaching content design" can be the existence of an e-platform which enables the design of teaching content according to the topics as well as the timing of a teaching process.

Ambient intelligence shall be developed throughout these contexts with the use of a neural network whose final result will be expressed through the fusion of neural network results, observed for each criterion individually. Basic connections within the system for the development of ambient intelligence are presented in Fig. 19.1.

This paper will explain the neural network model used for criterion C1 "Audio-video communication possibility." Video equipment should allow the monitoring of students' facial expressions. Based on these expressions, the system will propose corrective actions within the ambient environment. For this purpose, convolution neural networks models will be used, also known as non-cognitive models [17]. For the sake of analyses, these models will use data deployed in advanced databases [18].

Convolution Neural Networks

The term "neural networks of convolution" is used in explanations in image and signal processing in artificial neural networks [19]. We use convolution filters in the field of artificial intelligence and neural networks to sharpen or blur images likewise to detect edges in contrasting terms. Convolution neural networks are most frequently used when data represent images that are in pixel matrix and they are presented by their width, height, and pixel values. For color images, each of the three "RGB" channels is usually represented by value pixels within the range of 0–255. In convolution neural networks, the convolution filter represents a generalized linear model for the image region to which it is applied [20]. Apart from the filter name, the context of convolution neural networks, the name convolution kernel is also used. The filter is represented by a two-dimensional matrix of small dimensions, compared to the image to which it is applied and consisted of real values.

An important concept of convolution neural networks is displayed in maximal compression as a form of nonlinear pixel reduction [21]. Depending on a compression type, a pixel with a particular value in a particular region is selected. As for reduction operation, the pixel with the highest value is chosen for the maximum. The compression layer is used to reduce progressively the size of the image and therefore the number of features, thus leading to a decrease in the complexity of the calculation. An example of maximum compression is shown in Fig. 19.2.

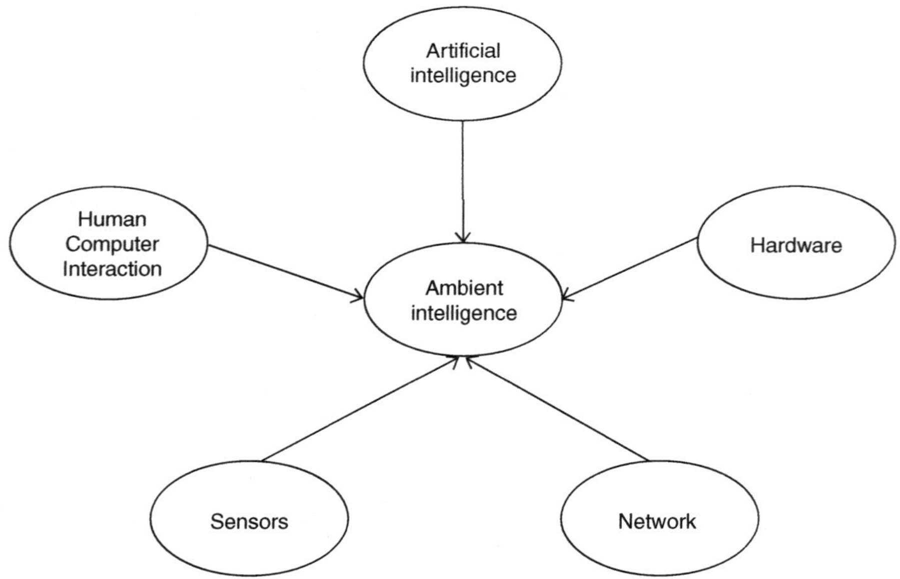

Fig. 19.1 Connections in the ambient intelligence development system

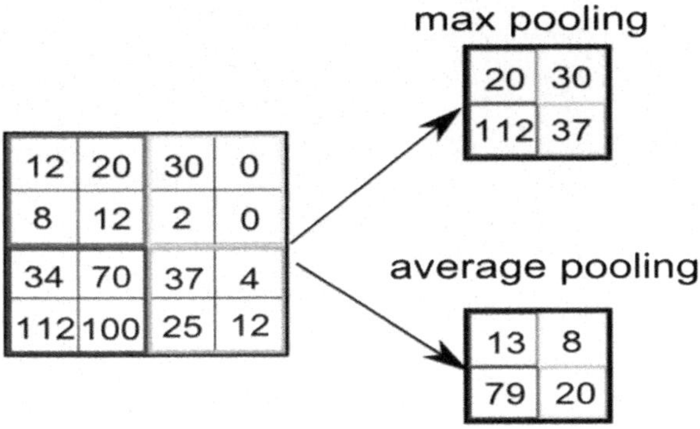

Fig. 19.2 An example of maximum compression

Description of the General Architecture of Convolution Neural Networks

The difference of convolution neural networks about other types of architecture neural networks reflects in the existence of a layer where convolution occurs. Convolution layer is essential for the functioning of a convolution neural network. It also performs various demanding calculations [22–24]. The essence of using convolution neural networks is to enable the response in real time regarding activities that take place in distance learning systems [25–28]. Figure 19.3 shows the architecture of convolution neural networks.

The initial author used to construct and recognize facial expressions with the use of her photos as input data toward processing within convolution neural networks. The initial author of the paper has also tried to present different states of facial expressions as shown in Fig. 19.4.

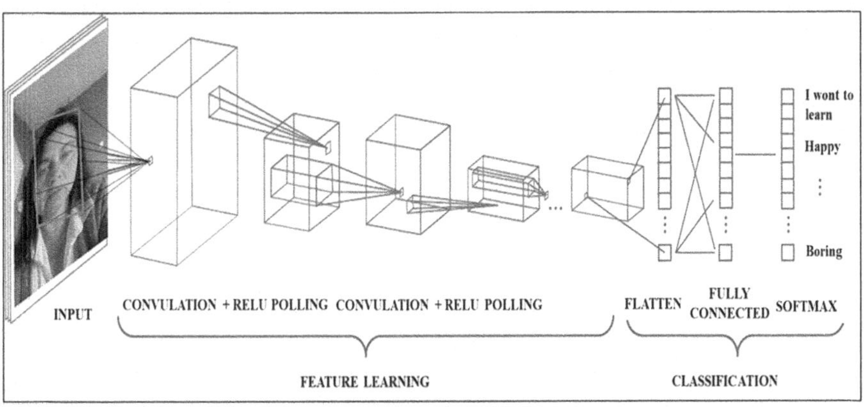

Fig. 19.3 The architecture of the convolution neural network

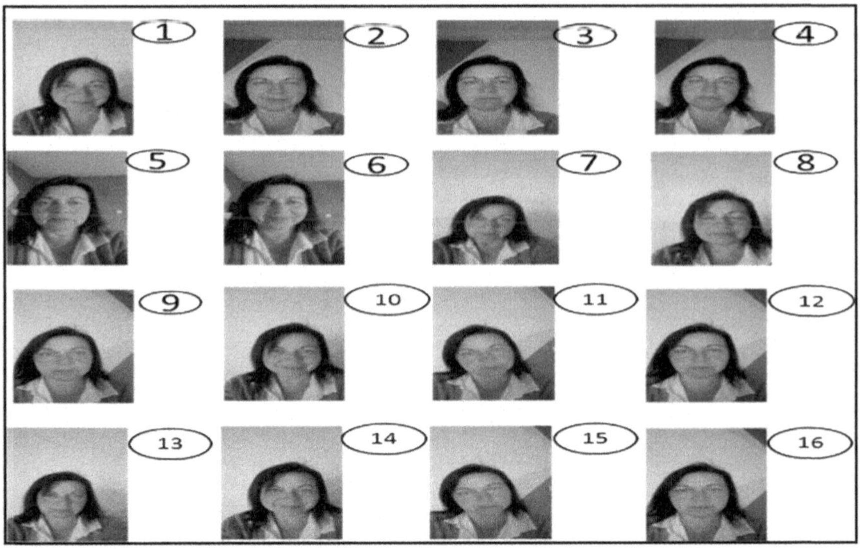

Fig. 19.4 Different facial expressions

Position 8 in Figure 19.4 shows the state in which the person expresses disinterest in teaching activities, manifested by eyes closed as well as lips folded. The lack of interest in teaching activities can be a consequence either of fatigue or poor ambient conditions. To compare the results of artificial neural networks, comparative analyses of the obtained values for the criterion C1 have been conducted by the author. The analyses are the result of image processing via convolution neural networks likewise through artificial "backpropagation" neural networks of which activation function is the sigmoid function. For the research, Neuroph Studio has been used in which neural network has been created as shown in Fig. 19.5.

After 10000 interactions, an error value of 0.01 has been reached, while in Fig. 19.6 a graph of the total network error after 136 interactions is shown.

19.3 Results

During the testing of the convolution neural network, facial expression detection has been performed, relating to disinterest in teaching in 98% of cases based on 16 different facial expression images. By testing artificial backpropagation of a neural network whose activation function is sigmoid recognition of facial expression, detection of facial expression relating to disinterest in teaching in 88% of cases has been conducted, based on 16 different photos of facial expressions. Further development of the e-learning platform should provide the teaching staff with information on which corrective measures should be taken to create ambient conditions to

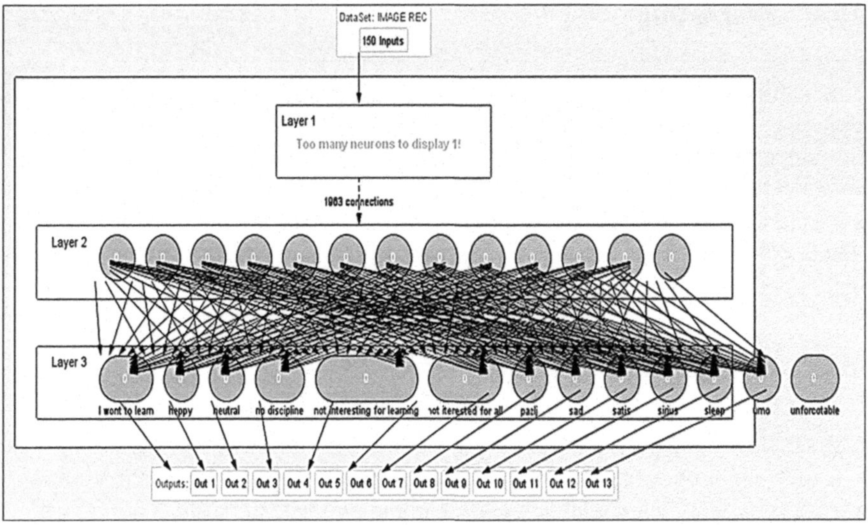

Fig. 19.5 A neural network created in a Neuroph Studio

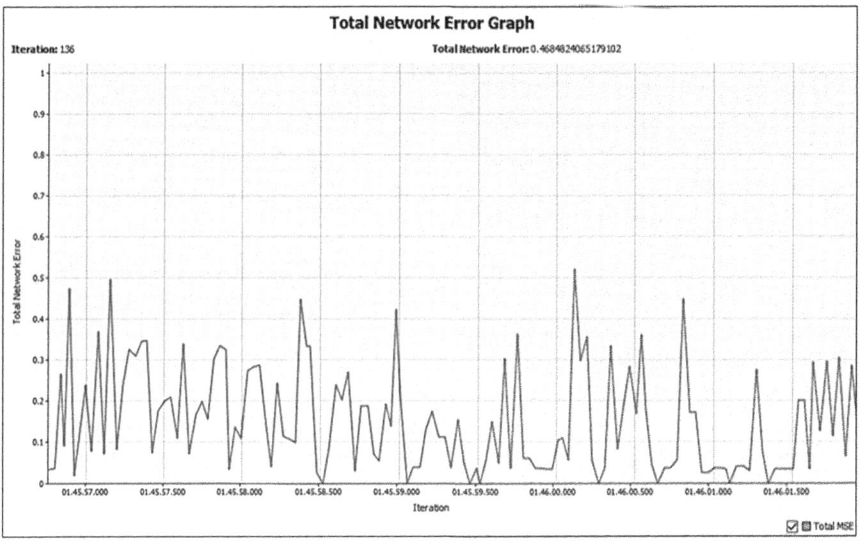

Fig. 19.6 Total network error graph after 136 interactions

achieve better results in mastering the curriculum. The initial author, mathematics professor, has applied corrective methods within the organization of e-teaching for the sake of mathematics course, which implied a higher level of achievement in ambient students' intelligence in comparison with students taught traditionally.

19.4 Discussion

Ambient intelligence evolves through contexts: e-learning environment, identification, and security. This paper presents the impact of the e-learning environment through criterion C1 "Possibility of audio-video communication" where the necessary conditions are: the existence of installed compatible audio-video equipment for students and teachers. Survey results indicate that with the use of ambient intelligence, better results are achieved, especially in elementary school's mathematics.

19.5 Conclusion

Contemporary trends, as well as lifestyles, are increasingly suggesting that e-learning will be applied on a much larger scale in comparison to the traditional way of schooling. Suppose that we see learning as a type of business or production, the final product represents knowledge, and we have to be objective and conclude that in case the final result is good, the production system will be simpler and cheaper. So there is no reason not to accept it. Research results indicate that the convolution neural network is more suitable for facial expression recognition in comparison with artificial backpropagation neural networks. Modern information technology in the creation of ambient intelligence has a key role in raising students' achievements as well as competency levels. It is also being executed through the concept of e-learning onto one of the convenient learning management platforms. Research results also indicate that with the use of ambient intelligence, better results are achieved, especially in elementary school mathematics education.

Contact Author Information Julijana Vasiljević was born on February 29, 1968, in Bitola, Republic of North Macedonia. She graduated from Faculty of Natural Sciences and Mathematics at the "Ss. Cyril and Methodius" University in Skopje in 1991. At the "Faculty of Technical Sciences" in Čačak, she enrolled a master's degree that ended in 2012. She is currently a PhD candidate at Pan-European University "Apeiron," Banja Luka, Bosnia and Herzegovina, Europe. (Address: Voždovac, V. Stepe 405 D, Belgrade 11000, Republic of Serbia)

References

1. J.C. Augusto, P. McCullagh, Ambient intelligence: concepts and applications. Comput. Sci. Inf. Syst. **4**(1), 1–27 (2007)
2. Siemens. http://networks.siemens.de/smarthome/en/index.htm (2006)
3. Nokia. research.nokia.com/research/projects/sensorplanet/ (2006)
4. ST Advisory Group, The European union report, scenarios for ambient intelligence in, 2001. ftp.cordis.lu/pub/ist/docs/istagscenarios (2010)

5. C. Augusto, D. Cook, Ambient Intelligence: applications in society and opportunities for AI. Lecture Notes, tutorial given during 20th International Joint Conference on Artificial Intelligence – IJCAI'07, Hyderabad, India (2007)
6. Д. Бјекић, Р. Крнета, Д. Милошевић, Компетенције за е-наставу у систему професионалних компетенција наставника основне школе, Иновације у настави, pp. 7–20 (2008)
7. D. Webster, *Knowledge Presenter: Learning about e-Learning*, http://www.knowledgepresenter.com/assets/freeguide.htm (Kookaburra Studios, 2006)
8. K. Nakajima, *Is "e-Teaching" Web Zero or Potentially Web 2.1?*, http://www.cccties.org/access/toukou/nakajima_20061117_2.pdf (2006)
9. M. Williams, E-Learning. History and prospects, http://apps.hct.ac.ae/ewb2005/2005/ewb_newsletter_04_10.pdf (2005)
10. H.S. Lin, H.L. Chiu, Using computers to support a beginning teacher's professional development. J. Sci. Educ. Technol. **4**, 367–373 (2000)
11. R. Sutherland, Designs for learning: ICT and knowledge in the classroom. Comput. Educ. **43**, 5–16 (2004)
12. T. Saaty, Decision making with the analytic hierarchy process. Int. J. Serv. Sci., 83–98 (2008)
13. D.Y. Chang, Applications of the extent analysis method on fuzzy AHP. Eur. J. Oper. Res., 649–655 (1996)
14. B. Srdjevic, Y. Medeiros, Fuzzy AHP assessment of water management plans. Water Resour. Manag., 877–894 (2008)
15. E. Triantaphyllou, Multi-criteria decision making methods, in *Multi-criteria Decision Making Methods: A Comparative Study*, (Springer, Boston, 2000), pp. 5–21
16. S. Tomić, D. Drljača, DDLM – quality standard for electronic education programs in higher education of Bosnia and Herzegovina. JITA – J. Inf. Technol. Appl. Banja Luka, Paneuropean University Apeiron, Banja Luka, Republika Srpska, Bosna i Hercegovina, JITA 9(2019) 2:67–79, (UDC: 004.735.8:621.39(497.6), (https://doi.org/10.7251/JIT1902067T), Volume 9, Number 2, Banja Luka, December 2019 (49-128), ISSN 2232-9625 (print), ISSN 2233-0194 (online), UDC 004
17. C. Padgett, G.W. Cottrell, *A Simple Neural Network Models Categorical Perception of Facial Expressions*, in Proceedings of the 20th Annual Conference of the Cognitive Science Society (Erlbaum, Mahwah, 1998), pp. 806–807
18. Z.Ž. Nedeljko Šikanjić, E.F.J. Avramović, Implementation of the Neural Network Algorithm in Advanced Databases, JITA – J. Inf. Technol. Appl., Paneuropean University Apeiron, Banja Luka, Republika Srpska, Bosna i Hercegovina, JITA 8(2018) **2**:54–63, (UDC: 004.738.5:551.588:551.506), (https://doi.org/10.7251/JIT1802054S), Volume 8, Number 2, Banja Luka, December 2018 (45–96), ISSN 2232-9625 (print), ISSN 2233-0194 (online), UDC 004
19. H. Waheed, S.U. Hassan, N.R. Aljohani, J. Hardman, S. Alelyani, R. Nawaz, Predicting academic performance of students from VLE big data using deep learning models. Comput. Hum. Behav. **104**, 106189 (2020)
20. J. Gu, Z. Wang, J. Kuen, L. Ma, A. Shahroudy, B. Shuai, T. Liu, X. Wang, G. Wang, Recent advances in convolutional neural networks. Comput. Vis. Pattern Recognit. https://arxiv.org/abs/1512.07108/ (2015)
21. C. Gulcehre, Convolutional Neural Networks (LeNet), http://deeplearning.net/tutorial/lenet.html/, Feb 2017
22. J. Johnson, A. Karpathy, Convolutional Neural Networks. Stanford Computer Science, http://cs231n.github.io/convolutional-networks/, Feb 2017
23. Y. Wang, Y. Li, Y. Song, X. Rong, The influence of the activation function in a convolution neural network model of facial expression recognition. Appl. Sci. **10**(5), 1897 (2020)
24. I. Goodfellow, Y. Bengio, A. Courville, *Deep Learning* (MIT Press, 2016) https://www.deeplearningbook.org/

25. T. Saito, Y. Watanobe, Learning path recommendation system for programming education based on neural networks. Int. J. Distance Educ. Technol. **18**(1), 36–64 (2020)
26. M.R. Syed, Methods and applications for advancing distance education technologies: international issues and solutions. Simulation **348** (2020)
27. A. Agrawal, N. Mittal, Using CNN for facial expression recognition: a study of the effects of kernel size and number of filters on accuracy. Vis. Comput. **36**(2), 405–412 (2020)
28. F. An, Z. Liu, Facial expression recognition algorithm based on parameter adaptive initialization of CNN and LSTM. Vis. Comput. **36**(3), 483–498 (2020)

Chapter 20
Reflections on the COIL Project Concerning Nuclear Power Generation, Sea Transportation of Radioactive Materials, and Renewables' Perspectives

Sanja Bauk and Edilene Gasparini Fernandes

20.1 Introduction

The COIL (Collaborative Online International Learning) projects are about using online tools for building bridges between students and professors from different countries in terms of developing soft skills as cross-cultural awareness, foreign languages, digital literacy, learning across disciplines, and critical thinking. The COILs allow mobility while students and professors are at home through virtual engagement. They make it possible for students from developing environments to have meaningful cross-cultural experience as part of their tertiary education [1–3]. The COILs facilitate more contacts between people, reach youth from different social backgrounds, and promote intercultural understanding [4]. Through the recent COIL between DUT and Fatec, we tried to acquire some basic knowledge about nuclear power cons and pros, uranium ore deposits in South Africa and Brazil, sea routes and shipments of nuclear materials around the world, sea transportation of irradiated nuclear fuel, plutonium and radioactive wastes, and renewables' potentials in comparison to nuclear power. Our initial plans were ambitious, but during this time we became aware of time constraints, our overwhelming regular workloads, and pressures due to the actual pandemic situation. It has many trials, hits, and misses around. However, regardless of the imperfection of the achievements,

S. Bauk
Maritime Studies, Faculty of Applied Sciences, Durban University of Technology, Durban, South Africa
e-mail: sanjab@dut.ac.za

E. G. Fernandes (✉)
Paula Souza Center – CEETESP, São Paulo State Technological College – FATEC, Sao Paulo, Brazil
e-mail: edilene@fatecriopreto.edu.br

S. Bauk, S. D. Ilčev (eds.), *The 1st International Conference on Maritime Education and Development*, https://doi.org/10.1007/978-3-030-64088-0_20

we decided to share them, since we believe this can assist future attempts in teaching, learning, and research in virtual space within the COILs and similar projects of virtual mobility in higher education.

The rest of the paper is organized as follows: (i) Section 20.2 gives overview of the COIL management and applied methodology; (ii) Section 20.3 contains some findings regarding nuclear power needs and resources, sea transportation of nuclear materials, and renewables' perspective in comparison to nuclear power generation in Brazil and South Africa; (iii) Section 20.4 gives some conclusion remarks and directions for further research in the field.

20.2 Project Management and Applied Methodology

Intensively involved in the COIL project were 14 students and 2 professors from South Africa and Brazil. Nationalities of the students were Brazilian, Zulu, and Xhosa. Prior to the project beginning, professors had several online meetings and made a preliminary plan of the activities. They decided about the topic, schedule, deliverables, and exchange of the students' contact details. Later on, teams had several online meetings, and students from both sides made short videos about their universities and their student life in South Africa and Brazil. They highlighted their different cultures and places of origin. All recordings, minutes of online meetings, instructional materials, and the like are uploaded at Trello platform (Fig. 20.1). Students were working in mixed groups. Initially it was five groups, but later on, this number was reduced to two. Students agreed it is easier to switch to two groups, and we supported them toward making their working environment more convenient.

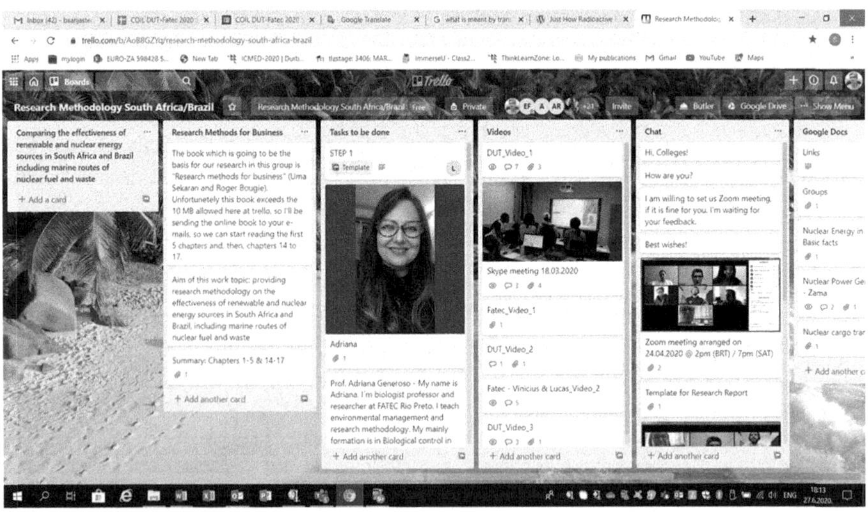

Fig. 20.1 The COIL Trello platform. (Source: Own)

Therefore, students created two comprehensive research reports in Google Docs from which we tried to pull out the essence, extend it, and present the outcomes in the form of this article. During the project realization, participants used mails, WhatsApp, and Zoom as online tools to present their ideas and accomplish tasks. Initially 2 months were planned for the project, but its duration has been extended to almost 5 months due to the actual pandemic issues and interruptions in the universities' academic calendars in South Africa and Brazil. When it comes to the issues of students' learning and teachers' collaboration, we were much more focused on students' learning because our research is being done on a basic level. Predominantly explorative and descriptive research approaches were used. As literature resources, we were using mostly non-print media as e-books, electronic journal articles, online documents, podcasts, streaming videos, etc. After several trials from both sides to resume the work after the interruptions, we came up with the findings, which are summarized in the following sections.

At the end of the project, Google survey has been conducted among the students. We sent the survey to all students via email, but we have received feedbacks from ten of them in due course. In the survey, we used Likert (S1–S16) and ordinal (S17) scales. The Likert scale examines how strongly the students agree with the offered statements. It is 5-point scale with the following anchors: 1, strongly disagree; 2, disagree; 3, neither agree nor disagree; 4, agree; and 5, strongly disagree. On another side ordinal scale helps us to determine the range among the transferable skills supported by the COIL (S17). The results are summarized in Table 20.1.

Even though the pool of surveyed students was small, we can conclude that all students were predominantly affirmative toward the statements in favor of the COIL. However, we can conclude students found work in an international group somehow challenging. In addition, they have faced some linguistics and technical difficulties. The first is understandable since for both teams English is not their native language. Native languages of the project members were isiZulu, isiXhosa, and Portuguese. When it comes to technical difficulties, some students did not have continuous access to the Internet. This also partly explains "imperfect" scores when it comes to the statements "Directions for group work were clear" and "Resources for research work were available." Both statements are connected (in)directly with students' Internet access. Besides, students recognized that the actual pandemic setting creates difficulties in group work. Regarding transferable skills, the students gave the highest priority to critical thinking and then to cross-cultural awareness, while team work and digital literacy share the third and the fourth positions.

20.3 The COIL Project Findings

Growing society's energy needs cause the development and utilization of nuclear power in combination with nuclear fuel recycling. Nuclear energy is very important in the sectors of medicine, agriculture, and industry. The countries with nuclear reactors need to transport radioactive materials to reprocessing plants prior of stor-

Table 20.1 Google survey results. (Source: Own)

No.	Statement	Average score
S1	COIL enhances internationalization (mobility) at home	4.6
S2	It enhances students' creativity	4.9
S3	It opens trans-disciplinary perspectives	4.6
S4	It boosts curiosity and motivates me to learn	4.7
S5	It supports informal learning styles	4.7
S6	My motivation in the university degree has been increased positively with COIL activities.	4.5
S7	"COIL further promotes internationalization in the curriculum opportunities for students to acquire the international perspective needed to success in global community" (L. Beth)	4.8
S8	Through COIL I have learned how to connect theoretical knowledge with real-world issues	4.6
S9	It is important for me to interact with colleagues from different geographical areas	4.3
S10	Working in an international group is somehow challenging	4.9
S11	Within COIL I have faced linguistics problems	3.5
S12	Within COIL I have faced technical difficulties	3.8
S13	Actual pandemic setting creates difficulties in group work	4.0
S14	Lecturers were supportive and resilient	4.8
S15	Resources for research work were available	4.1
S16	Directions for group work were clear	3.9
S17	Please rank COIL impacts on your transferable skills:	**Rank**
17.a	Critical thinking	1
17.b	Cross-cultural awareness	2
17.c	Team work	3/4
17.d	Digital literacy	3/4

age of residuals. Therefore, it is reasonable to expect that sea transportation of radioactive materials will be increasingly involved. The coastal countries as South Africa and Brazil, as well as its maritime industry in general, have to provide transportation of radioactive materials that include irradiated nuclear fuel, plutonium, and high-level radioactive wastes by ships in safe, reliable, and secure manner [5]. The comparisons among nuclear power and renewables are of considerable importance in this context, as well.

Nuclear Power Controversy and Prospective

Nuclear power is a controversial energy source. The accidents in Londonderry (Pennsylvania) in 1979, Chernobyl (Ukraine) in 1986, and Fukushima (Japan) in 2011 have raised serious safety concerns. Consequently, several countries moved away from it, citing safety risks and financial costs. On another side, nuclear power advocates say it has green credentials. Its carbon footprint is much smaller than other energy sources. The nuclear power plants can create a basket or a portfolio of different energy sources resilient from security point of view. In the early days of nuclear power, i.e., in the early 1950s of the past century, people did not know much about climate change or that this new low-carbon energy source could help address it. The thing is to steam climate change through producing far greater amounts of clean and reliable energy. For tackling climate change, 80% of all electricity will need to be low carbon by 2050 [6]. Worldwide, nuclear power saves more than two billion tons of CO_2 emissions each year. South Africa and Brazil are the fifth and seventh, respectively, in the world due to the deposits of uranium ore (Fig. 20.2). South Africa and Brazil have significant uranium deposits but are not at the top of nuclear power production. More precisely, Brazil is in the 18th place and South Africa in the 21st place in the world [6]. Brazil has one nuclear plant (Angra) and South Africa has also one nuclear plant (Koeberg). The nuclear generation was responsible for 1.2% of the electric production in Brazil in 2020 [7] and for 2.0% of electric generation in South Africa in 2018 [8]. On another side, according to the World Nuclear Association data from 2018, ten leading countries in the world in nuclear power generation are the USA, France, China, Russia, South Korea, Canada, Ukraine, Germany, the UK, and Sweden. On the basis of the aforementioned statistics, one may draw conclusion that developed and emerging countries take a larger share in the production of nuclear energy than developing ones. Developing countries should consider opportunities for building modern, efficient, and safe nuclear plants or consider implementing small nuclear power (modular) reactors to increase considerably their energy production capacities [10]. Nuclear energy can back up renewables ensuring round-the-clock clean power even when there is no sunshine or wind. Still, investing in a new nuclear technology has not been an easy task. The cost is a key concern. Developing small modular reactors can be an alternative solu-

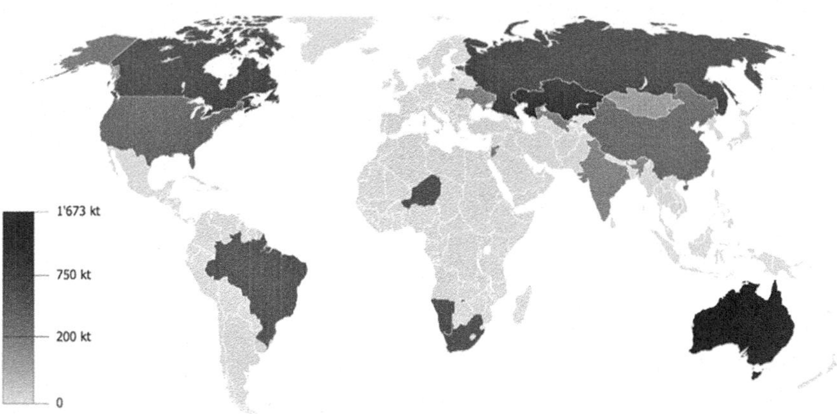

Fig. 20.2 Uranium reserves in 2010. (Source: [9])

tion. In addition, a combination of small modular reactors and renewables through so-called hybrid energy systems might be an alternative.

It is worth to mention that a one-of-a-kind floating nuclear power plant, designed and built by Russia, has become the most northerly nuclear facility in the world, as it begins commercial operations in a remote region in the Arctic (in a small town called Pevek in the far eastern region of Chukotka). The unique characteristic of this plant is its multi-functionality, i.e., the ability to provide faraway isolated regions of the country where the deficit exists with heat, electric energy, and drinking water. The new source of energy is extremely important for those living in the small Arctic town, as it will replace a coal power plant that had been providing Pevek with energy and heat for over 75 years. The facility polluted the air, with locals saying that they were worried about their children, as soot turned snow black and covered their windowsills. "We hope that the situation improves and ecology will be better," a local woman said. "[And] also there'll be more jobs, and new infrastructure will be developed" [11]. It might be concluded that the latest works more or less in general, over the world, but it is to be investigated in some more detail in different, specific settings.

Sea Transportation of Nuclear Materials

The main trade routes of nuclear material are between Japan, the UK, and France, including routes targeting Russia, Sweden, Canada, Argentina, Brazil, Chile, Ireland, South Africa, etc. Legislative framework for coastal states regarding sea transportation of irradiated nuclear fuel, plutonium, and radioactive wastes corresponds to the International Convention for the Safety of Life at Sea (SOLAS), the International Maritime Dangerous Goods (IMDG) Code, the United Nations Convention for the Law of the Sea (UNCLOS), the Irradiated Nuclear Fuel Code (INFC), etc. The International Maritime Organization's (IMO) Sub-Committee of Safety of Navigation (NAV) regulations covers the route planning, notification, and consultation with coastal states, including possible restrictions and exclusion of certain routes. Aids, equipment, and devices that would facilitate location and salvage of a ship and/or nuclear cargo in the case of emergency are covered by IMO's Sub-Committee on Carriage of Dangerous Goods (CDG) and Sub-Committee on Radiocommunications (COM) including both terrestrial and satellite channels [5]. The world's most experienced shipper of nuclear materials is Pacific Nuclear Transport Limited (PNTL). The homeport of this company is Barrow-in-Furness in England (UK). The company has successfully completed 180 shipments during the last 40 years [12]. PNTL has safely sailed over five million miles. This is the equivalent of going to the moon and back over ten times. More than 2000 casks of nuclear material have been safely transported by PNTL since its establishment. The company has the most experienced nuclear transport crew in the world. On average, each crewmember has more than 20 years of experience with PNTL. Today, three PNTL ships are in service: Pacific Heron, Pacific Egret, and Pacific Grebe. These vessels are capable of carrying spent fuel, mixed oxide fuel (MOX) assemblies, and vitrified high-level waste. The process of cask with nuclear material unloading is shown in Fig. 20.3.

PNTL ships have double hull construction, dual navigation monitoring and calling systems, twin engines, rudders and propellers, backup power generators, radioactivity monitoring, secured cargo, enhanced buoyancy, bow thrusters, backup generators, additional firefighting equipment, and weather routing system. It is important to emphasize that during the past 40 years and more of PNTL ships' operations, there has never been a single incident resulting in the release of radioactivity.

Brazilian Renewables' Potentials

When it comes to renewables, Brazil is considered the third globally in terms of installed capacity from renewable energy sources, being preceded by China and the USA. Besides, it is the first country by installed biomass capacity, second in terms of hydropower capacity, and eighth place considering wind capacity. This study was presented in the form of a report by the Brazilian Ministerial Secretariat from Mines and Energy, abridging a period from 2019 to 2023, and it was entitled "Brazil Renewable Energy Sector Report 2019-2023" [14]. According to this paper, the prospects for the evolution on the energetic consumption matrix by source show a tendency for maintaining the growing of the electrification of the country. The transportation sector shows that the by-products from sugar cane will probably keep its

Fig. 20.3 Unloading cask with nuclear cargo from PNTL ship's hull. (Source: [13])

importance throughout this period until 2023. Petroleum by-products, as well, will probably keep its importance as final energy source, growing 2.0% a year in this future period, however decaying its part in the final consumption matrix, due to the substitution of oil and gas by biodiesel and hydrated ethanol in the sector. Wood and coal are also losing position to sources presenting a better income.

From 2013 to 2018, the average of annual growth in the use of photovoltaic systems was 224.5% of centralized and 214.4% of distributed solar power installed capacity. That happened mainly because of the favorable geographic conditions in Brazil. Also in 2018, renewable sources reported 82.5% of total installed capacity in the country and 85.9% of the power generated during this year. Hydropower continues to be the most explored energy source. However, its part in the domestic electric energy matrix is slowly decreasing, as other renewable sources gain place, such as wind and solar. Besides, "almost 45% of primary energy demand is met by renewable energy, making Brazil's energy sector one of the least carbon-intensive in the world" [15]. However, the Brazilian electric system is a huge hydrothermal one with predominance of hydroelectric plants. This characteristic may turn it into very susceptible to climate changes, and this may possibly affect the restrictions in using the hydric resources. Some studies have mentioned a possible reduction of 15% in this type of generation in 2041 and ahead [16]. Wind energy source has become a very attractive investment for future parks when compared with other types of energy sources, considering that its cost has been reducing with time, according to the study from the Brazilian government. The offshore wind power is considered a candidate for expansion after 2027. However, costs are still high for this kind of plant. Photovoltaic technology has surprised by the reduction in costs. Bioelectricity, mainly the one generated by the sugar cane bagasse, is still a very competitive source of energy generation in Brazil. The following picture (Fig. 20.4) shows a plant that has been awarded the plant of the year in bioelectricity this year, by producing energy from sugar cane lignin, straw, and bagasse combustion in the boiler with a fluidized bed, which maximizes generation and uses the opportunities in the spot market [17]. Biogas is another attractive form of power source, and it is being considered a good candidate for expansion, due to the exceeding material coming from the agro-industrial production. Biomass coming from planted forests has suffered a management change. Plants have been installed closer to the forests, and this has reduced the costs for this type of source. With the pre-salt area exploitation, natural gas will be the main actor in the expansion of thermoelectric generation, according to the government plan for the period between 2020 and 2029.

Amplifying nuclear plants are not the main goal for generation, as one can read in this document. The investment on these plants, which next is called Angra 3, is expected to start only after this decade, as there are studies being conducted and licenses to be achieved before the construction begins [14] p. 60. However, "Brazil has the seventh largest uranium reserves in the world, but only a third of the country has been explored," according to the World Nuclear Association. Caetité is the only uranium mine in Brazil, but operations have been halted since 2014, and now, the government has announced it is targeting to resume production in this unique

Fig. 20.4 A bioelectricity plant in Brazil. (Source: [17])

Brazilian mine before the end of the year and, consequently, opening up the sector to private companies [18].

Considering the setbacks promoted by the buildup myths around making use of nuclear power by countries, in Brazil, one can overview a favorable horizon for the implementation of bioelectricity, due to the amount of crops and plantations the country holds. An elaborated study performed by SUCRE (Sugarcane Renewable Electricity), a project run by a group of researchers sponsored by the government, indicates that bioelectricity from sugar cane may supply all the residential demand for energy and still reduce 15% in the gas emissions [19]. Diversifying the use of biomass aiming at promoting a better utilization of all the parts of the sugar cane leftovers seems to be the recent news in the area.

South African Renewables' Potentials

In South Africa, renewables as solar and wind are in, while coal as fossil fuel is out [20]. Today, about 95% of the nation's power comes from coal-fired power plants, which do not comply with environmental standards. By 2030, more than 40% of total energy production should come from renewable sources. However, coil will continue to play a significant but declining role in electricity generation in the future, accounting for 59% of output by 2030. Nuclear power will contribute about 5%, hydropower 8%, solar 6% [21], wind 18%, gas 2%, etc. In May 2020, the

Department of Minerals and Energy Resources of South Africa has announced a work on a roadmap for a new nuclear bulb program of small modular nuclear power reactors. The extension of the life of Koeberg nuclear power plant beyond 2024 is announced as another opportunity for the industry. The Nuclear Industry Association of South Africa (NIASA) supports both plans [22].

20.4 Conclusion

The paper presents the results of a pilot research study within the COIL project between DUT (South Africa) and Fatec (Brazil) tertiary education institutions. It was a challenge for the professors to experiment with the potentials of cross-continental pedagogical and research collaboration. The involved students enjoyed sharing their national and cultural diversities, but they were not so eager when it comes to research part of the project. Majority of engaged students are working and studying at the same time, they may feel anxiety due to the current pandemic situation and economic uncertainty, some of them faced problems with Internet access, and the like. By taking into account these aggravating circumstances, we should be satisfied with the achieved outcomes and try to improve our performance throughout the following similar projects.

Since the purpose of our pioneer research was multifold, we highlighted some basic facts about nuclear power generation benefits and drawbacks. In addition, we considered basic features of sea transportation of nuclear materials. As this COIL involves counterparts from South Africa and Brazil, nuclear ore deposits in Brazil and South Africa were examined, as well. Both South Africa and Brazil are leading countries in uranium ore deposits, but they are not among the leading ones when it comes to nuclear power generation. In Brazil, hydro-potential is huge and its extensive exploitation is reasonable. Besides, bioelectricity production from sugar cane is an additional opportunity for electrical production in the future. In South Africa, coil is the main source of electricity production, but there are plans for intensifying renewables deployment. In parallel, there are some plans for deployment of small nuclear reactors and renewal of Koeberg nuclear power plant. For forthcoming research, more comparative analysis of nuclear power generation and renewables in both developing countries should be conducted, since both energy sources are components of future energy production portfolio.

Acknowledgments We would like to thank our students Zamangwane Hlongwane, Ntando Nqubeko Ndimande, Nhlanzeko Sbusiso Mathonsi, Thuladu Queen Mazibuko, Mxolisi Siphesihle Nxumalo, Amanda Pretty Ntombela, Lungisani Ndlovu, Alceu Chueire, Vinicius Francisco Camilo, Giovana Cruz Carlos, Lucas Bocchio, Marcelo Ito, Lara Pagliuso, and Vinicius Fantasia, although they might not agree with all of the interpretations and conclusion of the paper. Furthermore, this paper can be used for dissemination purposes of European Commission's Erasmus + EURO-ZA project (No. 598418), as an example of novel pedagogical method in maritime proposed by the project.

References

1. F. Helm, The long and winding road…. J. Virtual Exch. (1), 41–63
2. S. Bauk, S. Fajardo-Flores, Matching interaction design principles and integrated navigation systems in an electronic classroom. Trans. Maritime Sci. **9**(1), 90–98 (2020)
3. S. Bauk, Collaborative online international learning benefits vis-a-vis concerns: an empirical study. Montenegrin J. Econ. **14**(2), 207–216 (2019)
4. European Commission, Erasmus+ goes virtual. Press release. https://ec.europa.eu/commission/presscorner/detail/en/IP_18_1741. 15 Mar 2018
5. J. Wonham et al., Marine transportation of irradiated nuclear fuel, plutonium and radioactive wastes: the continuing debate on regulatory measures. Mar. Policy **24**, 287–299 (2000)
6. WNN – World Nuclear News, *IAEA Explains Nuclear's Vital Role in a Carbon-Free Future*. Retrieved 2020, May 14, from https://world-nuclear-news.org/Articles/IAEA-explains-nuclears-vital-role-in-a-carbon-free. 8 Jan 2020
7. World Nuclear Association UK, *Nuclear Power in the World Today*. Retrieved June 2020, 28, from https://www.world-nuclear.org/information-library/current-and-future-generation/nuclear-power-in-the-world-today.aspx. Mar 2020
8. World Nuclear Association UK, *Nuclear Power in the World Today*. Retrieved June 2020, 28, from https://www.world-nuclear.org/information-library/nuclear-fuel-cycle/nuclear-power-reactors/small-nuclear-power-reactors.aspx. March 2020
9. List of countries by uranium reserves, *In Wikipedia*. Retrieved 2020, June 28, from https://en.wikipedia.org/wiki/List_of_countries_by_uranium_reserves. 17 Mar 2020
10. IAEA, *Nuclear Power in the 21st Century* [Video file]. Retrieved from https://www.youtube.com/watch?v=8D1SVKjTJE4. 1 Nov 2017
11. RT, Russia. *World's ONLY Floating Nuclear Power Plant Project Goes Fully Operational in Russia* (PHOTOS, VIDEO). Retrieved from https://www.rt.com/business/489551-russia-floating-nuclear-plant/. 23 May 2020
12. PNTL UK, *40 Years of PNTL*. Retrieved June 2020, 29, from https://www.pntl.co.uk/
13. PNTL UK, *PNTL Ships* [Video]. Retrieved June 2020, 29, from https://www.pntl.co.uk/our-fleet/pntl-ships/ (n.d.)
14. PDEE – Plano Decenal de Expansão de Energia – February 2019. Ministério de Minas e Energia, Brasília, Brazil. Retrieved 2020, June 29, from https://www.epe.gov.br/sites-pt/publicacoes-dados-abertos/publicacoes/Documents/PDE%202029.pdf
15. Clean Energy Ministerial, *Advancing Clean Energy Together*. Retrieved July 2020, 4, from http://www.cleanenergyministerial.org/ (n.d.)
16. J. W. M. Lima, W. Collischonn, J. A. Marengo (eds.), *Efeitos das Mudanças Climáticas na Geração de Energia Elétrica* (AES Tietê, Hunter Books Editora, São Paulo, 2014) 360 p
17. Grupo Carlos Lyra, *Caeté é eleita usina do ano em bioeletricidade no MasterCana Nordeste* 2019/2020. Retrieved July 2020, 2 from https://www.usinacaete.com/2020/01/caete-e-eleita-usina-do-ano-em-bioeletricidade-no-mastercana-nordeste-2019-20/
18. Mining dot com, *Brazil to restart country's only uranium mine*. Retrieved July 2020, 2 from https://www.mining.com/brazil-to-restart-countrys-only-uranium-mine/
19. Laboratório Nacional de Biorrenováveis. O futuro da bioeletricidade da cana-de-açúcar. Retrieved July 2020, 2 from https://lnbr.cnpem.br/o-futuro-da-bioeletricidade-da-cana-de-acucar/
20. Renewable Energy World, Orlando, Florida, USA, *Renewables in, Coal Out: South Africa's Energy*. Retrieved July 2020, 4 from https://www.renewableenergyworld.com/2019/10/18/renewables-in-coal-out-south-africas-energy-forecast/#gref. 8–10 Dec 2020
21. 2OceansVibe News, *Using Solar Panels in Cape Town? You Have Three Days Left to Register*. Retrieved July 2020, 4 from https://www.2oceansvibe.com/2019/05/28/using-solar-panels-in-cape-town-you-have-three-days-left-to-register-videos/. 28 May 2020
22. ESI Africa, SA, *South Africa: Industry Welcomes Nuclear New Build Plants*. Retrieved July 2020, 4 from https://www.esi-africa.com/industry-sectors/generation/south-africa-industry-welcomes-nuclear-new-build-plans/. 8 May 2020

Chapter 21
Personalization of Teaching in E-Learning Systems

Boris Ribarić and Zoran Ž. Avramović

21.1 Introduction

Different types of intelligence described by the theory of multiple intelligences are the basic factors to be considered in correlation with the personalization and individualization of teaching methods in general [1–4]. Each person is an individual, characterized by specific traits and abilities that are different by their nature. Students differ in their ability to perceive, organize, conceptualize, recollect memories, and use information (holistic/analytical type, dependent/independent type, socially sensitive/insensitive type, etc.). Also, personality factors influence the development of the traits of an individual (extrovert/introvert type, social skills, inhibitions, and more) [5]. Numerous studies have been devoted to teaching/learning styles such as visual, auditory, and kinesthetic, as well as the strategies of teaching/learning such as cognitive, meta-cognitive, or social-affective [6]. In their research, Felder and Henriques concluded that students can learn in many different ways, by looking and listening, thinking, and acting, through logical reasoning or intuitively, memorizing, or visualizing. The teaching methods can also differ. Some tutors teach orally; others demonstrate or discuss. Some formally focus on rules, others focus on examples, some emphasize memorization, and so on. How much an individual student will learn in class depends in part on their innate ability and preparedness but also on the compatibility of his or her approach to learning and the tutor's approach to teaching [7]. The goal of personalizing teaching is to help students become aware of their learning strategies, analyze them, and identify which ones benefit them the most in different situations, as well as to develop new or to refine existing strategies. Developing awareness of such strategies would help students in becoming more effective learners and learn continually through their lives

---

B. Ribarić (✉) · Z. Ž. Avramović
Pan-European University APEIRON Banja Luka, B&H, Banja Luka, Bosnia and Herzegovina

S. Bauk, S. D. Ilčev (eds.), *The 1st International Conference on Maritime Education and Development*, https://doi.org/10.1007/978-3-030-64088-0_21

[8]. Usage of neural networks in data processing for personalized learning will provide the conditions for the adequate formation of classes, by respecting all the diversity of students, full understanding and adoption of the material prescribed by the curriculum, consistency with the general curriculum, and the constant insight into the progress of student achievement. This paper will use artificial neural networks as a method of processing and displaying data.

21.2 Method

Predicting student success is important for defining one's career, counseling, and guidance toward major choices. The development of predictive models is needed to become more aware of factors that could negatively affect student success (for instance, poor previous school experience, poor grades in an individual subject, etc.). Emphasizing students' basic characteristics is one of the most important items in predictive modeling, as well as analyzing the studies addressing different classification methods.

Artificial neural networks are systems of interconnected neurons, which relay messages between themselves. The connections between these neurons have numerical weights that may be subject to change depending on experience, which makes the neural networks adaptive and capable of learning [9].

The structure of artificial neural networks consists of an input layer, the output layer, and at least one hidden layer [10]. The data that is presented to the output layer are the desired values of the output variables. Before learning itself, it is necessary to define the model (input and output variables) and collect data from the school/student service and e-learning system on which the network will be applied.

Data should be divided into three samples: for training, for cross-validation, and for testing. There are no rules for the creation of this division, but it is recommended that most of the data points should be used for training of the network, while a smaller segment of data should be aimed at the testing and validation process.

A classification of data is defined to enhance the student's success. Such classification is determined by the minimum and maximum values related to student achievement during schooling. Its accuracy is increased during teaching by introducing variables such as records on student status, the average grade of students by current and past school years, number of subjects, and the average grade of students by subjects.

In the network of artificial neurons that we are presenting, a simple model of what a neuron does is used. The neuron is modeled as a logical unit [11]. Xt (Fig. 21.1) symbolizes the body of the neuron, while the input branches represent dendrites. Also, there is one output branch that represents the axon. In professional literature, such a unit is called a "perceptron."

A neural network with forwarding propagation was used in this paper. In such a network, information flows in one way; from the input units, data passes through

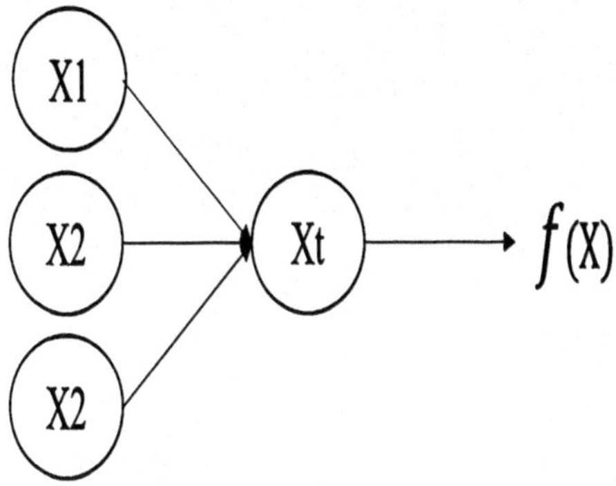

Fig. 21.1 Model of a neuron

hidden units, all the way to the output units. There are no cycles in the network, in contrast with recurrent neural networks [12].

The activation function used in the neural network is a sigmoid function:

$$h_\theta = \frac{1}{1+e^{-\theta^{T_x}}} \tag{21.1}$$

where "x" represents the input, "θ" weighting factors, and "T" a transposition function. The activation function is also referred to in the literature as a hypothesis or a model, which is a generic term referring to a solution to a problem of a certain data exploration algorithm. The general form of a sigmoid function is:

$$g_z = \frac{1}{1+e^{-z}} \tag{21.2}$$

and its graph is shown in Fig. 21.2. In addition to the sigmoid function, unit step function; part by part linearly function; Gaussian function, have a normal distribution; as well as some other functions [13–16].

To develop the neural network model, a database with variables was created containing records of student status, the average grade of students during current and past school years, number of subjects, and the average grade of students per academic subject. The database with said variables includes processed results for 101 students of elementary school "Vojvoda Radomir Putni" in Dedinje, Belgrade, Republic of Serbia.

The input attributes were:

- Student status records
- Grading
- Tests
- Average grade of students per current year
- Success in past school years
- Mathematics assignments

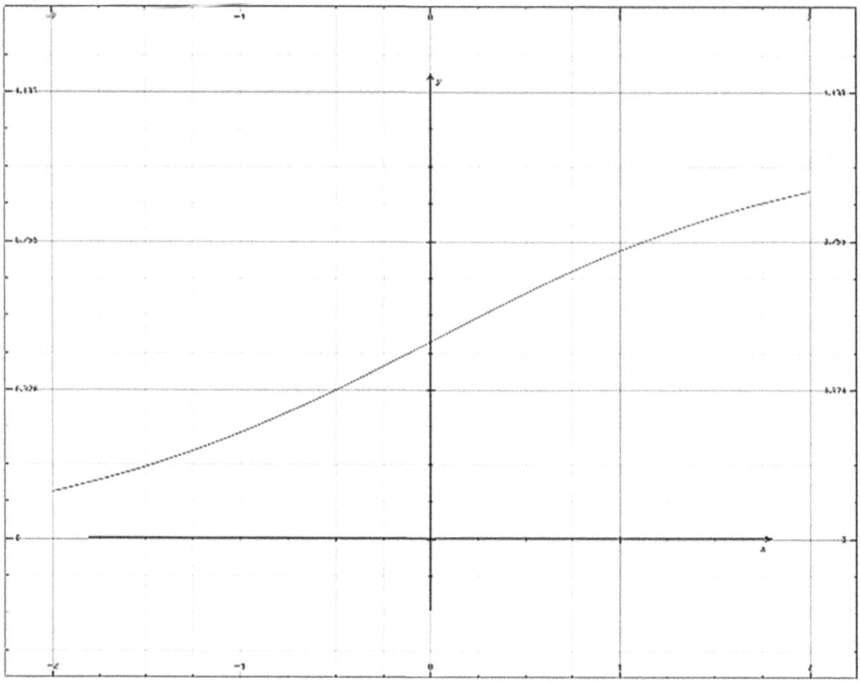

Fig. 21.2 Graph of a sigmoid function

The output or target variable was:

- The level of achievement – satisfactory

After arranging the data and transcribing it into CSV (comma-separated values) format, the data was compiled with the NeuroDesigner software application. Figure 21.3 shows the input variables and the output or target variable.

Furthermore, the data for training and testing were determined in such a way that 60.8% of the total data points were determined for training and 19.6% of the data points were determined for testing. Figure 21.4 shows the historical error data for the applied quasi-Newton optimization method by epochs.

The artificial neural network formed in this way has an input layer, three hidden layers, and an output layer as shown in Fig. 21.5.

The value of the target variable "result – level of achievement – satisfactory" is determined by the values (0 and 1). The value of the target variable "0" is a value that indicates that in dealing with students, it is necessary to take corrective measures, related to the individual characteristics and needs of the students. The corrective measures that need to be applied relate primarily to the revision of teaching methods, i.e., the adaptation of activities to the needs of the individual student, reflected in other learning styles.

Numerous studies are devoted to learning styles such as visual, auditory, and kinesthetic (abbreviation: VAK) and learning strategies such as cognitive, meta-

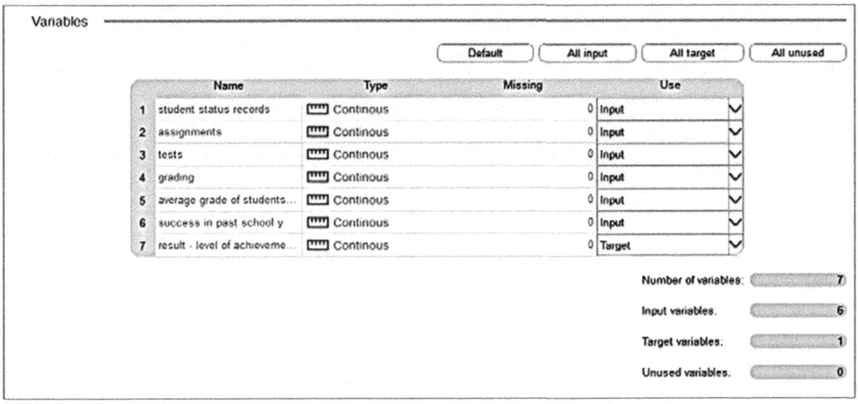

Fig. 21.3 Input variables and output or target variables

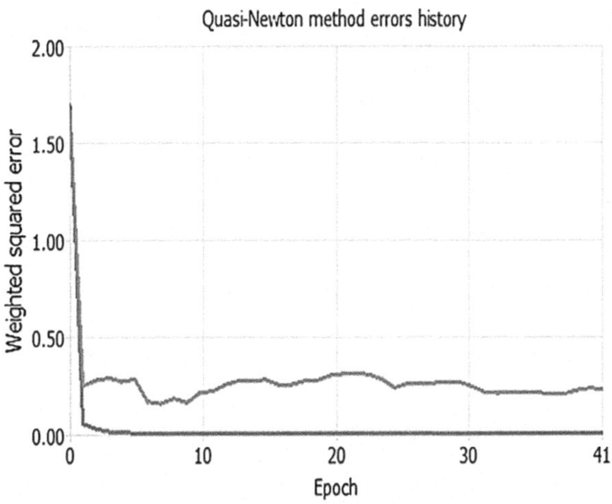

Fig. 21.4 Graphical representation of the quasi-Newton optimization method

cognitive, or social-affective. VAK concepts were originally developed by psychologists and teachers such as Fernald, Keller, Orton, Gillingham, Stillman, and Montessori, beginning in the 1920s. The VAK principles and theories today extend to all types of learning and developmental methods, well beyond its original fields of application. The VAK model provides a different perspective for understanding and the explanation of the preferred or dominant way of thinking and learning [17, 18]. To model an e-learning system that focuses on personalized learning, we have used the results obtained using artificial neural networks.

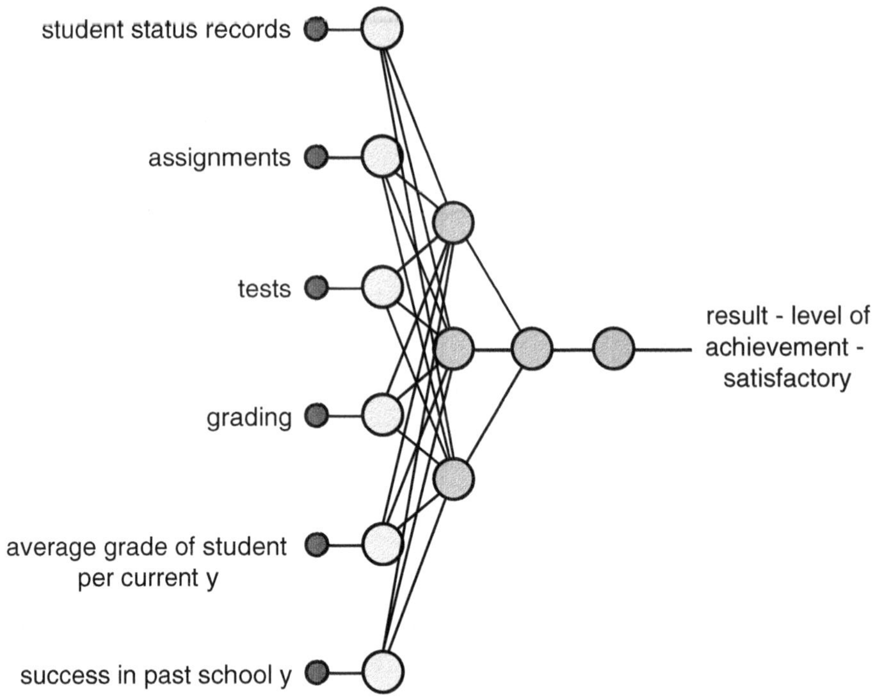

Fig. 21.5 View of the neural network

21.3 Results

The results of the classification accuracy of artificial neural network models represent the rate of correct classification on the training and testing samples [19]. The classification rate by output category applies to each grade separately. The average classification rate, in this case, is the rate of correctly classified students relative to the test sample. The average classification rate is 69.3%. As a model, we have used a multilevel perceptron with a forward propagation algorithm, containing three hidden layers, four neurons in hidden layers, a sigmoid activation function, and a forward propagation algorithm.

The obtained results show that in the test sample, 69.3% of cases were correctly classified, while 30.7% of cases were wrongly classified. This network represents the classification rate for the output variable – "student mathematical achievements."

Based on the research, we have learned that students will be 85% more likely to receive a grade of 3 or more in the subject of mathematics, depending on the number of access to the e-learning system, and with the active guidance of teachers, which primarily refers to the proper identification of an adequate learning style.

To determine the results related to the value of the output variable "result – level of achievement – satisfactory" in the "NeuroDesigner" software application, testing was conducted to find the extreme values for the obtained grades, in case no corrective measures related to learning styles were applied. Figure 21.6 shows the values of grades at the lowest level of achievement which is "1," while the cumulative value is expressed in percentages indicating the probability that a student will meet a satisfactory level of achievement and is 0.008%.

Contrary to the case above, when corrective measures are applied to learning styles and when the grades are in the highest achievement level of "5," the cumulative value is expressed as a percentage indicating that the student is likely to meet a satisfactory level of achievement which is at 0.987%. Figure 21.7 shows the values of the grades at the highest level of achievement which is "5."

The research presented in this paper indicates that the implementation of the proposed neural network model (Fig. 21.8) in the system of integrated e-learning platform can improve the rate of students' academic success and properly orientate them toward the implementation of corrective measures, related to the identification and the utilization of adequate learning methods.

21.4 Discussion

The basic contribution of this paper is reflected in several aspects. Firstly, the developed models of artificial neural networks can indicate to the teaching staff the success rate of the students. Models for predicting student achievements can indicate to students which classroom activities they need to improve. This primarily refers to certain activities in an adequate e-learning system, such as homework, tests, quizzes, etc., where the algorithms developed specifically for this task may indicate certain activities that need to be stepped up to increase the academic achievements among students. Secondly, models can also be used by students themselves, to assess their future development during schooling, based on existing learning and grading habits or patterns, assessing the time needed to organize themselves, or the amount of additional effort needed for achieving their desired success.

Fig. 21.6 Value of grades at the lowest achievement level

	Value
student status records	1
assignments	1
tests	1
grading	1
average grade of students per current y	1
success in past school y	1
result - level of achievement - satisfactory	0.00829400187

Fig. 21.7 Values of grades at the highest level of achievement

	Value
student status records	5
assignments	5
tests	5
grading	5
average grade of students per current y	5
success in past school y	5
result - level of achievement - satisfactory	0.98732141

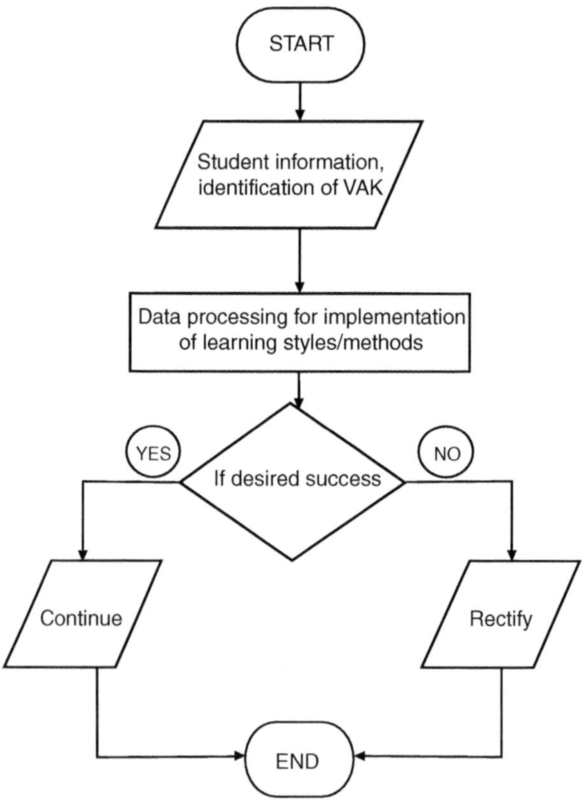

Fig. 21.8 The algorithm for applying artificial neural network results in the personalization of learning through e-learning systems

21.5 Conclusion

In terms of the achieved scientific goal of our paper, we can conclude that a good basis has been created for the improvement of the teaching/learning process, in accordance with the student traits obtained, by developing and analyzing models based on artificial intelligence and neural networks. The recommendation is based on the results of models that achieve predictive validity greater than 70% for average grades, grades for individual school subjects, as well as the identified factors and variables that are of great importance for the success of the learning process. The paper concludes that neural network models provide the highest accuracy in predicting student success. The research findings presented in this paper will help identify the requirements that prospective students are facing, as well as the ways they can achieve the best possible outcomes during school years. The results of this research may be used by other educational institutions.

References

1. H. Gardner, *Frames of Mind. The Theory of Multiple Intelligences* (Basic Books, New York, 1983)
2. H. Gardner, *Intelligence Reframed. Multiple Intelligences for the 21st Century* (Basic Books, New York, 1999)
3. M. Berman, *A Multiple Intelligences Road to an ELT Classroom* (Crown House Publishing, Carmarthen, 1998)
4. M. Berman, *ELT Through Multiple Intelligences*. NetLearn Publications, London. Retrieved Nov 2006 from http://www.netlearnpublications.com (2001)
5. R. Ellis, *Understanding Second Language Acquisition* (Oxford University Press, Oxford, 1986)
6. N.J. Anderson, Individual Differences in Strategy Use in Second Language Reading and Testing. Modern Lang. J. **75**(4), 460–472 (1991)
7. R.M. Felder, E.R. Henriques, Learning and teaching styles in foreign and second language education. Foreign Lang. Ann. **28**(1), 21–31 (1995)
8. G.M. Jakobs, T.S.C Farrell, Paradigm shift: understanding and implementing change in second language education. TESL-EJ. **5**(1), A-1, April 2001, http://www.cc.kyoto-su.ac.jp/information/tesl-ej/ej17/a1.html (2001)
9. В. Мишковић, Примена метода машинског учења у процесу развоја информационих система, Универзитет Сингидунум, Департман за информатику и рачунарство, Београд, Научни скуп са међународним учешћем, Синергија 2010, бр.стр.185–191 ISBN: 978-99955-26-20-7
10. M. Raissi, Deep learning of nonlinear partial differential equations. J. Mach. Learn. Res. Deep Hidden Phys. Models **19**(25), 1–24. http://www.jmlr.org/ (2018)
11. A.I. Galushkin, *Neural Networks Theory* (Springer, Berlin/Heidelberg, 2007)
12. K.-L. Du, M.N.S. Swamy, *Neural Networks and Statistical Learning* (Springer, London, 2014)
13. N. Ozgur, N. Taş, J.F. Peters, New complex-valued activation functions. Int. J. Optim. Control Theor. Appl. **10**(1), 66–72 (2020)
14. H. Afzaal, A.A. Farooque, F. Abbas, B. Acharya, T. Esau, Groundwater estimation from major physical hydrology components using artificial neural networks and deep learning. Water **12**(1), 5 (2020)

15. J. Socha, P. Netzel, D. Cywicka, Stem taper approximation by artificial neural network and a regression set models. Forests **11**(1), 79 (2020)
16. A.D. Jagtap, K. Kawaguchi, G.E. Karniadakis, Adaptive activation functions accelerate convergence in deep and physics-informed neural networks. J. Comput. Phys. **404**, 109136 (2020)
17. S. Wickramsinghe, S. Hettiarachchi, *Relationship among students' learning styles, assessment methods and students' performances*, in 3rd International Conference on Education and Distance Learning, at Sri Lanka, Volume: International Journal of education and distance learning. Online: ISSN 2424-6514, Print: ISSN 2424-6506 (2017)
18. S. Milinković, M. Maksimović (2013). Case study: using decision tree classifier for analyzing students' activities. J. Inf. Technol. Appl., 3(2), 87-95
19. H. Jaeger, Using conceptors to manage neural long-term memories for temporal patterns. J. Mach. Learn. Res. **18**(13), 1–43 (2017)

Chapter 22
The Importance of Information Technologies in Knowledge Management

Slavojka Lazić, Tijana Talić, and Dražen Marinković

22.1 Introduction

Education has always been an area that requires systematicity, responsibility, and exceptional dedication to work. It plays a leading role in acquiring the basic skills that provide an individual with a foundation for life and work. Education has always been exposed to the challenges. It is a continuous process that lasts, develops, and modernizes. In recent decades, the education of young people has become an increasingly important discipline. High demands continue to be made. Research and adoption of new knowledge are constantly sought. One of the most important goals of education is to raise the quality of teaching and learning. The best way to achieve this is to increase management in teaching.

Management in teaching is not one-way, only from teacher to students. It should also include guiding teachers toward the challenges of the new age. The traditional approach to teaching has long since become obsolete, and students and their parents today expect a more modern approach. Events, experiences, and information change very quickly. All this should be followed by the application of new achievements, especially the means of information technology. Teachers should recognize the importance and role of this area and apply it, in order to get closer to students and transfer knowledge from their area. How much do teachers use information technology in teaching today? If it is insufficient, is the application limited by inadequate technical support or lack of teacher's interest? There is no doubt that the modernization of teaching processes is always gladly accepted by students.

Educational institutions organize their work in accordance with the laws on education and the guidelines of the competent ministries. Their primary role is to

S. Lazić (✉) · T. Talić · D. Marinković
Pan-European University APEIRON, Banja Luka, Bosnia and Herzegovina
e-mail: tijana.z.talic@apeiron-edu.eu; drazen.m.marinkovic@apeiron-edu.eu

S. Bauk, S. D. Ilčev (eds.), *The 1st International Conference on Maritime Education and Development*, https://doi.org/10.1007/978-3-030-64088-0_22

provide students with the highest-quality knowledge, during the achievement of which, the teacher and the student will be mutually satisfied. If the process of knowledge transfer and acquisition has some shortcomings, educational institutions should identify them, define problems, and offer solutions to the competent institutions in order to eliminate the shortcomings.

22.2 Performance of the Educational Process

The educational process is very complex. It represents the continuous acquisition of new knowledge and skills and the constant monitoring of students in their learning and work. Modern and creative teaching enables the creation of conditions for the active participation of students, teachers, and parents in interaction and communication [1]. It is a process, which should be well considered and analyzed before its changes can be approached. Failed attempts at changes often cause negative consequences over an extended period of time.

Reform of the education system is inevitable. The changes that have affected the entire society in the last few decades have not bypassed this area either. The basis of any reform is to see the core of the problem that is being solved and to adopt measures that would eliminate the existing shortcomings and influence the creation of better conditions in teaching.

The education system and education are subject to continuous change. Every reform of the education system aims to modernize the content, especially of traditional subjects, and to introduce new ones. The main task of the program reform is to educate and train students for further education, work, and life.

Classical learning has long been modernized. The use of information and communication technologies (ICT) in the educational process is inevitable. By applying modern technologies and multimedia educational computer software, students are enabled to acquire new knowledge; acquire better skills, habits, and abilities; and apply them in practice. The application of technologies has brought society to such a stage that ICT skills, along with the knowledge of reading, writing, and arithmetic, are already considered an element of literacy [2]. Primary education, followed by secondary and then higher education, should equip the young person with the basic skills necessary for his work. This is followed by lifelong learning. It is also difficult to predict what kind of training needs we will need in the coming decades. Evaluation of the results achieved in the teaching process encourages the modernization and improvement of teaching content, enrichment and completion of methods for the transfer and acquisition of knowledge, and expansion and use of new teaching aids.

In order to provide the highest-quality service in education, the education system must adapt to the changes that have occurred, especially in the field of application of information technologies. The modern education system today is dynamic, adaptable, and open.

As modern technologies are applied in almost all aspects of social life, their use in education is inevitable. Their application enriches education and improves the teaching process. Education accepts the changes brought by the new age. By adopting the technical possibilities of using different but increasingly available sources of knowledge, education takes a new form in relation to the traditional one. Knowledge, the only resource that is increased by sharing [13], needs to be constantly directed and increased.

With the development of telecommunication technology, teaching aids, teaching methods, and forms of work are more intensively developed and improved, in the function of raising the efficiency and effectiveness of the teaching process. The role of teachers in the entire educational process remains in the first place. That is the starting link in direct work with students. This is especially pronounced in the lower grades of primary school. Teachers are certainly required to monitor, adopt, and transfer new knowledge in the field of information and communication technologies in their work with students.

The education of young people begins at their earliest age. It starts from kindergarten through primary and secondary schools and continues through higher education institutions. It is transferred to the workplace, where the need for lifelong training is most likely to arise today. Therefore, when the educational process, improvement, and reforms of that process are improved, it would be included the entire young population. No step in education should be skipped in order to get the best results.

Education is regulated by the laws on education. Educational work in schools is realized on the basis of plans and programs. It is very important to make good plans and programs in primary schools.

The role of the teacher is to help students find the easiest way to work and learn. It should not be forgotten that the teaching staff is always key and leading in the implementation, especially in the lower grades of primary school. Pupils are small and at this age they are still ready to accept a teacher-leader. Their knowledge of new technologies is not negligible. The teacher is there to guide them through the teaching process, pass on the necessary knowledge, and give instructions and explanations, developing their freedom of expression. During schooling, over time, this role changes more and more. The role of teachers is now to create more and more freedom of expression in students; initiate appropriate forms of work and learning; encourage social and personal motives in students; help them in terms of providing the necessary knowledge, explanations, and framework instructions. In this way, young people are created from young students, ready to absorb the offered information, through systematic work and improvement. Over time, they acquire freedom of expression and opinion, individuality, a creative approach to teaching, and to modern means of information technology. Educational technology includes knowledge of students and, accordingly, planning the content of education and the choice of teaching methods [3].

The traditional form of teaching dominated until almost the end of the last century. It mainly represented the frontal form of teaching. The teacher is the central person, performs the role of the lecturer, and decides on the content of the lecture. The student is part of a group and often does not have the opportunity to progress at their own pace. Motivations are often sought elsewhere, in other areas.

Because of that, didactic triangle, the teacher, the teaching content, and the student, had to be transformed into a quadrangle, with technology (ICT) as the fourth corner.

The application of information technologies has today become an inevitability, which is more or less involved in the educational processes of schools, countries, and geographical areas. The speed of application depends on the funds invested in this area. Teacher education mainly follows the development and application of new technologies in teaching.

22.3 Modern Information Technologies in Teaching

Communication technologies have transformed modern society into an "information society" [12]. Improving and modernizing the teaching process implies the introduction of modern information technology into the educational process. Harmonization with the modern needs of the technological society implies modernization of teaching and rationalized management of the teaching process. This achieves a better effect. At the end of the past and the beginning of this century, teaching aids, teaching methods, and forms of work with students are intensively developed and improved. Today, more and more opportunities are being created for children to use modern technologies in teaching from the earliest period of their education, and significant steps are being taken to prepare young people to actively and continuously continue their self-education later. With the application of ICT, it is easier to identify individual differences of students and continuously monitor their development [4].

The application of information technologies in the educational process should enable new ways of approaching the teaching process, from classical to modern, where an analysis of the educational process, students, means used, teacher functions, and evaluation of achieved results can be performed. In this way, it is possible to use programmed teaching, educational television, computers, operating systems, application software, multimedia presentations, etc. (Fig. 22.1)

It should be pointed out that in order to modernize the technological process of services in education, technical support is also necessary, modernization of teaching aids. Most of our classrooms are still not equipped for conducting and organizing modern classes, with the application of information and communication technologies. In the Republika Srpska, BiH, there are certainly a large number of schools that have such a form of teaching, which in some way indicates their advantage over others. However, regardless of the equipment of individual classrooms with modern ICT, students do not always use this space. Due to the large number of students and classes in schools, and the need to harmonize class schedules, ICT-equipped classrooms often remain inaccessible for certain subjects. It should not be forgotten that the reform of the education system aims to modernize the entire teaching process,

Fig. 22.1 Teaching aids. (Source: web)

and not just some of its parts, i.e., subjects in schools. An important application of ICT in education has been achieved in the EU. The diversity of education systems present in the Member States also requires different approaches to this area. The most developed countries in the field of ICT are Sweden and Finland, where about 90% of the population has access to the Internet [14]. The education systems of these countries have the longest application of ICT. BiH is certainly not at the top of the scale, but in the last 10 years, the application of these funds in the teaching process has intensified.

22.4 The Role of Teachers in Modern Education

If the teacher does not progress, the students go back [15].

The modernization of teaching and the application of IT significantly change the relationship between teachers and students. While the leading role of teachers is very pronounced in the traditional way of education, it is now increasingly losing such a form, and the student is given more and more freedom to express their ideas, concepts, and solutions. Certainly, the role of the teacher should not be neglected or marginalized at any time, because he is the one who is ultimately responsible for the teaching process. He should work skillfully with students so that they achieve full freedom, independence and initiative. The role of leader is retained only in the lower grades of primary school, while other teachers are gradually losing that role, especially in student education.

Teaching staff, certainly professional in their field but insufficiently competent in the field of modern information technologies, should be trained and prepared for a modern way of teaching. A multimedia textbook can help the teacher in the function of quality teaching, in a way that he is just adopting [5]. Educators go through various forms of training (ECDL, etc.), after which they need to pass an appropriate test. This includes knowledge and use of computers, and file management: MS Word, word processing; MS Excel, spreadsheets; MS PowerPoint, presentations;

and MS Outlook, IE - Internet and email. The role of each teacher is to continue with their own training, according to their own needs and to apply the acquired knowledge in the field of information technology for their subject, the scientific field to which the subject belongs. In this way, a better-quality teaching staff is created, which is ready to accept the changes in the modern environment.

Integrating ICT skills into the teaching process should be a challenge for both teachers and students. Student activity has a primary role in solving the set problems. The teacher seems to move to the side, creating space for students to be able to engage in solving specific life problems. The IT conception of teaching changes the role of the frontal form of teaching into interactive, individual, or group [6]. The ways in which students learn new content affect the quality and durability of their knowledge [7]. Certainly the teacher carefully uses selected educational activities that are interdisciplinary, student-focused, and applied to real issues and practice.

22.5 Forms of Teaching Modernization

Modernizing teaching requires a serious approach. School education is the basis for lifelong learning [8]. It is the application of information technologies that enables a comprehensive analysis of other people's ways of education and the observation of the best. In that way, one can see exactly what the education system in the country is like and what it is that needs to be eliminated. Based on that, strategic directions of development in the field of education are determined, with an implementation plan. The influence of modern information and communication technologies in the realization of the educational process should not be neglected [9].

One way to compare the education system with other countries is to monitor the results of the PISA survey. PISA is the world's largest survey in education, which has been organized every 3 years since 2000 under the auspices of the Organization for Economic Cooperation (OECD), among the target age group – 15-year-olds.

It is a "program for international assessment of students, so far the most comprehensive international assessment of knowledge and skills of fifteen-year-old students in member countries of the Organization for Economic Cooperation and Development (OECD) and partner countries" [16]. The aim of the PISA research is to determine how much 15-year-old students are prepared to enter the world of adults and to continue their education or involvement in the work process. The age of 15 was chosen because in most participating countries, students at that age are nearing the end of compulsory education, so the assessment can provide insight into their knowledge and skills accumulated over a period of approximately 10 years of schooling.

Since the OECD launched the PISA program, with the aim of developing a reliable and relevant indicator of student achievement, to date, 80 countries have joined the PISA survey.

The PISA survey conducted aims to assist participating countries in making strategic decisions in the field of education based on empirical, statistical data on student achievement. In other words, PISA points to weaknesses in education systems and indicates the direction in which these weaknesses need to be addressed and the education of a country improved. PISA results do not affect the assessment of students and teachers, but indicate the situation in education systems and education policy. The results of students and schools are not evaluated individually, but are presented as the overall achievements of students at the level of participating countries.

Certainly, the PISA research could not be conducted without the support of modern information technology resources for this research. This means that today it is impossible to determine and perceive one's own place and quality of education without the application of information technologies.

By modernizing teaching and applying IT, faster and better results in learning are achieved, and it can be performed in different ways: programmed teaching, the use of computers and the use of the Internet, the use of so-called electronic diaries, e-learning, distance learning, distance education with the help of Skype programs, the use of blogs in teaching, the use of virtual classrooms and multimedia simulators in teaching, etc.

The use of computers and the use of the Internet can be applied at the earliest age of education [10]. Through educational games, children can be given the opportunity to reason, guide, retain attention, make logical inferences, and make decisions. In the later course of students' education, other forms of using modernized teaching can be introduced. Modernization and innovation in education put every teacher in a situation to act as a creative, in order to achieve the best possible results.

Interactive distance learning is based on a systems approach using multimedia electronic sources of information. In lower grades, the presentation of multimedia textbooks from individual subjects that are included in the curriculum can be used. It is the application of multimedia teaching content that changes the position of teachers. The focus of his work shifts to the detailed preparation of teaching, motivating students, and control and evaluation of the teaching process. The teacher creates more space for educational work with students [11] (Fig. 22.2).

Fig. 22.2 Modern teaching/learning. (Source: web)

22.6 Conclusion

The harmonization of the educational process with the modern needs of the technological society implies the modernization of teaching and rationalization during the teaching process. It should not happen that the introduction of new teaching technology in schools is delayed in relation to other segments of society. It is necessary to find a module on how to equip, if not all, then as many schools as possible, with the appropriate means necessary for conducting modern, often multimedia teaching. The concepts of the so-called e-learning are increasingly being adopted today. The modern school, therefore, must turn more to ICT technology. Teacher preparation is very important for conducting this form of teaching. The teacher certainly retains a very important role, with a conscious emphasis on the student. By using multimedia computers and computer-generated teaching contents, students follow the lessons more carefully, often remembering the teaching contents better, especially those that are more difficult to learn just by listening or reading. This form of teaching provides students with the opportunity to think, analyze and conclude, more actively participate in the process of learning the teaching content, as well as finally dedicate themselves to learning and research.

References

1. M. Ilić, *Teaching Different Levels of Complexity* (Faculty of Teacher Education, Belgrade, 1998)
2. A.W. Bates, *Managing Technological Change: A Strategy for University Leaders* (CARNet, Zagreb, 2004)
3. P. Mandić, D. Mandić, *Educational Information Technology* (Faculty of Teacher Education, Belgrade, 1996)
4. D. Mandić, *Didactic-Information Innovations in Teaching* (Faculty of Teacher Education, Belgrade, 2003)
5. I. Solaković, Multimedia textbook in the function of quality teaching. Bijeljina (2007)
6. M. Vilotijević, Informatics concept of teaching. Belgrade (2002)
7. D. Mandić, Information Technology in Education. Srpsko Sarajevo (2001)
8. J. Đorđevic, *School, Lifelong Learning and the Importance of Modern Information* (Faculty of Philosophy, Belgrade, n.d.)
9. M. Knežević, *Influence of Modern Information and Communication Technologies in the Implementation of the Educational Process* (Faculty of Pedagogy, Sombor, n.d.)
10. G. Milosavljević, Internet and Education. FON (material from specialist studies). Belgrade (2009)
11. Authors, *Proceedings* (ITeO, Banja Luka, 2009)
12. J. Vasiljević, Ž. Stanković, *Ambient Intelligence and e-Learning*, vol 10 [1] (Panevropski Univerzitet Apeiron, Banja Luka, JITA, 2020)
13. https://iserbia.rs/novosti/miodrag-zec-znanje-je-neograniceni-resurs
14. http://www.ns.ac.yu/stara/doktorske/ivanTasic/disertacija.pdf
15. http://europa.ed.int/comm/education/plicies/pol/policy_en.html#programme
16. http://www.mpn.gov.rs/priprema-za-pisa-test/

Part IV
STCW Concerns

Chapter 23
Qualifications of MET Instructors Case Study: Republic of Croatia

Marija Pijaca and Ana Gundić

23.1 Introduction

In the selection of maritime education and training instructors (hereinafter: MET instructors) for conducting the lectures and education of seafarers in the subjects at the management level, maritime higher education *institutions (hereinafter: MHE institutions) in the Republic of Croatia usually* require that MET instructors have appropriate qualifications. Qualifications which are most frequently required are the possession of an appropriate certificate of proficiency or capacity, such as officer, master or chief engineer. Calls for vacancies for a MET instructor position at an MHE institution exhaustively list additional requirements which the course candidates must meet. Very often this is the reason for a low number of candidates for the course as only those who sail, i.e. seafarers, meet these additional requirements and it is difficult for them to "get away" from their seafaring practice and dedicate themselves exclusively to teaching and lecturing positions for the education of seafarers at an MHE institution.

Key requirements for carrying out the education of seafarers in the Republic of Croatia are regulated by the Ordinance on Ranks and Certificates of Qualifications for Seafarers of the Republic of Croatia (hereinafter: Ordinance) [1]. The Ordinance was adopted based on the provisions of the Maritime Code [2] and the Act on Security of Sea-Going Vessels and Ports [3] which are important legal sources of maritime law in the Republic of Croatia. Also, the Ordinance is harmonised with the provisions of the European and international law, including the provisions of the International Convention on Standards of Training, Certification and Watchkeeping for Seafarers (hereinafter: STCW Convention) [4]. Furthermore, if we take a look at

M. Pijaca (✉) · A. Gundić
Maritime Department, University of Zadar, Zadar, Republic of Croatia
e-mail: mpijaca@unizd.hr; agundic@unizd.hr

S. Bauk, S. D. Ilčev (eds.), *The 1st International Conference on Maritime Education and Development*, https://doi.org/10.1007/978-3-030-64088-0_23

the contents of the Ordinance in the subject of requirements which must be met by MHE institutions and MET instructors, provisions of Part Five of the Ordinance, from Article 79 to 89, under the title "Education and Training of Seafarers", are the most important. As stated in the provisions of Article 79 of the Ordinance, requirements for obtaining certificates of competence of the seafarers are carried out by MHE institutions authorised in accordance with the special regulations and with the permission of the ministry of the Republic of Croatia responsible for maritime affairs.

Also, according to Article 79 of the Ordinance, MET instructors in the subjects of a narrower seafaring profession in the education and training processes, in addition to requirements prescribed by special regulations, must also possess a certificate of competency for a dedicated capacity which is, at the minimum, relevant to the educational programme. In addition to the referenced provisions of the Ordinance, the provisions of "Annex G" of the Ordinance, titled "Requirements for Carrying Out the Education Programme", are also important. It is stated in "Annex G" of the Ordinance that an MHE institution "must have at its disposal a sufficient number of MET instructors for the envisaged education plan and programme and an envisaged number of attendees in accordance with the pedagogical measures". An especially important part of the provision of "Annex G" is the one according to which the following is the duty of an MHE institution:

"to ensure that all MET instructors possess necessary experience in their vocation and subject, that they are familiar with the entire education programme in which they take part, and that in their teaching activities they conscientiously and consistently adapt to the technological and other changes in shipping". [1]

However, in the continuation of the provision, the Ordinance does not explicitly prescribe whether the MET instructors who teach the subject of narrower seafaring vocation should possess certain certificate or profession, for example, officer, commander or chief engineer. But MHE institutions interpret the provisions of the Ordinance by assuming that a MET instructor possesses certain certificate or vocation and exhaustively prescribe the described additional requirements in the content of the call for vacancies for the position in the subject at the management level.

Further to the above, the following questions arise: are the obtained competencies of MET instructors at the time of employment at MHE institutions sufficient within the meaning of the obligation of an MHE institution to ensure that MET instructors "conscientiously and consistently adapt to technological and other changes in shipping"; can a MET instructor, while employed at an MHE institution, while being the beneficiary of mandatory insurance on the basis of such employment, join the vessel and sail and thus keep pace with the "technological and other changes in shipping"; and why should MET instructors, who have a smaller workload at an MHE institution during the semester, be prohibited from joining a vessel, sailing and having a seafarer status while they are employed at an MHE institution?

Considering the obligations of an MHE institution from the Ordinance, we believe that while they are employed at an MHE institution, MET instructors should be given an opportunity to board a vessel and sail, i.e. to have refresher sailing periods, and to work as a seafarer. Regardless of their employment at an MHE institution, such instructors should be given an opportunity, through lawful legal sources, to board a vessel, and it should be prescribed that, during their time aboard a vessel, they have a seafarer status and all the rights and obligations from their status as a seafarer. Therefore, here we are not referring to practical navigation training which MET instructors conduct with students, for example, on a training vessel, but under the terms of "boarding a vessel and navigating", we refer to the boarding and navigation of a MET instructor in the capacity of a seafarer and crew member in order to perform duties aboard a ship (transport of cargo, passengers, etc.). We believe that boarding and navigation during the employment at an MHE institution surely means "obtaining experience in the profession and subject" taught by MET instructors and adaptation to "technological and other changes in shipping", as stated in the Ordinance.

However, in realising this idea, MET instructors and MHE institutions are limited by the legal provisions in the field of labour law and mandatory insurance (health and pension insurance).

The aim of this paper is to present the legislation in the Republic of Croatia which prevents MET instructors from sailing while performing their duties at MHE institutions and which we believe does not support the provisions of the Ordinance in the part referring to the required qualifications for MET instructors. For that purpose, in this paper we will primarily point out the challenges for MHE institutions with regard to MET instructors, considering the legal framework for seafarers' education in the Republic of Croatia. Within the scope of the same title, we shall also examine the seafarers' education programmes. Furthermore, in the paper we will set out all potential solutions for boarding a vessel and navigation by MET instructors during their employment at MHE institutions in accordance to legal regulations in force. Then we shall analyse all legal sources from the field of labour and social legislation in the Republic of Croatia which, as they stand now, prevent MET instructors from simultaneously having a seafarer status and being employed at the MHE institution. We shall primarily analyse regulations pertaining to labour law, mandatory insurance schemes (health and pension insurance), basic acts of MHE institutions and other sources. Finally, based on the carried-out analysis, we shall propose the incorporation of a legal provision in the legislation of the Republic of Croatia which would enable MET instructors to board vessels and sail while they are performing their duties at MHE institutions. Incorporation of such a provision in the legislation would contribute to the qualifications of MET instructors and quality of education of seafarers at MHE institutions.

23.2 Challenges for MHE Institutions with Regard to MET Instructors Regarding the Legal Framework for the Education of Seafarers in the Republic of Croatia

The Republic of Croatia is a party to the International Convention on Standards of Training, Certification and Watchkeeping for Seafarers (hereinafter: STCW Convention) [4] by virtue of which, among other things, provisions of the Ordinance have been harmonised. The STCW Convention is the first international instrument legislating minimum basic requirements on the training, certification and watchkeeping for seafarers on an international level, and the education of seafarers in the Republic of Croatia is based on the competencies prescribed by the STCW Convention. The STCW Convention was adopted in 1978; it came into effect in 1984 and has been significantly amended in 1995 and 2010. Considering that the content of the STCW Conventions from 1978 was primarily focused on the obtainment of knowledge by seafarers, in 1995 the competence-based approach was introduced. In the same year, Seafarers' Training, Certification and Watchkeeping (STCW) Code was adopted (hereinafter: STCW Code), and it is an integral part of the STCW Convention. The STCW Code consists of two parts: Part A, which is mandatory and comprises minimal requirements which must be met in order to ensure the meeting of provisions from the STCW Convention, and Part B, which is a recommendation and contains guidelines which serve as help when applying the provisions from the STCW Convention.

Minimal standards pertaining to seafarers' education are divided by duties aboard a ship which are included in the performance of ship activities: deck department, engine department and radiocommunication. Ship activities include duties, responsibilities and work which ensure safe operation of the ship, safety of human lives and protection of the marine environment. Required competencies are prescribed for each activity, and knowledge, understanding and skills are set out for each competency and must be obtained following the conclusion of the educational process. Also, ship activities can be performed at a management, operational and support level. The management level pertains to the function of a master, chief officer, chief engineer and second engineer. The operational level of responsibility pertains to the function of an officer responsible for navigational watch or engine room watch, electrical office and radio operator. The support level of responsibility pertains to the function of a ship crew member who is a member of navigational watch or engine room watch, senior crew member who is a member of navigational watch, i.e. senior crew member who is a member of navigational watch and ship electrician.

Programmes which comprise the competencies set out by the STCW are carried out at maritime education and training institutions (hereinafter: MET institutions) and can be divided into:

1. Programmes oriented only to obtaining competencies prescribed by the STCW Convention – as a rule, these types of programmes are carried out at MET institutions.
2. Programmes which, in addition to competencies set out by the STCW Convention, also include wider knowledge of shipping industries – these types of programmes are usually carried out at universities, i.e. maritime higher education institutions (MHE institutions).

Implementation programmes carried out at an MHE institution are oriented towards obtaining certificates of competency at management level but are more demanding than the programmes carried out at MET institutions, i.e. they are more demanding than the programmes that are based solely on the STCW Convention. The aforementioned stems from the fact that the programmes which are carried out at an MHE institution are oriented towards acquiring competences prescribed in the STCW Convention but also include competencies from, for example, maritime law, maritime economy, logistics, etc. In this way it is ensured that the students will acquire competencies they need in order to perform duties on land in addition to performing duties on board a ship. The average duration of seafarers' service aboard a vessel is 10 years, which means that the programmes based only on the STCW Convention do not meet their long-term needs.

In the Republic of Croatia, as in most countries of the European Union (hereinafter: EU), the education of seafarers for the management level that awards certificates of competency is based on the provision from the Bologna Declaration from 1999 [5] and usually lasts for 3 years. Study programmes for obtaining a certificate of competency are, as a rule, divided into continuing education systems and the so-called sandwich systems. The difference between the continuing education systems and sandwich education systems is that, in the "sandwich systems", on-the-job training aboard a ship is combined with classes at an MHE institution. In the continuing education system, after completing the study programme, the students themselves search for companies in which they will perform their navigation training for 1 year (as deck or engineering trainees). As a rule, this system lasts from 3 to 8 years, i.e. from the undergraduate to postgraduate level [6]. Standards prescribed by the STCW Convention are most often covered on the undergraduate level. Very frequently, training centres operate as part of an MHE institution, which offers training programmes (short courses) for obtaining certificates of proficiency, which are oriented on acquiring practical skills for the performance of certain tasks on board. Such programmes are offered to students, seafarers or companies.

All programmes carried out at an MHE institution must meet the requirements set out by the ministry competent for education and science and the ministry competent for maritime affairs. This means that MET instructors who carry out the study programme at an MHE institution must, according to the requirements of the ministry competent for education and science, have scientific and teaching professions and, according to the requirements of the ministry competent for maritime affairs, they must also possess seafaring experience. In addition to the formal requirements which the lecturers must meet, it is essential that they are up to date with the new technologies on ships to ensure that the students obtain the competencies they need for the performance of tasks on sophisticated ships.

The implementation of new technologies represents a great challenge in almost all industries and organisations [7]. Increase in the number of systems, changes of parameters in the system and their connections to other parts of the system and other elements in the process make it more difficult to understand how the system operates [8]. Due to this, in most cases, the introduction of new technologies includes the development of new competencies of crew members and/or improvement of existing competencies. The adjustment of existing competencies of the crew to new technologies on ships represents a key element in the processes of safe running of a ship. Successful implementation of new technologies on a ship, in addition to crew competencies, also depends on the competencies of qualified MET instructors. Each country which is a party to the STCW Convention must ensure that MET instructors are appropriately qualified. However, the STCW Convention does not clearly define what "appropriately qualified" means, and it is left to each country to define the meaning of this term, which leads to very different interpretations of the term in various countries which are parties to the convention [9].

MET instructors [4]:

- Should have an understanding of the specific training objectives for the particular type of training
- Be qualified in the task for which training is being conducted
- Should have a full understanding of the training programme and the specific objectives for each type of training being conducted
- Should have appropriate knowledge of instructional techniques
- Should have appropriate knowledge of training methods and practice
- Should have a full understanding of the assessment system, assessment methods and practice

For the purpose of conveying knowledge to students, the importance of practical on-board experience of MET instructors is especially stressed [10], and in the comparative literature, it is often stated that they should have on-board experience at the management level, i.e. "Captain or Ch. engineer's CoC" [9]. In addition to the above, MET instructors should be up to date with new technologies on ships and modern-day ship operations [11]. Also, according to the researches carried out by shipping companies, MET instructors also show a lack of practical experience in ship operations [9], and it is therefore believed that refreshment sailing periods are the best way for the instructors to improve their knowledge and ensure that students obtain up to date competencies [12, 13].

23.3 Possible Solutions for Boarding a Ship and Navigation by MET Instructors During Their Employment at a MET Institution

Considering the legislation in force in the Republic of Croatia, MET instructors have two possible solutions for boarding a ship and sailing during their employment at an MHE institution.

The first option is for a MET instructor to request, in writing or verbally, for a sabbatical or free part of the academic semester from the MHE institution to board a ship and navigate. This option is based on the provision on the "idling" of employment as stipulated by the Labour Act [14] and internal acts of the MHE institution (e.g. the statute) and provisions of the title "Sabbatical", if the internal act of the MET institution does indeed contain such a provision. Some internal acts of MHE institutions allow sabbatical to instructors with a scientific-teaching title after they spend 6 years at the MHE institution and provided that the MHE institution ensured the carrying out of the education in subjects for which the curriculum is carried out by the subject MET instructor [15]. A downside of this provision is that the MET instructor has a right to another sabbatical only after the expiry of 6 years from the exercise of this right.

The second option is that a MET instructor terminates their labour relation at an MHE institution. This also means that the MET instructor must be deregistered from mandatory insurance schemes (health and pension insurance) of which they are a beneficiary under their employment at the MHE institution. However, the MET instructor may only enter into employment at a certain shipowner after the termination of labour relation and deregistration from the mandatory insurance schemes. After the MET instructors come ashore, they can once again be employed at a MET institution and registered for mandatory insurance schemes. Labour and social rights, as well as other obligations, such as an obligation to pay income tax prescribed in the Income Tax Act [16], start or stop with the establishment or termination of employment at a MET institution. This solution is based on the provisions of the Labour Act [14], Health Insurance Act [17] and Pension Insurance Act [18].

According to the legal regulations in force in the Republic of Croatia, these are the only solutions available to a MET instructor for boarding a vessel and navigation during their employment at an MHE institution.

In addition to the above, we would also like to point out that boarding a ship and navigation by a MET instructor using the first or second solution is most often due to their personal reasons, usually for the opportunity to earn additional income, and is not linked to acquiring competencies or any provisions and obligations of a MET institution from the Ordinance. Considering the inadequate and insufficient solutions from the legal regulations on the options for boarding a vessel and navigation by MET instructors in the capacity of a seafarer, we believe that legal regulations from the field of labour and social legislation should be harmonised and linked to those which prescribe education and competency requirements for MET instructors, etc. and which are regulated by the Ordinance.

23.4 Starting Points for the Regulation of Labour Law and Social Security Relations in the Legal Regulations of the Republic of Croatia

Labour and social legislations in the Republic of Croatia are generally regulated by the Labour Act, the Mandatory Health Insurance Act and the Mandatory Pension Insurance Act, as well as branch collective bargaining agreements concluded individually for certain activities. Specifically, according to Article 1 of the Labour Act, this Act regulates employment in the Republic of Croatia, unless specified otherwise by another act or an international agreement which has been concluded in accordance with the Constitution of the Republic of Croatia [19] and confirmed with it [14]. Also, according to Article 1 of the Mandatory Health Insurance Act, this Act regulates mandatory health insurance in the Republic of Croatia, the scope of the right to healthcare and other rights and obligations of mandatory insurance beneficiaries, the terms and manner in which they are exercised and financed as well as the rights and obligations of a mandatory health insurance institution [17]. Also, according to Article 1 of the Pension Insurance Act, the said Act regulates mandatory pension insurance based on generational solidarity and payment of pensions from the total amount of the capitalised payment of contributions from the mandatory and voluntary pension insurance of personal capital savings [18].

All the above-mentioned acts represent fundamental legal framework for the creation, realisation and termination of labour and social rights of employees in the Republic of Croatia, but there are distinctive characteristics and issues related to labour and social rights of employees in certain activities. Therefore, we shall determine whether there are any distinctive characteristics with regard to labour and social relations of MET instructors and seafarers and, if any, what kind they are.

Regulating Labour Law and Social Security Relations of MET Instructors

With regard to labour law and social security relations of MET instructors, in addition to the Labour Act, the Mandatory Health Insurance Act and the Pension Insurance Act, Collective Bargaining Agreement for Science and Higher Education [20] is also a part of the legislative framework for these relations. The Collective Bargaining Agreement for Science and Higher Education is a branch collective bargaining agreement for activities in the system of science and higher education and is in effect for all public institutions which form a part of the system of science and higher education in the Republic of Croatia, and it also applies to employees of an MHE institution – MET instructors.

MET instructors are primarily employees of an MHE institution, and, according to the legal sources on labour, mandatory insurance schemes (health and pension insurance) and the Collective Bargaining Agreement for Science and Higher

Education, they cannot be contemporaneously employed by two employers or be beneficiaries of mandatory insurance schemes on more than one legal basis. This rule is not a "prohibition" which applies only to MET instructors but is a provision that applies generally to all employed persons. It is based on the fact that every employer is an independent entity, sole trader, company or craftsman of other natural or legal person that establishes employment with a person who can be a subject of such employment in the capacity of an employee [21].

Employment of MET instructors at an MHE institution is established by virtue of an employment agreement. This is a contractual relationship, most commonly fixed-term, with the requirement of obtaining a PhD or a scientific-teaching title of a MET instructor, working either full-time or part-time. Labour and social rights of MET instructors, such as the right to participation in mandatory insurance schemes, are based on concluded employment agreements, and liabilities, such as income tax, are based on the provisions of the Income Tax Act [16]. According to the provision of the Mandatory Health Insurance Act, "persons employed at a legal or natural person with registered seat in the Republic of Croatia have the status of an insured person" [17]. Therefore, according to the provision of the Mandatory Health Insurance Act, MET instructors are beneficiaries of the health insurance scheme. According to the provision of the Pension Insurance Act, "insured persons are all persons on the basis of employment" [18]. Therefore, MET instructors have the status of an insured person on the basis of their employment at a MHE institution. Also, a MET instructor is also a taxpayer and pays taxes on salary received as remuneration for their work at an MHE institution pursuant to the Income Tax Act [16].

Based on the above, we can summarise that labour law and social security relations of MET instructors are equal to those of the majority of employed person and that the legal framework which regulates them, in addition to legal sources (Labour Act, Mandatory Health Insurance Act and Pension Insurance Act), also comprises autonomous legal sources (employment agreement and Collective Bargaining Agreement for Science and Higher Education). Moreover, internal fundamental acts of an MHE institutions (e.g. the statute) and all regulations on which these acts are based are also relevant to their employment (regulations of the ministry competent for education and the ministry competent for maritime affairs).

Regulating Labour Law and Social Security Relations of Seafarers

Numerous special provisions are applicable to seafarers and regulate their labour law and social security relations. This is based on distinctive characteristics of seafaring that cannot be set within the standard employment framework equivalent to, for example, employees who work on land. We had previously stated that the Labour Act generally defines employment relations in the Republic of Croatia, unless set out differently by a law or international agreement concluded and confirmed in accordance with the Constitution of the Republic of Croatia [19] that is published and in force [14]. However, the Labour Act does not contain a single special provision on labour law and social security relations of seafarers, so the same has been replaced by a lex specialis legal source that regulates different areas of maritime law (e.g. maritime transport, navigational safety, etc.). Therefore, the Maritime Code is the most direct legal source of labour law and social security status of seafarers. However, the Maritime Code is not the only source which defines the status of seafarers. In addition to the Maritime Code, seafarers' status is regulated by a series of by-laws. Their status is primarily regulated by the Ordinance [1], the regulation on maritime booklets and authorisation for boarding, and procedures and means of registration and termination for mandatory pension and health insurance [22]; the regulation on the content and manner of registering employment contracts of seafarers and fisherman [23]; the regulation on performing duties and keeping watch by seafarers on sea-going vessel of merchant navy of the Republic of Croatia [24]; the regulation on the minimum safe manning required for safe navigation of sea-going vessels, floating objects and stationary offshore objects [25]; the regulation on the application of the Maritime Labour Convention from 2006 [26]; etc. In addition to the above, there is room left for autonomous regulation by employment agreements and collective bargaining agreements for seafarers [27, 28].

By virtue of an employment agreement, special records and other seafarers' rights, obligations and responsibilities are agreed. For that, provisions of the Maritime Code, by-laws, collective bargaining agreements and other legal sources are used. The main characteristic of collective bargaining agreements for seafarers is the normative character of such agreements and their contents, which constitute the main rules on labour law and social security relations in the field of maritime affairs. Collective bargaining agreements can be implemented only through individual employment agreements, in accordance with international and national legal sources [27, 29].

It is especially important to highlight international legal sources, due to the fact that seafarers participate in both international and domestic competition and international conventions, along with national legal regulations, also play an important role in regulating labour law and social security relations of seafarers. As a member of the International Labour Organization and other international organisations, the Republic of Croatia incorporated many universal and other international sources pertaining to labour law and social security relations of seafarers. The most important international legal source in this subject is the Maritime Labour Convention of 2006, which was amended in 2014, 2016 and 2018 [30] and which succeeded in establishing uniform rules on a global level with regard to the minimum requirements for seafarers' work, employment, accommodation aboard a vessel, health protection, medical care and social security in a single text [31].

From the extensive analysis of the contents of different national, international, legal and autonomous legal sources on the seafarers' status, we note numerous special and specific requirements for employment as a seafarer. Primarily, requirements such as professional qualifications, medical fitness, nationality, special knowledge and capacities, possession of necessary certificates and permissions, etc. are stipulated [1, 2, 30]. The specific quality is in that a certain legal relationship is established between a seafarer and a shipowner even before the signing of an employment contract, such as undergoing an additional medical examination at the request and expense of the shipowner, preparation and preliminary check [29]. Also, from the analysis of contents of these legal sources, we note special provisions on social security of seafarers. According to the provisions of the Maritime Code and the Mandatory Health Insurance Act, seafarers employed by a natural or legal person registered in the Republic of Croatia are insured and have the insured person status [2, 17]. Also, all seafarers who boarded vessels in international voyage, regardless of the vessel's nationality, are equated in their rights with regard to pension and health insurance. Exceptionally, seafarers who are insured under the pension or health insurance scheme of a country with which the Republic of Croatia has a social security agreement do not file an application for mandatory pension and health insurance in the Republic of Croatia. Instead, they deliver a certificate issued by a foreign social security institution to the competent port authority. As prescribed by the Maritime Code, a seafarer in international voyage is a payer of contributions for mandatory insurance schemes on a monthly basis for the calculation of contributions prescribed and promulgated by the Minister of the Sea, Transport and Infrastructure. The amount of the monthly base for contribution calculation is determined according to the salary that a seafarer would earn for the same, i.e. similar jobs on a vessel in international voyage. Also, in the context of the social security relations of seafarers, the provision of the Maritime Code which lists the cases of achieving 183 days of navigation necessary for the exemption from the payments of seafarers' income tax based on work on a vessel in international navigation is also important. The actual number of days of navigation can be increased by, for example, days spent travelling to the port of embarkation, return travelling days, sick days due to illness, days spent in education and training and days occurred due to the termination of the employment agreement due to business-related reasons [2].

Pursuant to the above, we can summarise that labour law and social security relations of seafarers are accompanied by the application of maritime legislation. The starting points are primarily in the Maritime Code, and we should look in addition for solutions for this subject matter in general labour and mandatory insurance laws. In addition, it is necessary to respect international sources, and there is room left for autonomous regulation of these relations.

23.5 Proposal on the Means of Regulating the Rights of MET Instructors for Boarding a Vessel and Sailing While They Are Employed at an MHE Institution

We have separately presented all legal sources which regulate labour and social security relations for MET instructors and for seafarers, and we took note of specific characteristics of the way in which they are regulated. MET instructors are employed in a way that, according to the Labour Act and laws on mandatory insurance, they cannot simultaneously hold a seafarer status within the meaning of the Maritime Code. On the other hand, seafarers have their specific characteristics with regard to labour law and social security status. Therefore, we can describe the presented legal sources as sources that preclude MET instructors from boarding a vessel and sailing while they are employed at an MHE institution, thus preventing them from taking "refreshment" sailing periods during their employment at an MHE institution. However, the presented legal framework should at the same time take into consideration the wider context of the activities of MET instructors and seafarers, and it should be harmonised with the provisions of the whole legal framework which regulates labour and social security legislation currently in force with regard to MET instructors and seafarers.

As this paper has repeatedly pointed to the inconsistencies between the legal framework regulating the labour law and social security matters and the provisions of the Ordinance regarding the requirements for carrying out seafarers' education, we shall hereby propose to add a provision to the legal framework which would correct the observed deficiencies. The legislator should unambiguously legislate that MET instructors are allowed to board a vessel and sail and prescribe that, as long as a MET instructor is aboard a vessel, they shall have the seafarer's status within the meaning of the Maritime Code, by-laws and other national and international regulations that regulate the seafarers' status.

Therefore, for the purpose of securing the obligations of an MHE institution from the Ordinance, de lege ferenda, we propose adding a provision to the legal framework which would allow MET instructors to board a vessel and sail while employed at an MHE institution. The provision should prescribe that, during the employment at an MHE institution, the employment status of a MET instructor is not interrupted if the MET instructor boards a vessel for refresher sailing periods within the meaning of the Ordinance. The provision should also regulate that the MET instructor's employment at an MHE institution should be temporarily "idle" until they disembark the vessel. It is important to prescribe that, during the boarding of a vessel and sailing, a MET instructor has all the rights and obligations under their seafarer status. According to the proposed provision, the service period of a MET instructor would be continuous, i.e. uninterrupted, and their rights to mandatory insurance schemes would not be terminated. The most important thing is to prescribe the continuity of service period and mandatory insurances regardless of the place of work.

Adding this provision to the legal framework of labour law and social security relations would also affect other laws, such as those that regulate the obligation of paying income tax. Therefore, we believe that MET instructors should pay their obligations, such as those from the Income Tax Act, depending on the length of their employment at an MHE institution, i.e. depending on the number of days sailing in a capacity of a seafarer.

The proposed provision should be added to the general laws regulating labour law and social security legislation in the Republic of Croatia, as well as to the contents of the Ordinance, the Maritime Code (based on which the Ordinance was adopted) and the internal acts of an MHE institution. It is possible to refer to the same provision in the autonomous legal sources (MET instructors' employment agreement and collective bargaining agreements).

It is up to the legislator to consider the above proposal. By accepting this proposal, the legislator would demonstrate an entirely novel point of view with regard to the seafarers' education.

23.6 Conclusion

Considering the inadequate solutions from the legal regulations currently in force and current potential solutions of boarding a vessel and sailing by MET instructors during their employment at an MHE institution, this paper proposes adding a provision which would enable MET instructors to board a vessel and sail as seafarers while they remain employed at an MHE institution. According to the proposed provision, MET instructors' employment at MHE institutions would not be terminated, and they would not have to demand a sabbatical by invoking the Sabbatical provision from the internal acts of an MHE institution and Labour Act in order to board a vessel and sail. We presented all the legal regulations in force that preclude MHE institutions from acting in accordance with the provisions of the Ordinance, especially regarding the obligation requiring them to ensure that all MET instructor continue to obtain experience in the vocation and the subject they teach and continue to conscientiously and consistently adapt to the technological and other changes in shipping.

Considering that general labour law and social security regulations, as well as autonomous legal sources, apply to the employment of MET instructors, the proposal for adding a new provision primarily pertains to the following regulations: The Labour Act, the Mandatory Health Insurance Act and the Pension Insurance Act. However, the proposal for adding a new provision also applies to the Ordinance because it determines the obligations of an MHE institution and the competency requirements that MET instructors must meet. Because the Maritime Code directly regulates the labour and social rights of seafarers in national regulations, we propose the adding of the same provision to the Maritime Code as well, in order to achieve mutual synthesis of all regulations regulating the labour and social status of MET instructors and seafarers.

The idea on the potential refresher sailing periods of MET instructors during their employment at an MHE institution would mean that the MET instructor holds the seafarer status within the meaning of the Maritime Code while aboard a vessel and sailing.

Considering the issues presented in the introductory part of this paper, we believe that the acceptance of the proposed provision would enable a MET instructor to board a vessel and sail, thus acquiring vocational experience and experience in the subject they teach and conscientiously and consistently adapting to the technological and other changes in shipping; during the semester in which there is a smaller workload at an MHE institution, a MET instructor may board a vessel and sail and hold the seafarer status parallel with their employment at an MHE institution.

For the purpose of conveying knowledge to the students, the importance of practical sailing experience is especially emphasised for MET instructors. MET instructors should be "up to date" with new technologies, and therefore refresher sailing periods are the best way for MET instructors to improve their knowledge and competencies.

References

1. The Ordinance on Ranks and Certificates of Qualifications for Seafarers of the Republic of Croatia, Official gazette of the Republic of Croatia, Narodne novine no. 130/13, 45/14, 124/15, 72/16
2. Maritime Code, Official gazette of the Republic of Croatia, Narodne novine no. 181/04, 76/07, 146/08, 61/11, 56/13, 26/15, 17/19
3. Act on Security of Sea-Going Vessels and Ports, Official gazette of the Republic of Croatia, Narodne novine no. 108/17
4. International Convention on Standards of Training, Certification and Watchkeeping for Seafarers, *STCW Convention* (IMO, London, 2010)
5. Bologna Declaration., http://bolonjska.50webs.com/Bologna%20declaration.pdf
6. D. Ivanišević, A. Gundić, Đ. Mohović, Difficulties in Understanding the COLREGs among the Students from Different Systems of Education for Seafarers. TransNav **13**(4), 869–873 (2019)
7. P. Monga, A system dynamics model of the development of new technologies for ship systems. Diss. Virginia Tech (2001)
8. M.R. Endsley, *Automation and situation awareness. Automation and Human Performance: Theory and Applications*, vol. 20, pp. 163–181 (1996)
9. T. Gamil, Upgrading MET Instructors: the Skills and Knowledge Enhancement Demanded by MET Stakeholders, WMU (2008)
10. Semjonovs, D., Bogdaņecs, A., Fernández González, MJ.: *Instructors' Competence for Enhancing Quality of In-House Training in Maritime Education*, in International Scientific Conference, Society, Integration, Education-SIE2015 (2015)
11. The Nautical Institute, The International Maritime Human Element Bulletin, Alert, Issue No 22 January 2010, ISSN 1747-5015
12. S. Vujičić, N. Hasanspahić, A. Gundić, N. Hrdalo, Assessment for Ensuring Adequately Qualified Instructors in Maritime Education and Training Institutions. Athens J. Sci. **7**, 115–126 (2020)
13. The Nautical Institute. The International Maritime Human Element Bulletin, Alert, Issue No 29 May 2012; ISSN 1747-5015
14. Labour Act, Official gazette of the Republic of Croatia, Narodne novine no. 93/14
15. Statute of the University of Zadar, consolidated text from December 17th 2019, class: 012-01/11-01/02, no. 2198-1-79-06-19-34, https://www.unizd.hr/o-nama/propisi-i-dokumenti/osnivanje-dopusnica-i-statut
16. Income Tax Act, Official gazette of the Republic of Croatia, Narodne novine no. 115/16, 106/18, 121/19, 32/20
17. Health Insurance Act, Official gazette of the Republic of Croatia, Narodne novine no. 80/13
18. Pension Insurance Act, Official gazette of the Republic of Croatia, Narodne novine no. 115/18
19. Constitution of the Republic of Croatia, Official gazette of the Republic of Croatia, Narodne novine no. 56/90, 135/97, 08/98, 113/00, 124/00, 28/01, 41/01, 55/01, 76/10, 85/10, 05/14

20. Collective Bargaining Agreement for Science and Higher Education, Official gazette of the Republic of Croatia, Narodne novine no. 9/19
21. Ž. Potočnjak, I. Vukorepa, Labour Act and other sources of labor law, Narodne Novine, Zagreb (2011)
22. Regulation on maritime booklets and authorisation for boarding, and procedures and means of registration and termination for mandatory pension and health insurance, Official gazette of the Republic of Croatia, Narodne novine no. 3/08
23. Regulation on the content and manner of registering employment contracts of seafarers and fisherman, Official gazette of the Republic of Croatia, Narodne novine no. 32/15, 109/19, 13/20
24. Regulation on performing duties and keeping watch by seafarers on sea-going vessel of commercial navy of the Republic of Croatia, Official gazette of the Republic of Croatia, Narodne novine no. 5/03
25. Regulation on the minimum number of crew members needed for safe navigation of sea-going vessels, floating objects and stationary offshore objects, Official gazette of the Republic of Croatia, Narodne novine no. 63/07, 76/11, 46/13, 104/15, 31/16, 83/16
26. Regulation on the application of the Maritime Labour Convention from 2006, Official gazette of the Republic of Croatia, Narodne novine no. 122/16
27. A. Lutteberger, Challenges of the Autonomous Regulation of Seafarer's Status in the Republic of Croatia. Pomorski zbornik **43**(1), 307–313 (2005)
28. A. Lutteberger, Pomorsko upravno pravo, Maritime faculty, University of Rijeka, Rijeka, p. 30 (2005)
29. M. Đ. Učur, Radnopravni status pomoraca, Faculty of Law, University of Rijeka, Rijeka, pp. 5–27 (2003)
30. Maritime Labour Convention., https://www.ilo.org/wcmsp5/groups/public/%2D%2D-ed_norm/%2D%2D-normes/documents/normativeinstrument/wcms_090250.pdf (2006)
31. D. Petrović, Global rights for seafarers: A new maritime labour convention. Naše More **53**(1–2), 29–33 (2006)

Chapter 24
A Survey of Maritime English Textbook Activities for Marine Engineers

Sandra Tominac Coslovich

24.1 Introduction

Selecting, creating, and applying activities in teaching English as a foreign language (EFL) and especially teaching English for specific purposes (ESP) is one of the most significant issues in foreign language teaching. Research into the selection of teaching materials to be used in foreign language classroom is relevant due to two important assumptions. First assumption is that teaching activities used in class should never be separated from the cognitive dimension of learning. In other words, a choice of certain types of activities is going to support the acquisition of a certain type of knowledge. The second assumption is that all teaching materials have a potential to foster learning. Thus, there is a spectrum of activities ranging from those focused mainly on form and acquisition of structural aspects of language to those focused on the development of communicative competence, i.e., the four language skills – reading, listening, speaking, and writing. Naturally, there are also activities that are focused both on language structures and communicative competence [13].

The following paper analyzes English language practice activities aimed at marine engineers that can be found in some of the currently available ME (Maritime English) textbooks. There are many aspects of the teaching materials that can be examined and various ways in which they can be categorized [6, 13, 17, 25]. Selected activities analyzed in this paper are categorized according to Neuner's typology of exercises, their form, aim, and content, i.e., according to whether they promote guided or unguided (re)production of language, fluency or accuracy, acquisition of language systems (grammar, vocabulary, pronunciation), and/or development of communicative competence. The analysis of activities is also carried out

S. Tominac Coslovich (✉)
Faculty of Maritime Studies, University of Rijeka, Rijeka, Croatia
e-mail: stominac@pfri.hr

S. Bauk, S. D. Ilčev (eds.), *The 1st International Conference on Maritime Education and Development*, https://doi.org/10.1007/978-3-030-64088-0_24

within the framework of the IMO Model Course 3.17 Maritime English and its recommendations regarding the type of teaching activities and approaches to be used in teaching Maritime English.

Since teaching materials are managed and administered by the foreign language instructor, EFL teachers have the responsibility to provide learners with activities that promote balanced learning of EFL. "This argument addresses the fact that, with a biased implementation of focus-on-form versus focus-on-meaning activities in the class, learners might not develop the necessary communicative competences (that is, the ability to produce or comprehend language in a natural and fluent way) or accuracy in language use (the achievement of grammatical commitment" [13, pp. 166].

24.2 Aim and Methodology

The textbooks for teaching ME to marine engineers were selected according to several criteria: time criterion, according to which the textbooks were selected with dates of publication ranging from the 1980s to the present to examine the potential influence of the communicative method; availability; target learners, i.e., the textbooks aimed (at least in part) at marine engineers as target learners were selected; and the inclusion in the IMO Model Course 3.17 recommended textbook list. Several textbooks used in the analysis are not included in the IMO Model Course 3.17 list but were chosen because of their availability and official usage in a number of MET institutions. Activities were extracted from one sample unit in each of the selected ME textbooks and scanned in a separate document for further qualitative and quantitative analyses. From the ME textbooks intended both for marine engineers and navigating students, only the exercises from sample unit covering marine engineering topics were extracted. Once the activities were compiled in a single document, qualitative and quantitative analyses were performed. Both the quantitative and qualitative analyses were done manually to discern the features and nuances which made an activity focused either on developing communicative skills (reading, listening, speaking, and writing) or language systems (grammar, vocabulary, and pronunciation), fluency or accuracy and whether it was aimed at language reception, strongly guided (re)production, less guided production, or communicative production. The collection of samples encompassed 143 examples of activities.

The focus was especially placed on the activities aimed at teaching Maritime English to marine engineers in order to see to which extent they adhere to the tenets of the communicative approach to teaching, recommended for teaching Maritime English by the IMO Model Course 3.17. Thus, the first several paragraphs of this paper will be dealing with the characteristics of Maritime English as a special type of ESP, followed by a brief outline of the teaching approaches recommended by the IMO Model Course 3.17 for teaching Maritime English, namely, communicative approach, task-based approach, and content-based instruction. The central part of the paper will give an overview of various exercise typologies and apply some of their perspectives to the analysis of ME textbook activities aimed at marine engineers. The final segment of the paper will provide a brief discussion on the results and give further recommendations.

24.3 ESP: The Case of Maritime English

The main distinction between English for specific purposes (ESP) and English as a foreign language (EFL) pertains to learners and their reasons for learning English. ESP learners are mainly adults or learners at secondary school level who are already to some extent familiar with English and are "learning the language in order to communicate a set of professional skills and to perform particular job-related functions" [ESP site]. As a result, an ESP program is based on learners' needs analysis and built on an assessment of purposes and functions for which English is required. Whereas in EFL all four language skills, listening, reading, speaking, and writing, are treated equally, in ESP a needs analysis ascertains which language skills are the ones most required by the students, and the program is adapted accordingly. ESP incorporates both subject matter and English language instruction, thus increasing learners' motivation because they are able to apply what they learn in their English classes to their main field of study. The knowledge students possess in their subject matter fields reinforces their ability to acquire English. Maritime English is considered one field of ESP with its unique characteristics which combine plain language knowledge with specialized and professional knowledge of English [8]. It subsumes five different sub-varieties according to the specific purpose they serve within the maritime context: English for navigation and maritime communications, English for maritime commerce, English for maritime law, English for marine engineering, and English for shipbuilding [3]. This paper focuses on English for marine engineering, i.e., the textbook activities used to teach it.

Generally speaking, in order for seafarers to communicate efficiently, they need to be able to use and understand English in a variety of situations. Using English properly implies the seafarers' ability to use grammar, vocabulary, and phonology to express themselves clearly and suitably in speech and writing, whereas understanding English implies that the seafarers can successfully interpret messages that they hear and read and respond to them appropriately and intelligibly.

According to the IMO Maritime English Model Course 3.17, language can be divided into two broad categories for teaching purposes: systems and skills. The first language category refers to language systems that cover grammar, vocabulary, and phonology, and this is what constitutes the majority of the input about language, or the new material that the students are taught. Seemingly, if the foreign language learning process is to be successful, the input about language should always be based on a "common core" of the language required by all learners, irrespective of their specific profession. However, in maritime context, this common core input should be integrated with the maritime input, i.e., the terminology and phrases that seafarers require for communication [16, pp. 207]. The second language category encompasses language skills or, more precisely, four communication skills: listening, speaking, reading, and writing. Although people do not usually need to think about how they use these skills in their first language, research into the way adults learn foreign languages shows that learners do not automatically transfer the skills they use from their first language to the new language. Therefore, it is important to teach ways of listening, speaking, reading, and writing in addition to teaching structure, vocabulary, and phonology. Naturally, learners have to be competent in each of these skill areas in order to integrate and apply language systems efficiently. Therefore, teaching activities should, whenever possible, be adapted from authentic maritime situations, topics, or materials. However, effective communication requires more than the ability to integrate language systems and skills. The students also need to understand how social contexts and specific situations influence the choice of language and the type of communication. For instance, they should be able to recognize which register of language is appropriate in a given social context or know the conventions for using the language in specific situation, as well as the strategies for coping with misunderstandings in cases when communication breaks down.

"Knowledge and technical know-how are clearly important, but these must be presented with an excellent standard of communication skills. After all, success is not only based on what you know but also on how you can communicate it. Indeed, communication skills are considered one of the best career enhancers. In fact, evidence suggests that employers in all occupational fields place greater value on employees' communication skills than they do on their technical skills and rate it as a top priority for both securing and retaining employment" [15, pp. 13].

This paper is particularly concerned with activities in ME textbooks aimed at marine engineers for several reasons. Most Maritime English syllabi focus on deck officers, whose English proficiency is addressed by the STCW Convention, a collection of seaspeak manuals, Standard Marine Communication Phrases, and a number of books commonly used in Maritime English training, whereas engineers, apart from a few rather general requirements set out in the STCW Convention, until recently, have had no internationally accepted teaching standards for developing communication skills. IMO language competence requirements (Table A-III/1 of the STCW Code) suggest that engineers should know how to read and write in English so that they may use engineering publications and perform their duties and clearly understand and correctly interpret all technical information. However, the IMO Model Course 3.17 Maritime English (2015 edition) has recently provided a detailed specialized Maritime English teaching syllabus for officers in charge of engineering watch. Thus, according to IMO Model Course 3.17, 2015 edition, and investigation [19] into which engineers' duties require some communicative competence and studying parts of well-organized company standing orders, one quickly realizes that there are considerably more jobs that require written than oral communication. In fact, whatever happens in the engine room has to be recorded and reported in writing, and only a few routine duties are associated with oral communication, which of course does not mean that developing speaking skills should be neglected. This must surely be taken into account when creating appropriate language teaching materials. Keeping these facts in mind, we will investigate the available ME textbook materials to see whether the practice activities focus on developing the aforementioned written communication skills and in which manner. The paper will also investigate whether the available textbook activities are indeed communicative in nature, whether they focus on fluency or accuracy, and whether they are reproductive, guided, and closed or productive, unguided, and open.

24.4 Communicative Approach

Communicative approach refers to the ways in which teachers can focus the teaching of foreign language in such a way so as to enable the students to engage in real-life conscious and meaningful communication in the target language. Communicative language teaching – CLT (sometimes also referred to as notional-functional approach and functional approach) – originated in the late 1960s in the British language teaching tradition as a result of growing dissatisfaction with the audiolingual and grammar-translation methods used in foreign language teaching. The main reason for dissatisfaction was that teachers felt that students were not learning enough realistic language and that they did not know how to communicate using appropriate social language and expressions in a particular real-life context of the language they were trying to acquire. The British applied linguists pointed out a significant dimension of language that was insufficiently addressed in the language teaching approaches at that time, i.e., the functional and communicative capacity of language. Thus, they began to promote the development of communicative competence in the target language as opposed to the knowledge of its structures. Since language interaction is an interpersonal activity that has a distinct relationship with society, the study of language has to observe the function of language in context, both in terms of its linguistic and situational context [2, 12]. Although the communicative approach to language teaching started in Great Britain, its scope has expanded since the 1970s and is now seen as "an approach (and not a method) that aims to (a) make communicative competence the goal of language teaching and (b) develop procedures for the teaching of the four language skills that acknowledge the interdependence of language and communication" [20]. As a result, authentic language use and classroom exchanges where students engaged in real communication with one another have become quite popular.

The development of the CLT was based on a wide range of language theories, ranging from those of the British linguists Firth [12] (view that language is interaction) and Halliday (theory of language functions) to those of the American sociolinguists Hymes [14] (theory of communicative competence in contrast to Chomsky's [7] theory of competence), Gumpertz (interactional sociolinguistics), and Labov to those of the language philosophers Austin and Searle and their speech act theory. Another important contribution came from Wilkins [27], who proposed a notional syllabus based on a functional and communicative definition of language that was incorporated by the Council of Europe to aid the teaching of major languages in the European Common Market.

As opposed to the language theories which serve as the basis of CLT's approach, its theory of learning does not have such a sound basis, and relatively little has been written about it. However, according to Richards and Rodgers [25], three different elements can be discerned from CLT practices: the communication principle, activities involving real communication that promote learning; the task principle, learning supported by activities in which language is employed for performing meaningful tasks; and meaningfulness principle, language meaningful to the student in order to enhance the learning process. Thus, activities are chosen according to how well they engage the student in meaningful and authentic language use rather than just automatic use of language patterns.

The topic that is considered central to CLT is syllabus design. One of the first models to be proposed is that of Wilkins' notional syllabus [27], specifying notional and semantic-grammatical categories (such as frequency, motion, location, etc.) and communicative function categories (e.g., complaints, denials, requests, etc.) that students need to express. The syllabus was further developed by the Council of Europe that included objectives of foreign language courses, situations in which students might typically use a foreign language (for instance, business or travel), topics they might need to talk about (e.g., shopping, education, etc.), functions they needed language for (e.g., requesting information, complaining, etc.), and notions used in communication (such as time, frequency, duration, etc.). The results were published in Van Ek and Alexander's *Threshold Level English* [27].

The syllabi of the type mentioned above are realized through various types of learning and teaching activities. There are two types of activities. One type includes functional communication activities, which require the use of communicative processes such as information sharing and negotiation of meaning based on the information-gap principle [22]. The other type includes interaction activities which are focused on "learning to communicate through interaction in the target language" [18, pp. 279]. Typical examples of this type of activities include dialogues, role plays, discussions, debates, etc.

Likewise, instructional materials fall into the three main categories which can be labelled as text-based (typically referring to textbooks), task-based (based on jigsaw or information-gap principles), and realia (different types of authentic materials from real life around which communicative activities can be built, such as manuals, graphs, pictures, magazines, charts, etc.) (Richards and Rodgers 2001).

Communicative language teaching is based on real-life situations that require communication. The language teacher constructs a situation or a context that the students are likely to encounter in real life. Thus, the students' motivation to learn is derived from their willingness to communicate efficiently and meaningfully about relevant topics. They must cooperate, negotiate, and interact with each other. In this way they become pivotal members in the learning process, i.e., the teaching becomes student-centered. On the other hand, the teacher takes on several roles. According to Breen and Candlin [4, pp. 99], "the teacher has two main roles. The first is to facilitate the communication process between all the participants in the classroom, and between these participants and the various activities and texts. The second role is to act as an independent participant within the learning-teaching group. The latter role is closely related to the objectives of the first role and arises from it. These roles imply a set of secondary roles for the teacher; first, as an organizer of resources and a resource himself, second as a guide within the classroom procedures and activities … A third role for the teacher is that of researcher and learner." Other roles the teacher also assumes are that of a counselor, needs analyst, and group process manager.

Two additional approaches to foreign language teaching also recommended by the IMO Model Course 3.17 in regard to teaching ME are content-based instruction and task-based learning. These two approaches have also originated from communicative language teaching and are briefly outlined below.

In task-based learning lessons, students solve a task that involves the authentic use of language rather than just complete simple language questions about grammar or vocabulary. Task-based approach can be defined as "how a learner applies his or her communicative competence to undertake a selection of tasks" [6, pp. 183]. According to Littlewood [18, pp. 320], "definitions of 'task' range along a continuum according to the extent to which they insist on communicative purpose as an essential criterion." Furthermore, according to Estaire and Zanon [21, pp. 13–20], tasks can be oriented toward communication (communication tasks), where the learner's attention is focused on meaning rather than form, or toward learning the linguistic rules (enabling tasks) that enable them to perform the communicative tasks; thus the main focus is on linguistic aspects (grammar, vocabulary, pronunciation, functions, and discourse). Tasks involve meaningful communication, i.e., problem-solving activities with a single solution or multiple solutions; they are built around a specific genuine sociolinguistic situation (for instance, attending a job interview). Since there exists a wide range of definitions of what a task can be, perhaps the definitions could be clarified further by stating what tasks are not. "They are not completion exercises involving transformation or meaningless repetition, or question-and-answer strings with the teacher, and they are not oriented towards analysis of linguistic structures, though enabling tasks focus on language" [26, pp. 95].

Content-based instruction (CBI) is an approach to language teaching which integrates language instruction, i.e., teaching of English in the content areas. This approach is focused on learning language through learning about specific subject matter content. Content-based instruction (CBI) is "an approach to second language teaching in which teaching is organized around the content or information that students will acquire, rather than around a linguistic or other type of syllabus" [25, pp. 204]. The language learning objectives are thus achieved through content learning. In short, CBI is a method of teaching language and content simultaneously. As a result, CBI requires language teachers to be knowledgeable in content areas and capable of eliciting knowledge from students. In addition, language teachers should use and adapt suitably selected authentic materials and make them comprehensible and meaningful to the learners at their level of proficiency. Also it must be noted that CBI has proven very effective and useful in ESP.

In order to evaluate the type of activities found in textbooks used for teaching ME to marine engineers in accordance with the communicative approach postulates, the following paragraphs will cover some basic notions about exercise typologies that have been developed throughout the history of foreign language teaching. The focus will be especially placed on the type of activities and their characteristics which are seemingly associated with the communicative language teaching.

24.5 Exercise Typologies

Foreign language practice activities are at the very center of a language learning process. The ideal sequence of foreign language practice activities should help learners advance from closed strongly guided exercises to open exercises and eventually to independent reception and production of the target language. So far, various exercise typologies have been designed in foreign language teaching in order to incorporate a range of different exercise types into a well-organized and meaningful whole. However, despite being based on insights from educational psychology, developmental psychology, or linguistics and seemingly presented as logical organizing principles on which teaching foreign languages can be based, they have often been subject to criticism. Some of it is aimed at the fact that exercise typologies vary according to different methods and approaches. Also, they have to be adapted to the characteristics of the learners and the learning context. Furthermore, the level of difficulty of some exercise types and the relative degree of translatability of particular exercise typologies into actual teaching practice are affected by the learners' motivation, previous knowledge, relative degree of complexity of the contents that need to be acquired, etc. [5]. "Despite these reservations, when observing the principle of meaningful learning as defined by Ausubel [1], observing De Corte et al.'s [10] taxonomy of cognitive objectives, and putting Craik and Lockhart's [9] concept of levels of processing into practice, exercise typologies are helpful instruments for curriculum development, course evaluation or for planning a sequence of practice activities" [5, pp. 243]. Firstly, the principle of meaningful learning implies that teaching materials and exercises can only promote learning when they are meaningfully related to what students already know. Secondly, the underlying principle behind taxonomy of cognitive objectives is that of an increase in the complexity of cognitive operations and in the degree of independence in information processing, thus resulting in the types of exercises supporting the gradual development from lower to higher levels of involvement, from receptive to productive tasks, from teacher-guided to learner-independent learning. Finally, according to the notion of levels of processing, incoming information is processed by various operations, which can be termed perceptual-conceptual analysis. The level of processing reflects an individual's attention. In other words, if an individual considers incoming material deserving of long-term recall, they will process it differently from material they consider relatively irrelevant. Furthermore, the level at which the stimulus will be processed depends on the nature of the stimulus, i.e., the exercise type, the time available for processing, subject's own motivation, attitudes, feelings, goals, and knowledge base [5].

A wide array of exercise types has been developed over the years in foreign language teaching reflecting the changes in the concept of foreign language competence.

There are several ways in which exercise types can be distinguished. It is possible to classify them based on their formal characteristics, such as gap-filling, matching, classification, completion, true or false, multiple-choice, and other exercises. Another principle according to which exercise types can be classified is content. Thus, exercises can focus on grammar, vocabulary, pronunciation, or development of the learners' listening, speaking, reading, or writing skills or special competencies, for instance, intercultural competence. A third perspective categorizes exercises according to where they can be placed on the reception-reproductive-productive scale. Also, exercise types can be classified as those aiming to develop learner's accuracy and fluency [5]. Although none of the criteria of exercise classification should be used in isolation, one should be aware of the orientation of exercises and its effect on language learning.

The sample of exercises in this paper will be analyzed by using a combination of perspectives. First, we will grade the exercises according to their formal characteristics and the abovementioned reception-reproductive-productive scale based on Neuner's typology, which is in line with the taxonomy of cognitive objectives [10], mentioned earlier in the paper and thus based on the principle of an increase in the complexity of cognitive operations and in the degree of independence in information processing, resulting in the types of exercises supporting the gradual development from lower to higher levels of involvement, from receptive to productive tasks, from teacher-guided to learner-independent learning. The second perspective will focus on whether the exercises aim to develop learners' accuracy or fluency, and the third perspective will categorize exercises according to whether they focus on development of learners' listening, speaking, or writing skills. The language skills can also be grouped into the receptive ones (listening and reading) and productive ones (speaking and writing), thus corresponding to different types of activities that can also be placed on the abovementioned reception-reproductive-productive scale.

The ideal sequence of exercises that should be available in a good foreign language lesson should be organized in a logical way and progress from closed to open exercises, from receptive to productive ones, and from predictable to unpredictable use of language. Therefore, based on the work of Neuner et al. [21], exercises can be classified into four categories. The first category of exercises (type A exercises) is designed for checking and organizing information. They are associated with the phase in which material is presented to the learner and comprehension is checked. The exercises in this category are aimed at understanding and are usually created around written text. At this point no production in speaking or writing of the target language is required of the learner. The emphasis is placed on recognizing and understanding the material, usually by engaging students in reading or listening activities. The second category of exercises (type B exercises) is reproductive, strongly guided, closed, and controlled activities and is aimed at implanting language skills. These types of exercises are strongly guided exercises where the learner is encouraged to reproduce elements or building blocks of language (sounds, words, grammatical structures, phrases, idioms, functions, etc.) in the same context as the one in which they are provided. Typical examples are reciting or copying texts; spelling exercises; and translation, matching, and the reproduction of meaning-directed exercises, such as selecting given words or sentences and using them for gap-filling on the basis of their meaning in the context provided. The third category of exercises (type C exercises) is aimed at developing skills. This category provides guided speaking and writing exercises where students can practice using the building blocks or elements of language from the previous stage. These exercises are less guided, half-open, productive tasks that allow the room for variations, i.e., the language choices are not as predictable as they are in the previous stage. In this stage, the learner is expected to (re)produce the language elements in a different context than that which was previously provided. Typical tasks include finishing sentences or texts, answering questions about a text, describing or writing what is shown in diagrams, role plays with information gap, etc. Although these exercises are still focused on form, they may be considered more communicative since situational meaning plays a bigger role. The fourth category of exercises (type D exercises) is unguided, open, and productive exercises in which learners use the skills they have learned in the previous phases. These exercises or tasks are always focused on meaning and can be considered fully communicative. They are hardly guided or not guided at all, i.e., open. At this stage, learners are expected to produce pieces of language in speech or writing in a context that has not been predefined. Lessons are usually centered around a (real-world rather than language-focused) task.

The main activity types that reflect the principles of a communicative methodology and are considered the outcomes of CLT are the ones which focus on fluency instead of accuracy, communicative and meaningful practice instead of mechanical practice, and functional communication that requires learners to use their language resources to overcome an information gap (such as in jigsaw activities or reasoning-gap activities), solve a problem, complete a task, engage in role plays, gather information or transfer information that is presented in one form, and represent it in different form [24].

A combination of perspectives on which the exercise typologies presented above are based will be used in the following section to analyze and categorize the selection of ME activities extracted from ME textbooks in order to investigate how well they fit within the communicative approach framework.

24.6 Results of the Analysis and Discussion

The selected textbook activities are first classified based on their formal characteristics. According to this criterion, the following exercise types were identified: gap-filling, true or false, matching (pictures and terms, terms and definitions), multiple-choice questions, identifying words or grammatical structures in texts, indicating/naming the elements (in diagrams, pictures, etc.), finding common features, finding odd one out, answering questions, completing sentences (or conversations, diagrams, tables, worksheets), describing/explaining processes (or pictures, diagrams, parts, functions), comparing and contrasting, unscrambling sentences, rewriting sentences, asking questions from statements, turning active sentences into passive ones, making one's own sentences using specific structures/vocabulary items, translating terms and sentences into English/mother tongue, recording unknown terms and finding their meanings in a dictionary and learning them by heart, taking notes during listening/reading, reading text and tracing a process on a diagram, studying a diagram/meaning of a word, writing out formulae to show relationships between symbols, writing an essay/description, doing a presentation, troubleshooting, and acting out a dialogue. Various exercise types that have been recorded in the collected samples are discussed in the following paragraphs in terms of their frequency, purpose, content, placement on the reception-reproduction-production scale, focus on fluency or accuracy, and communicative nature or lack thereof. Some suggestions on how to make some activities more communicative are also provided.

According to the formal characteristics, the most frequently used type of exercise appears to be gap-filling (also found in the collection of samples under the titles "filling in the blanks" or "writing/inserting missing words"), which appears 34 times and accounts for 24% of the total of 143 samples in our collection. Gap-filling exercises are one of the most widely used techniques for testing both vocabulary and grammar and are considered one of the traditional reading comprehension tasks, together with yes/no questions, true/false questions, and multiple-choice questions. In our collection of samples, gap-filling activities are mainly used to practice or test the acquisition of new vocabulary and to a lesser extent grammar and are typically aimed at developing accuracy. Gap-filling is also considered a strongly guided (re)productive task. True/false questions are recorded 11 times or in 8% of the cases, whereas multiple questions appear in 5% of the cases or 8 times. Almost all of the multiple-choice question exercises found are aimed at reading comprehension of written texts and only one at listening comprehension. Yes/no questions have not been found among the collected samples.

Other exercise types recorded with notable frequency are matching exercises, which have been found in 11 instances. Matching tasks found in our collection of samples are almost exclusively used to check vocabulary items by inviting students either to match pictures or labels in a picture or diagram with terms or to match terms with definitions. It should also be noted that matching is considered a typical example of a receptive task, where the emphasis is on recognition and understanding of the material that is offered.

Another activity similar to this one is also a completion task in which learners were asked to complete a diagram with the appropriate terms. This type of exercise appears four times in the collected samples. Other types of completion exercises which appear in the collection of samples require students to complete a conversation based on listening (only one instance), a worksheet based on watching a video (only one instance), and a table based on reading a text (three instances). Table completion is also considered a good alternative to comprehension questions. For instance, in one of the examples found in the collection of samples, learners are given a table to fill in with pieces of information from a text on two-stroke and four-stroke diesel engines in which they have to compare and contrast the two items. Simpler version of a table completion activity is also found where a table is provided with different features and items which learners have to tick according to which feature applies to which item. This exercise works well with factual texts, with texts which compare two or more items, systems, or methods and with texts which refer to classifications of different objects. It also gives students a clear framework for practicing note-taking skills, which in turn provides a solid base for eventually performing tasks that build writing skills. Incidentally, taking notes during listening and reading is a task that appears twice.

Answering questions is an exercise that appears 12 times or in 8% of the samples. Majority of questions are reading comprehension questions which appear right after the text, and only one exercise is found with a set of questions which serve to elicit learners' already existing knowledge about a topic prior to reading a text. Answering questions is a less guided productive task that typically focuses a bit more on fluency and content than accuracy (of form). Comprehension questions found in our collection of exercise samples are all about a text, and apart from being aimed at fluency, they also seemingly serve to elicit the accurate vocabulary items introduced in a text.

Explaining or describing concepts contained in a text or a visual is a task that appears in ten instances and accounts for 7% of samples. In this type of exercises, learners are asked to explain or state how something works or describe pictures and diagrams. Describing in writing or in speech what is, for instance, shown in a diagram is considered a less guided productive task in which language control is still present (e.g., learners are given a list of prepositions and vocabulary items), but they still have to make meaningful choices while performing the task because they have to respond according to the location of different items in a diagram. In other words, the tasks of this type are still focused on form, but they can be called more communicative because situational meaning plays a more significant role. Additionally, a suggestion on how to make this activity more communicative is to perhaps invite students to compare their descriptions of a diagram in pairs.

Identifying, indicating, or naming elements or concepts (without provided list of word choices) is a type of exercise that appears seven times or in 5% of samples, and it serves to elicit or practice vocabulary items.

It is interesting to note that only one activity aimed at practicing word combinations or multi-word units is found among the samples. The exercise in question asks learners to form nominal compounds (which are very important for marine engineers) according to example. Traditional grammar-translation methods stressed the importance of structures which individual words could be inserted into to produce meaningful sentences. However, major breakthrough in teaching vocabulary came about in the 1990s with the emergence of the lexical approach, which reverses this view and says that native speakers carry around in their vocabulary thousands of units called lexical chunks, ranging from words, polywords, and collocations to institutionalized utterances and sentence frames and heads [13, pp. 238].

Finding a common feature or finding the odd one out and explaining one's choice is an exercise that appears in four samples and provides controlled practice which allows the learners to use the new words for communicative purposes within clear contexts. In turn this allows the teacher to check the learners' understanding of the new items and helps them consolidate the meaning of the new word. This is not an entirely unguided and open communicative activity, but it is still considered more communicative since it does not simply imply a mechanical practice but an activity where language control is still provided but students are required to make meaningful choices when carrying out the practice.

One instance is also found of an exercise where learners are asked to make their own sentences using new words. Although the presentation of new vocabulary should be followed up immediately with controlled practice, "this should not involve asking students to 'make sentences with new words' on their own as this is not an activity which native speakers would do and can be daunting for the students. Controlled practice should let the students use the new words for communicative purposes within clear contexts [16, pp. 244–45]."

Unscrambling sentences is an exercise that appears two times in the samples, just as writing wh-questions and asking questions from statements. Rewriting sentences according to a model by using specific grammatical structures and turning sentences into passive are recorded once each. Unscrambling sentences, rewriting sentences by using specific grammatical structures, and creating wh-questions out of statements are all guided productive exercises dealing with syntax and aimed at developing accuracy. Although accuracy tasks are not authentic communicative tasks, they serve to consolidate new language for learners and are thus providing them with the confidence they need to use language that they have just acquired as well as ensuring that they are able to produce new forms correctly. Controlled and semi-controlled accuracy exercises encourage shy students to speak, leading on to tasks that build fluency.

Acting out a dialogue is as an activity that appears only once. Learners are asked to complete a conversation (presented with gaps) with the phrases already given to them and then take roles and act out the dialogue. This is again an example of a guided (re)productive practice.

Translation exercises were recorded four times in the collection of samples, i.e., one exercise required translation into mother tongue and three exercises required translation into English. Traditional approaches to teaching, such as grammar-translation method, placed great importance on the acquisition and application of grammar rules. Students were required to memorize structures by drilling and had to translate complex structures into their first language. Vocabulary was also learned in isolation by using lists of words in any two languages to learn and memorize by repetition. However, "translation to the students' first language is not encouraged; even at lower levels, seafarers should be encouraged to think about vocabulary items in English in order to promote their understanding of English as a seafaring language worldwide." Although, grammar-translation and direct method gave way to language in use, with the focus placed on the learner, translation and the need for dealing with cases of first language interference have not completely disappeared. Even today, when translating a lexical item, especially when it comes to highly specialized terminology, translation sometimes leads to better comprehension and acquisition and is more efficient. However, students should be made aware that in most cases a word-for-word translation will not work. "There are a number of reasons for this: there may not be an exact equivalent or the word may not even exist in their first language; the closest equivalent in their language may be used quite differently and could even be a different part of speech; the word may also collocate quite differently in their language" [16, pp. 257]. If translation exercises are aimed at presenting new vocabulary, they could be replaced by other techniques, such as contextualizing, guessing meaning from context, using visuals (which is used quite extensively in our collection of samples), paraphrasing, using monolingual dictionaries, etc. For practicing new vocabulary, a number of information-gap activities could be adapted, such as describing and drawing practice, in which learners work in pairs, i.e., one describes the diagram to the other who listens and tries to draw the diagram from the description. Finally, each pair compares drawings at the end of the task. Other techniques for practicing new vocabulary can be also found in IMO Model Course 3.17 [16, pp. 249–50].

Another exercise which appears six times in our collection of activities is recording unknown terms from the text, checking their meaning in a dictionary (not specified whether monolingual or bilingual), and learning them by heart. The use of dictionaries (preferably monolingual, followed by a bilingual) and memorizing word lists and their translations are both considered determination strategies for vocabulary learning. The use of dictionary is a legitimate method of learning new vocabulary. However, it could be supported by other techniques of presenting learners with new vocabulary which are mentioned in the previous paragraph, such as contextualization, paraphrasing, or, where this is not possible due to complexity of a term, using visuals. It should also be noted that visuals, such as drawings, diagrams, real and symbolic pictures, etc., are of tremendous importance in the process of knowledge acquisition in general, especially information processing, storage, and retrieval. They may also have important implications for the quality of visual communication within a specific profession. Thus, it is clear that every engineering profession, including marine engineers, relies substantially on the use of visual forms as a means of non-verbal communication [23].

One type of exercise which is associated with communicative language activity types is information transfer. Activities of this type require learners to take information that is presented in one form and represent it in a different form. One example of this type of activity found in our collection is an exercise where learners have to transfer written down warnings and good practices on board into visual signs and vice versa and compare them with a partner. This type of activity appears only once in our collection of samples, but it is still worth a mention because of its communicative potential and task-based nature. Other examples of information-transfer activities may require learners to read instructions on how to get from A to B and then draw a map showing sequence, or they may read information about a subject and then represent it as a graph, etc. [24].

Another example of a simple task-based activity used for reading comprehension is an exercise in which learners have to read a text describing the fuel oil system of a diesel engine and trace the circulation of heavy oil and the diesel oil on a diagram and use a different color for each. This is an example of a one-way tasked-based activity based on input. For further extensive reading on task-based language teaching, see [11].

Troubleshooting activities appear seven times in the collection of samples and are all dealing with different cases of engine failure. Incidentally, all of the activities of this type appear only in two textbooks. For instance, in one exercise learners are asked to work in pairs and use the information given to them in faults checklist, service reports, and fault diagnosis chart to find out what is wrong with the engine. In the other exercises of this type, learners are also asked to perform fault diagnosis, but they are also asked to offer possible solutions. These exercises are good examples of task-based communicative activities because the students have to rely on using one's language resources to complete the task. This activity requires learners to deduce possible causes of events and establish and explain reasons for faults. In this case, it is very important to choose and pre-teach appropriate words for the learners since marine engineers need to know more verbs to describe mechanical faults than, for instance, navigational terms, so it would be useful and advisable to consult specialized manuals or maritime trainers from other departments to ensure that appropriate vocabulary is selected for the classes.

Correct sequencing of jumbled paragraphs is an activity that appears twice. In both cases, learners are asked to put the sections of text in the correct sequence. This activity draws students' attention to the organization of information within the text and can also serve as one of the practices that can help develop writing skills at a later stage.

Guided writing as an activity appears twice, and in both of the cases, it involves descriptions. Learners are asked to write a description by following instructions presented in three stages; first, they are asked to join the groups of sentences to form paragraphs and use any of the linking words they have learned so far. In the second stage, learners are presented with a diagram and asked to sort out the paragraphs into a logical order, and finally, in the third stage, learners have to write out paragraphs in a logical order to form a description. Guided writing is considered a less guided productive task. While it is of great importance that students are given opportunities to write discursively, the lack of guidance in writing can make the task difficult for the student to produce. "Furthermore, essay writing alone does not prepare students for the range of different text types they need to produce in the world of work. Traditionally, the main concern regarding writing was the accuracy of the finished product. This form-focused or 'bottom up' approach to writing can be described as controlled writing or product writing. Product writing tasks, like accuracy focused speaking tasks, are necessary for helping students to manipulate and practice the structures they learn. However, communicating in writing involves more than producing a string of English sentences which are grammatically accurate. Effective written communication in 'real life' requires us to be able to read effectively, choose an appropriate means of response, plan the content of our response before writing it, have a clear idea of the purpose of the text we are writing, organize our thoughts in a logical, linear sequence, in keeping with the rhetorical pattern of English, signal the organization of information by using paragraphs and cohesive markers, use a layout appropriate to the type of text, use an appropriate register of language, draft, check, edit and revise our own writing (both at sentence level and at discourse level). In other words, classroom instruction needs to take account not only of the written product but also of the processes involved in writing. Writing should not be given as solitary practice of grammar exercises or reserved for exclusively for homework. If students are to be able to produce the range of written work that is expected of them as officers, they will need guided practice in each of the sub-skills outlined above. Process writing provides a way of doing this by means of classroom activities which take a global or 'top down' approach to writing. This requires ongoing input, supervision, support, revision and regular feedback, as with the teaching of any communication skill [16, pp. 301–302]."

Doing a presentation is an activity that is recorded six times. However, no instruction is given as to how to tackle this activity. Learners are simply asked to do a presentation on a specific topic. IMO Model Course 3.17 encourages language trainers to ask pairs or groups of students to give presentations to the class in English about a subject matter content they have learned about in subject matter courses. This provides learners with the opportunity to consolidate their knowledge from specialized subjects and revise technical terminology in English. Learners should also be encouraged to use visual aids to assist their presentations, such as drawings, diagrams, or notes that they can project on screen. Presentations and talks are demanding activities that require a lot of preparation, so learners should be given enough time to research, plan, and rehearse their presentations. "In the preparation stage, include discussion of what makes a good presentation and remind students that, in order to sound natural and to keep their audience's interest, they should use notes and look at the audience instead of reading from a script. Give controlled practice to help students memorize some useful functional phrases. Encourage students to practice by recording themselves and assessing a partner's performance … Give the students a while-listening activity to do and encourage comments and questions after the presentations. Mark each student according to his/her spoken accuracy (pronunciation, grammar, vocabulary), choice of register and fluency. If possible, give individual feedback by talking to each student about the strengths and weaknesses of his/her presentation [16, pp. 351]." Presentations are an excellent opportunity for learners to practice all language systems areas (vocabulary, grammar, discourse, and phonology) and skills (speaking, reading, writing, and listening). Typically, a presentation activity comes toward the end of a lesson or a sequence of lessons that focus on a particular language area or skill. Students should be prepared throughout the course by gradually being engaged in plenty of controlled and semi-controlled practice activities, such as gap-filling, drills, etc., before they are asked to deliver a presentation, which involves much freer use of the target language.

When it comes to the placement of activities on the reception-reproduction-production scale, based on Neuner's typology, it becomes quite apparent that a large majority of exercises fall into the category of receptive and reproductive, closed, strongly guided, and controlled practices aimed at checking and organizing information and implanting skills. These types of activities account for 65% of all the collected samples, followed by less guided productive, half-open exercises which allow room for variations and serve for developing skills, which amount to 27% of the samples. Unguided, productive, fully communicative, and open tasks seem to be scarce or completely absent. Even some of the activities identified in the collection of samples as potential candidates for this category, which account for 8% of samples, could be considered more as half-open less guided communicative activities than as examples of unguided, productive, and open tasks that are genuinely communicative. Also, majority of exercises are aimed at accuracy, vocabulary, and grammar practice and reading comprehension, whereas activities aimed at developing writing skills appear less than one would expect, i.e., only in 12 samples, where learners are asked to write descriptions of different pieces of equipment. Speaking activities are quite frequent, and they generally appear in exercises where learners have to answer comprehension question or explain or describe a certain concept or process or do a presentation.

It is quite obvious from our collection of activity samples that great emphasis is placed on the acquisition and practice of specialized vocabulary. Indeed, vocabulary is the most obvious feature that distinguishes any specialized use of language from general English. While it is essential to teach specialized maritime vocabulary, this needs to be balanced with other areas of language systems and skills practice. The majority of exercises associated with vocabulary acquisition found in the textbooks are typically guided (re)productive activities, such as gap-filling, or even receptive, such as matching, in which learners are expected to (re)produce or in certain just recognize the language elements in the same context as the one they have been previously provided. Of course, this does not imply that activities of this type are not to be used. They are necessary because they provide solid basis for communicative activities and fluency-oriented activities that follow or should follow later on. Descriptions in writing or in speech and answering questions about a text are solid activities that seem to be a step in the right direction toward developing fluency and are considered more communicative because they are less guided and the language choices are not entirely predictable. However, majority of sampled units usually stop at this level of activities and unfortunately do not follow them up with genuinely communicative and task-based activities based on real-life scenarios that could truly enhance learner's communicative skills.

Overall, most of the exercises in our collection of textbook activities lack the communicative component, and only a few of them can be considered fully communicative and/or task-based. However, one consistent feature of the activities extracted from the ME textbooks is that they are based on specialized content. As was mentioned previously, content-based instruction (CBI) incorporates language instruction with instruction in the content areas. Still, CBI can present a substantial cognitive challenge for the learners, who might feel overwhelmed and even frustrated at times, which can lead to overuse of learners' native language. CBI is also challenging for the language instructors since they have to master the knowledge of the subject matter content in addition to target language knowledge. Another issue also pertains to insufficient amount of materials available on the market that are adapted to the learners' level [25]. All of the abovementioned issues associated with CBI surely complicate things even further when taking into account the fact that majority of activities in the ME textbooks are focused on accuracy and controlled (re)production of vocabulary and grammatical structures and only few activities engage learners into communicative and/or task-based activities which encourage them to make meaningful language choices relying on their own language resources in a context that has not been predefined. Thus, faced with specialized content-based materials and activities mainly focused on accuracy and usually strongly guided (re)production of language, learners are often left unable to express themselves fluently and communicatively both in speech and in writing.

24.7 Conclusion

Selection of adequate teaching activities in foreign language acquisition is an important issue which becomes even more complex in teaching ESP such as Maritime English. It is extremely important that EFL and ESP trainers are aware of the teaching potential of the materials and activities they use in their classroom in order to provide learners with a balanced approach which would enable them to acquire both language structures and communicative competence. IMO Model Course 3.17 provides useful guidelines and recommendations which promote the use of communicative approach and task-based learning in teaching Maritime English. Thus, this paper set out to investigate the typology of activities sampled from a selection of ME textbooks aimed at teaching Maritime English to marine engineers in order to determine how well they fit within the communicative and task-based framework. The results of this survey suggest that there is still a much stronger presence of more or less strongly guided (re)productive activities that focus on accuracy, vocabulary, grammatical structures, and reading comprehension than those focused on meaningful use of language and development of communicative competence. A number of task-based and partially communicative activities were identified, but they only accounted for a small number of samples and seemingly not developed to their full communicative potential. Although reproductive, closed, and strongly guided activities are an indispensable element in successful achievement of accuracy and proficiency in a foreign language, it is also important to balance those activities with the ones that develop learners' communicative competence.

References

1. D. Ausubel, The facilitation of meaningful verbal learning in the classroom. Educ. Psychol. **12**, 162–178 (1977)
2. M.S. Berns, Functional approaches to language and language teaching: Another look. Stud. Language Learning. **4**, 4–22 (1983)

3. A. Bocanegra-Valle, Maritime English. (2013). https://doi.org/10.1002/9781405198431.wbeal0746
4. M. Breen, C.N. Candlin, The essentials of a communicative curriculum in language teaching. Appl. Linguis. **1**, 89–112 (1980). https://doi.org/10.1093/applin/1.2.89
5. M. Byram (ed.), *Routledge Encyclopedia of Language Teaching and Learning* (Taylor & Francis, London, 2017)
6. F. Cerezal, Foreign language teaching methods, in *A Handbook for TEFL*, ed. by N. McLaren, D. Madrid, (Marfil, Alcoy, 1996), pp. 161–185
7. N. Chomsky, *Aspects of the Theory of Syntax* (MIT Press, Cambridge, MA, 1957)
8. C. Cole, P. Trenkner, Whither Maritime English? – 2012, in *Proceedings of IMEC 24, The International Maritime English Conference 1st- 4th OCTOBER 2012*, (Myanmar Maritime University & Uniteam Marine, Myanmar, 2012), pp. 3–18
9. F.I. Craik, R.S. Lockhart, Levels of processing: A framework for memory research. J. Verbal Learn. Verbal Behav. **11**, 671–684 (1972)
10. E. De Corte, C.T. Geerligs, N.A.J. Lagerweij, J.J. Peters, R. Vandenberghe, *Beknopte Didaxologie*, 5th edn. (Wolters-Noordhoff, Gröningen, 1981)
11. R. Ellis, *Reflections on Task-Based Language Teaching* (Multilingual Matters, Bristol, 2018)
12. J.R. Firth, *Papers in Linguistics* (Oxford University Press, London/New York, 1957)
13. J. Gris Roca, Categorization of activities used in English as a foreign language: A corpus based study with pedagogical materials. Procedia - Social and Behavioural Sciences **198**, 165–173 (2015)
14. D.H. Hymes, On communicative competence, in *Sociolinguistics. Selected Readings*, ed. by J. B. Pride, J. Holmes, (Penguin, Harmondsworth, 1972), pp. 269–293
15. E. Iakovaki: English in Maritime Educational system: The Greek Example., https://hellanicus.lib.aegean.gr/bitstream/handle/11610/18250/English%20in%20Maritime%20Educational%20System.pdf?sequence=1&isAllowed=y (2018)
16. IMO Model Course 3.17, *Maritime English*, 2015th edn. (International Maritime Organization. International Maritime Organization, London, 2015)
17. A. Littlejohn, The analysis of language teaching materials: Inside the Trojan horse, in *Materials Development in Language Teaching*, ed. by B. Tomlinson, (Cambridge University Press, Cambridge, 2011), pp. 179–211
18. W. Littlewood, The task-based approach: Some questions and suggestions. ELT J. **58** (2004). https://doi.org/10.1093/elt/58.4.319
19. J. Luzer, A. Spincic, *Reports on Level of Competence in Communication. Marcom Project, European Commission Research Programme into Waterborne Transport, Institute für Sichercheitstechnik/ Verkehrssichercheite. V(ISV) im Technologiezentrum, Warnemünde* (Final document. Part of project Level of Competence in Communication Among Marine Engineering Officers, Workpackage 5.2.1/2, 1998)
20. Ludescher, F.: Communicative Language Teaching, http://www2.vobs.at/ludescher/Alternative%20methods/alternative%20methods.index.htm
21. G. Neuner, M. Krüger, U. Grewer, *Übungstypologie zum kommunikativen Deutschunterricht* (Langenscheidt, Berlin, 1985)
22. Nunan, D.: Communicative Tasks and the Language Curriculum. TESOL Quarterly 25, 2, pp. 279–295. (1991)
23. Z.J. Pudlowski, Visual communication via drawings and diagrams. Int. J. Appl. Eng. Educ. **4**(4), 301–315 (1988)
24. J.C. Richards, *Communicative Language Teaching Today* (Cambridge University Press, 2006)
25. J.C. Richards, T.S. Rodgers, *Approaches and Methods in Language Teaching: A Description and Analysis* (Cambridge University Press, Cambridge, 1986/2001)
26. P. Skehan, *A Cognitive Approach to Language Learning* (Oxford University Press, Oxford, 1998)
27. D.A. Wilkins, *Grammatical, Situational and Notional Syllabuses* (Julius Gross Verlag, Heidelberg, 1972)

Part V
Voice of Seafarers

Chapter 25
What Seafarers Say

Quentin N. Cox

25.1 Introduction

The objective of this paper is to give an insight into the mindset of seafarers with regard to recruitment and retention. There were 22 interviewees in all, but there is no value in extrapolating the findings, for example, suggesting that these views represent a majority of seafarers. These are merely examples of responses. For that very reason, a series of caveats should be established with regard to the views expressed in this paper. All participants were willing volunteers, the significance of this statement being that there were many delegates of the same human factor training courses from which these participants were sourced, who chose not to participate in this study. Their views were not garnered. The participants were professional enough to be interviewed outside working hours, whilst at college, which is to their credit. They were each motivated enough to give up their own recreational time for the purposes of this study. Purely as an observation, they all came across as serious about their careers, which is in keeping with their willingness to participate in the first place and for which they deserve my thanks at very least.

Q. N. Cox (✉)
Teekay Shipping, Glasgow, UK
e-mail: Quentin.cox@teekay.com

S. Bauk, S. D. Ilčev (eds.), *The 1st International Conference on Maritime Education and Development*, https://doi.org/10.1007/978-3-030-64088-0_25

25.2 Almost All Had Close Family Connections Already in the Merchant Navy or, If Not, the Offshore Industry

This connection gave the participants an indication of what the career entailed, especially in terms of working rotas and time spent away from home. Each of the research participants had served for at least 4 years in the industry so at the time of the interview were by no means looking toward the end of their seagoing careers, more the middle years. There was an implicit family expectation on the participant to at least start a career in the MN, and given the fact they were still in the industry after a number of years and in most cases after some negative experiences, it appeared they planned to stay for a while yet.

The fact that many of the participants spoke of a family connection already having worked in the maritime industry, was more than likely the result of the location of the research interviews, in Glasgow. However, Gekara [1] maintains most British seafarers come from existing traditional seafaring communities or families, a point which concurs with the observation of Barnett, Gatfield, Overgaard, Pekcan and Graveson [2, p. 134]. Many participants grew up in small fishing villages where generations of different families lived closely together. This meant that "news" spread quickly. On a more contemporary level, without breaching the realms of social science, it would be fair to say that "gossip" spreads quickly [3, p.149]. A number of participants declared clearly that such an environment was not one they felt comfortable within, and perhaps their aversion to this helped motivate them to find gainful employment in the wider world. These factors may have influenced the data provided by the participants purely due to the convenience sampling method used, namely, a nested group of participants each having undertaken Human Element Leadership and Management (HELM) training all at the same maritime educational institution in Glasgow. This HELM training is the UK's manifestation of the management level of training mandated by the IMO, via the 2010 Manila Amendments. "The characteristics of the STCW Convention adopt that the human element is of a crucial significance in any institution and that education and training are vital to improve the skills and competencies of the human element" [4, p.282].

Not all participants were from the same background, but those that were from similar backgrounds told similar stories. Aligned to the objectives of the study, there were also several participants from different countries [UK, India, Greece, Philippines], with vastly different backgrounds which therefore lent a variety of different cultural values to the study. Five of the 22 participants were women, so some gender diversity was involved.

These participants may also have witnessed the benefits associated with the career, including long leave periods. However, it was a common observation that the MN was a good career up until the point the participant wanted to start a family. It is not uncommon for spouses of seafarers to suffer from the long absences imposed by one partner being away from home on a ship for weeks, if not months, at a time. Divorces are not uncommon [5, p. 68], and indeed one of the participants had witnessed his parents' divorce at a young age, and he claimed his were not the only parents he knew of who had separated.

Perhaps that had an impact on the participant's aspirations to leave the seagoing career in order to retrieve a failing relationship. A number of participants emphasised the importance of internet accessibility on board ship. This claim was largely on the basis of communication with home and spouses, rather than just the opportunity to "surf" the net. The implication from some participants was that divorce rates would drop as a result of more frequent communication.

> "We are getting a bit better at that at sea because [communications] are a lot better now so people are speaking to home a lot more. Especially now everyone's sSkyping or getting more time".

Recruitment and Retention Point Far from nepotism, family or social connections from within the industry prepare, to a degree, the mindset of new joiners. So, these inductees may be more prepared than others, though that is no guarantee and of course that should not dictate a recruitment policy or obviate the recruitment of any other candidate. The mention of internet access is one that repeatedly features in seafarer satisfaction surveys, so this requires no further exploration here.

25.3 No Participant Indicated That a Career in the Merchant Navy Was Their First Choice of Career

Aspirations varied between military, air force, police detective or simply going to a university before having to make their mind up. This appeared to be in part because they harboured more ambitious careers whilst at school. Whilst they knew of close relatives in the MN, they did not necessarily discuss it as a career choice whilst at school. It was only one's choices had to be made, that they were advised in more detail of the pros and cons of an MN career, from these relatives or connections.

Yet those participants interviewed found satisfaction in their work, perhaps because the elements of the careers they once aspired to had been replicated, to some degree, by their roles on board within their current MN career. The elements which they found satisfying, such as travel and the taking on of responsibility, formed part of the working environment to which they had originally aspired. Most significant was the avoidance of a 9 to 5 job in the same location they grew up in.

Another common response was that "I didn't know exactly what I did want to do, I only knew what I didn't want to do". In each of these cases, the career they found themselves in at the time of the interview was the right one, despite being cognizant of the perils [6, p. 158] as far as they were concerned at the time.

At the time of the interview, few F participants indicated any intention of leaving the industry . This was perhaps not surprising when they had invested so much time, study and, in many cases, their own money in progressing through the ranks. Those that did mention future plans ashore suggested they'd prefer to remain in the industry but work in shore-based offices of shipping companies, rather than change careers altogether. Certain participants chose university as a matter of course, as soon as they had completed their schooling, with the intention of maximising their earning potential [7, p.2].

Another not unique problem encountered, as certain participants pursued an alternative career path, was that their university studies did not go according to plan. This appeared to hasten their change of direction and certainly presented them with pressing decisions.

> All the way in school I never really considered the Merchant Navy at all, it didn't even cross my mind, despite the fact that on both sides of my family a few of my relatives had been fishermen and Merchant seafarers. I finished school and went off to university and did three years of biology at Edinburgh University. Didn't get good enough grades to stay on the course so I had to find something to do with myself without.

There are two possible interpretations to be made when considering the factor of graduates or undergraduates (not finishing their degree course). Firstly, the participants realised by the time they'd given the MN career a chance, they were of an age that a definitive choice had to be made, or they would risk never holding down a job. Secondly, these individuals are often cited as being easier to train, since they have been used to absorbing knowledge. Yet even this characteristic could be double-edged, if they are old enough to have become set in their ways.

"Yes again it's up to the individual but sometimes it does [make them easier to train]. Yes, of course it does because they're more settled down. Whereas at other times it doesn't because they're more set in their ways".

Recruitment and Retention Point Even candidates with close family connections do not often cite the merchant navy as their first choice of career. In many cases, the option is not even recognised, yet many sturdy and reliable officers have been nurtured from such beginnings. However, this should not absolve potential candidates from researching information about prospective employers.

25.4 Having Started Their First Trip at Sea, Expectations Were Not Always Met

The first trip away to sea is a traumatic experience for anyone, at any age. Even those who started in the MN at a later age, perhaps late 20s, were apprehensive. Most participants who had stared in a different career or who had started a university course prior to joining the MN, unusual as rare as this was, agreed that whilst they were still apprehensive their earlier experiences had helped them. This was because their spell at university or in other employment, was beneficial in terms of developing non-technical skills, such as communication skills or personal etiquette.

Reaction to the duties given on the first trip varied, though the sense of apprehension was almost ubiquitous and entirely to be expected. It was the worst of both worlds; they would worry about the technicalities of the job. It was not just the work in their particular department that required familiarisation but the ship as an entity, even something as fundamental as how to find their way around. This would have been especially difficult on a cruise liner where the accessible parts of ship are extensive. On a cargo ship, most of the deck area covers the cargo area, so there's only an accommodation block to get used to, which would be much smaller. Then there were the non-technical skills, how to speak to others and how to treat them.

> "I think when you first go on board you're just totally intimidated because you don't know how anything works, you don't know how to behave.
> "Yes, especially the first one [trip]. I would say the toughest for me honestly speaking I came from a background where I never picked up a broom at home".

Some were puzzled at the contrast between their pre-sea study and the first few weeks on board. Their studies had covered, perhaps not surprisingly, many of the technical details of their work including the operation of equipment, both bridge and cargo equipment for deck officers and engine room equipment for engineer officers. Yet on board they would often be given menial tasks, such as cleaning, chipping and painting.

> No it wasn't what I expected. I'm sure it's an age-old cliché, from the cadets going into the classroom, you're doing all these theoretical classes, learning advanced mathematics and dynamics. Then you go onto the vessel and all of a sudden you're pumping bilges, you're starting pumps, you're stopping pumps. You kind of feel like it's slightly different
>
> .

> You just take it as it comes. I remember the first time on board, the ship was massive and I was just walking around and I was thinking that there was no way I was going to find my around this ship, never mind do the job.

This concern actually raised an important issue related to their training – just about all cadets given a training task book before they join their first ship. During their cadetship they are expected to have technical tasks "signed off", to ensure they are competent in that task. There may be two or three signatures from a qualified officer required for each task. Yet many cadets are given very little help to complete their task book by the senior officers.

> Nobody would teach you anything. You were there as just another person on board. "Go and work on deck". "But I've got to do this in my book". "Go and work on deck and do it later." By the time later comes he's signed off and then you'd get the next guy.

Even when cadets did receive assistance toward their progress, such as senior officers following training plans and tasks, the career did not necessarily give them what they might have expected before they joined.

> And the phrase was always, superstar wages, superstar wages. I should buy this car, you'll get all the most attractive women that you wish for. For the first two years of my career after I qualified it was just a massive reality shock. This is not the money I was expecting and it was certainly not the life I was expecting.

> Honestly my brother was against me going into the Merchant Navy because of what he'd seen [The participant's brother had been in the Merchant Navy]. It's a tough life, you have to all understand it because the initial perception of the Merchant Navy is that it's still very lucrative. From the outside people see the money and they see the travelling everywhere across the world but they don't realise what's actually behind it. It's not all those things, it's the work that you do on board.

> And when I landed and the training was done in such a way that they made you feel that you were an officer. And you should go on board and behave like an officer. But when you actually go on board it's a totally different thing.

In some cases there was a tangibly hostile atmosphere on board, which meant the first trip was even more challenging than they might have expected.

> it was actually quite negative because the guys on board at the time, they didn't work very well together. They were always drinking and fighting with each other and I just thought, this really isn't for me. Plus, even when I was trying to help some of these guys, they'd always say, "What the 'expletive' are you doing?" basically, so I thought this is not for me.

Another not uncommon observation was from a participant whose early experiences on board were not good. He was unable to make much progress on his training record book and was largely used merely as an extra pair of hands. However, this did not diminish his resolve.

> I think apart from that I just kind of told myself that you're in that situation and there's no point in dwelling on it, just of more ways how you're going to do it. And that's kind of what happened. It wasn't as bad as I thought it was going to be.

A lesson to draw from this, in terms of my own analysis, is that if new joiners are treated badly, they will leave. Bearing in mind the cost of training a junior officer [8, p.142], employers may wish to reflect on this. It cannot be taken for granted that new joiners will endure the hardships and stay [9 p 91]. As has been made clear earlier, I happened to be speaking to 22 young officers who did endure the hard times and have since enjoyed a more tolerable future. The study did not scrutinise groups of former junior officers or apprentices whom have left the industry in light of their difficult experiences. The work of Elizabeth Gould [10] speaking to cadets illustrates this all too clearly, as well as that of Victor Geraka [1, p.224]. It was not all doom and gloom. There were participants who opined how favourable early stages were.

> It surprised me by the point of view that I was expecting more as concerned about the Masters and Chief Officers to be more strict. But all of them were very friendly. The climate and the atmosphere on the vessel was very friendly. So that was the only thing that surprised me because I had an idea from my father and some years ago I joined the vessel to visit my father so I saw the vessel when I visited there, so I had an idea about that.

The above citation may seem incongruent to the views of many counterparts, so it should be remembered that it came from a Greek officer, who had been to a university, served in National Service and had a short spell employed in retail prior to joining his first ship. The life skills gained during these periods, especially during the militaristic spell in National Service, perhaps shaped some of his views. On that basis it is perhaps little wonder that he found the merchant navy senior staff not quite strict as the senior staff in a military domain.

Reassuringly, not all British starters endured such trying environments. This one did have to deal with heavy weather, which he described as "interesting" but he appeared to keep the long-term benefits of the career in mind and put a positive light on his early experiences.

> When everyone sees a ship they think it's high tech and interesting as it is. I knew offshore was high tech because of the challenges I had to face, the weather, the high tech instruments they use, the gear like that. And I knew that the guys were highly paid as well, which made sense because they were carrying out hard stuff. When I went on board it was the first time I'd ever been on a ship. So that was interesting, out of Peterhead [North East Scotland] in a gale. So that was interesting but I took to it straight way. I'm not sure if my expectations were met because I didn't really know what to expect but I enjoyed every bit of it. Yes, it was good.

Recruitment and Retention Point It is easy for shipowners and operators to effectively forget what is happening on their ships and look only at the finances of operating it. The contract by contract nature of some ship management employment terms does not necessarily lend themselves to close scrutiny of the shipboard environment. Some agencies agree to offer cadets sea time on ships owned by enterprises with whom they have a tenuous connection. In such cases, neither the shipowner nor the agency appears willing to accept the cadet as an employee, leading to an implicit (if not legal) absolution of responsibility. It should not be forgotten by either that careful nurturing of these cadets could produce a long-term and reliable asset. The accounts of certain participants suggest these matters have at times been neglected by those enterprises who should take responsibility.

25.5 Going Away to Sea Changed the Mindset Somewhat

Another common factor observed by many seafarers is how their early experiences, if not too intolerable, help mature them as individuals. This became clear at an early stage of their seagoing careers, comparing themselves with their former peers. A number of participants talked of how they felt they had matured in a relatively short space of time and articulated this specifically. Others talked of being able to grow in confidence which is closely related in terms of being willing to take on more responsibility. The benchmark coted was their "drinking pals" from their home town. The change was commonly explained in a kind of "before and after" comparison. Before leaving home for their initial trips to sea, they felt they may have been very similar to their peers, which is understandable if they sent their teenage years together up to the point of leaving school or college and having similar experience of life. The day-to-day existence and experiences of the peers of the participants continued in much the same way as they had previously . However, the experiences of the participants themselves grew increasingly different, mixing and interacting with older and more experienced seafarers, gaining confidence as they learned on the job not just routine menial tasks but more complex operations dealing with cargo and navigation or engineering. The overall impression of this sudden change in environment largely contributed to what the participants interpreted as an increase in confidence. The "after" scenario was going home back and interacting on a social basis with their former peers who had gained very little in terms of experience, confidence or maturation, by comparison.

> There was a fairly major difference and I used to sit down with friends my age, schoolmates – I couldn't talk to them anymore because the level of, I'd say, maturity kind of changed. The way I looked at things, the way they looked at things totally different. I was on a more safe way of life. At that age everybody wants to go wild and do crazy things but my perspective on life was that I wanted to stay safe, to be on the back foot and watch.

> And I came back after my first trip and I'd been working and drinking with older people. And I came home at the time and my friends were all… I couldn't believe the difference. I'd matured beyond my years and they were all doing the same thing.

> You worry that you won't get on well with folk, I suppose, you're going to really miss being away, are you going to be able to do the job itself. As the weeks went by, you got to know everybody and you're getting along well with everyone. You start going ashore, you do the job. You have a bit more responsibility and start to grow in confidence.

> Naturally as you progress through your cadetship, you're given more and more responsibility and learning more and more. You get a better idea of what's actually going on.

For what it's worth, my own experiences echo these sentiments, though this paper is technically about the interpretations of the participants rather than my own opinions. It is also perhaps worth remembering that these are recollections of these participants a few years after the experiences of which they speak. They are benefiting from the wisdom of hindsight, and as explained earlier, they are commonly thinking of periods of apprehension and tension. In most cases, the participants had persevered thorugh the difficult periods and were rewarded with working environments which they found more enjoyablke and during which their career progression appeared brighter.

Other than the first citation above, which was from an Indian participant, the remainder were from Scottish seafarers, generally from small coastal towns. To put a more culturally diverse perspective on the issue, the Greek participant who, remember, served in National Service [11 p 72], as well as a spell at a university and in retail, had this to say about the transition:

> I mean the people from the village, from the countryside usually are – how can I say that – they are not used to be very socialised. You know people from the town and all this. I am from the village in Greece but I live in the city, so I can make the comparison with. Of course, in the village they are more friendly but they are more friendly if they know you. It's more a closed society but when someone's coming from outside they are always coming back. They say, 'Who is he now, what is he doing? Is he here to do something bad to me?

Recruitment and Retention Point A good sign that cadets are adapting to life at sea is this kind of reaction. It may not be easy to garner at the time, but as a reflection it may be a useful pointer. It also highlights the benefit of maintaining a personal type of contact with young officers and not just treating them as a plug in a hole.

25.6 Likes and Dislikes

To round off these interviews, participants were asked what they liked and disliked about their careers so far and the work involved. The responses were so similar it is hardly worth attributing the comments to any particular participant.

The responses can be listed as follows:

Favourable:
- Long leave periods at home with friends and family.
- Generally, above-average salaries.
- Responsibility that comes with the work especially at junior levels.
- Diversity of colleagues.
- Travel.
- Clear career progression.

Unfavourable:
- Long periods away from home.
- Not being relieved on time.
- Loss of communication with home.

Of course, the participants involved in this study did find "better times ahead" which enabled them to peruse their ambitions within the industry. What this paper does not take into account are the individuals who found their daily life so hard to tolerate that they did leave their career.

Slišković and Penezić [12] refer to this as:

> the "healthy worker effect", which is explained by self-selection and adaptation: seafarers who cannot adapt to work on board, as well as those who suffer from the effects of occupational stressors, leave the occupation.

This should be taken into account so that the data does not become biased in favour of those that did remain in their career. Invariably it was the prospect of "better times ahead" and the existing favourable factors of the work that motivated the participants. By no means all complained of having to endure harder times earlier in their careers. Several participants envisioned a clear career path, and again this is not a unique discovery [2, p.134].

One very common response, in these days of advancing technology, was the expectation that ships transiting the oceans would have reliable internet access. In some cases, according to the participants, this factor alone could be the difference between a participant deciding between one prospective employer and another. Existing research into these issues indicates that internet access is a common concern amongst seafarers, for example:

> The analyses of the results have shown that respondents perceive the separation from family and the inadequate communication with family and friends while at sea as the most important issues hindering retention in the seafaring profession. [13, p.159].

> "Seafarers were also asked if there were any facilities they would like onboard that were not currently available. By far the most frequent answer, suggested by 66% of seafarers, was that they would like access to the internet/Wi-Fi onboard". "When seafarers were asked if they had internet access onboard, nearly two thirds (61%) indicated that they had no internet access at all". [14, p.105].

The use of internet at sea provides another double-edged sword since it has been observed that "The result is that seafarers isolate themselves in their cabins, rather than interacting with their co-workers" (Carol-Dekker 2017, p. 166) though this is another debate altogether.

Recruitment and Retention Point These findings are not especially different from existing research, which bothers to seek the opinion of actual seafarers, yet it should be incumbent on prospective employers to point each of these factors out to candidates. No job or career is perfect, but if reasonable expectations are made clear at the outset, then both parties will be better equipped to cope with the ups and downs.

References

1. V. Gekara, Understanding attrition in UK maritime education and training. Glob. Soc. Educ. **7**(2), 217–232 (2009)
2. M. Barnett, D. Gatfield, B. Overgaard, C. Pekcan, A. Graveson, Barriers to progress or windows of opportunity? A study in career path mapping in the maritime industries. WMU J. Marit. Aff. **5**(2), 127–142 (2006)
3. T. Kohn, Becoming an islander through action in the Scottish Hebrides. J. R. Anthropol. Inst. **8**(1), 143–158 (2002)
4. M. El Ashmawy, *The maritime industry and the human element phenomenon* (Proc. The 13th Annual General Assembly of the IAMU, 2012)
5. L. Wu: Sailing on a Neoliberal Sea: Multinational Seafarers on Container Ships (Doctoral dissertation, Chinese University of Hong Kong) (2016)
6. M. Lipowski, M. Lipowska, A. Peplińska, M. Jeżewska, Personality determinants of health behaviours of merchant navy officers. Int. Marit. Health **65**(3), 158–165 (2014)
7. A. Chevalier, G. Conlon, *Does it Pay to Attend a Prestigious University?* (IZA Discussion Paper No. 848. Available at SSRN: https://ssrn.com/abstract=435300, 2003)
8. L.D. Caesar, S. Cahoon, J. Fei, Exploring the range of retention issues for seafarers in global shipping: Opportunities for further research. WMU J. Marit. Aff. **14**(1), 141–157 (2015)
9. Cahoon, S., Caesar, L. and Fei J, Human Resource Practices in Seafaring: Opportunities for Improving Retention (2014)
10. E.A. Gould, *Towards a Total Occupation: A Study of UK Merchant Navy Officer Cadetship* (Cardiff University, 2010)
11. D. Tarabar, J.C. Hall, Explaining the worldwide decline in the length of mandatory military service, 1970–2010. Public Choice **168**(1–2), 55–74 (2016)
12. A. Slišković, Z. Penezić, Testing the associations between different aspects of seafarers' employment contract and on-board internet access and their job and life satisfaction and health. Arch. Ind. Hyg. Toxicol. **67**(4), 351–363 (2016)
13. A. Papachristou, D. Stantchev, I. Theotokas, The role of communication to the retention of seafarers in the profession. WMU J. Marit. Aff. **14**(1), 159–176 (2015)
14. N. Ellis, H. Sampson, Corporate social responsibility and the quality of seafarers accommodation and recreational facilities, in *IMISS2013–Proceedings of the International Scientific Meeting for Corporate Social Responsibility (CSR) in Shipping*, (2013), p. 91
15. H. Dickey, V. Watson, A. Zangelidis, Job satisfaction and quit intentions of offshore workers in the UK North Sea oil and gas industry. Scottish J. Polit. Econ. **58**(5), 607–633 (2011)

Part VI
Business vs Sustainability

Chapter 26
An Economic Development Policy Implementation Approach to Creation of Sustainable Maritime SMMEs in the KwaZulu-Natal Province

Bheka Clive Zulu

26.1 Introduction

This paper explores the concept of the ocean economy policy framework for SMME development relevant within the maritime sector in KZN province. The concept of the special-purpose vehicle (SPV), its relevance and possible applications within the blue economy context will all also be explored in this section.

The blue economy, also called ocean economy, has gained global prominence in both developed and developing countries due to its significance in promoting sustainable economic development and growth [16]. The ocean economy is also referred to as the blue economy hereafter. According to Pauli [18], the blue economy is a sustainable economic development model which aims to produce 100 million jobs globally, providing significant opportunities for businesses linked to the sea and shipping industry, port transport and logistics and marine activities. In the African context, the blue economy is based on the African Union's (AU) Agenda 2063, and the AU's 2050 Africa's Integrated Maritime Strategy (2050 AIM Strategy) presents significant economic development contribution through SMME's development and sustainability [25]. The 2050 AIM Strategy focuses on the use of Africa's ocean economies to foster inclusive growth for the maritime sector and create sustainable economic development using oceans and sustainable maritime industries, respectively. Arguably, the ocean economy is high on the South Africa's economic growth agenda, as is on the African Continent's Agenda.

South Africa's blue economy's interests range over 3000-km coastline across the Indian, Atlantic and Southern Oceans with extensive offshore interests with the jurisdiction of an exclusive economic zone (EEZ) ranging over 1.5 million km^2 of ocean space [20]. Given the latter, the maritime sector has sizeable potential

B. C. Zulu (✉)
Maritime Studies, Durban University of Technology, Durban, South Africa

S. Bauk, S. D. Ilčev (eds.), *The 1st International Conference on Maritime Education and Development*, https://doi.org/10.1007/978-3-030-64088-0_26

contribution to the South African economy. In 2013, the sector contribution to the GDP increased to 50%, and the sector employment reached 800,000, and sector investments by Transnet alone totalled R11.2 billion, most of it aimed at the creation of new ocean economy SMMEs and sustainable SMME development. However, despite the notable economic contribution, the sector remains at nascent level, is fragmented and under-resourced and continues to operate in a virtual vacuum since democratic dispensation in 1994 [5, 9]. In addressing the sector's deficiencies, the government has prioritised strengthening the sector's economic contribution through the country's Operation Phakisa launched in 2014. Operation Phakisa aims to within the maritime context aims to promote entrepreneurship development and creation of maritime SMMEs and is regarded as high-impact sector with a potential to contribute to the country's economic development.

Globally, the SMME sector remains significant towards national output, and combating inequality, unemployment and poverty and through creation of job opportunities, reduction of poverty and income inequalities in several economies (Organisation for Economic Co-operation and Development [17]). In the South African context, the government has initiated several policy initiatives such as the National Development Plan (NDP) which aims to develop the SMME sector to create 90% of 11 million jobs needed in South Africa by 2030 and to increase job creation from 13 million to 24 million and the GDP by 5.4% [15], thus reducing the unemployment rate to 6% (Department of Small Business Development) [6]. However, business environment specialists SBP [21] also underscores that despite government's commitment through policy interventions aimed at promoting entrepreneurship, SMMEs still encounter sustainability challenges.

The KwaZulu-Natal (KZN) province through its 2030 visions aims to be a prosperous, skilled province acting as a gateway to the African continent and the world [12]. Several economic policy frameworks have been enacted towards the realisation of the province's economic development; these include the Provincial Growth and Development Strategy (PGDS) and the Provincial Growth and Development Plan (PGDP) which are aligned with the Millennium Development Goals (MDGs), the National Development Plan, the Nine-Point Plan and the Medium-Term Strategic Framework (TSF) (Provincial Planning Commission, 2012). The KZN Integrated Maritime Strategy (KIMS) focuses on stimulating the ocean economy in the province.

However, the traditional form of government interventions is unlikely to fulfil the economic development role; as such special-purpose vehicles (SPVs) have become a "third way" of development, an approach that involves a process of "guided development". The key mechanism for development is the use of special-purpose vehicles (SPVs) that represent the partnerships between industry, academia, government and labour. The KwaZulu-Natal (KZN) province has also identified special-purpose vehicles (SPVs), concurrently with the utilising tools such as clustering, incubators, value chain development, competitiveness research and benchmarking, to analyse and develop priority sectors in the province.

26.2 Purpose of the Study

The aim of the study is to establish an economic development and policy implementation approach for the creation of sustainable maritime SMMEs in the KZN province.

Despite many elaborate government programmes, economic and policy frameworks that have been put in place to assist with small business development, SMMEs in the country continue to experience a high failure rate and stagnancy. Literature cites that SMMEs continue to lack developmental instruments to expand entrepreneurship; currently, a mismatch exists between the various development models of the SMME sector used by South African policymakers and its reality. Addressing these challenges will require researchers, practitioners, government institutions and all relevant stakeholders to start taking a holistic approach. This is because everyone must realise that SMMEs exist in an ecosystem and operate within a particular framework. There is a need for a holistic approach to the industry needs in the form of a one stop SPV, if proposed solutions are to be effective [8].

When effectively crafted, SPVs can be a very effective conduit to deal with system gaps. Current literature shows little or no information on the current existence of any SPV model or structure within the South African ocean economy and specifically in KZN. Also, the similar current models in other sectors countrywide, such the financial sector, have not been tested on the emerging and developing countries like South Africa. In addition, current studies on special-purpose vehicles have been focusing on estate and venture capital financing. The concept has not been explored within the maritime studies and emerging context.

The study has the following objectives:

- To determine the challenges faced by SMMEs within the ocean economy in KZN province.
- To explore the impact of economic policy on the development and sustainability of maritime SMMEs in the KZN province.
- To identify the need and explore the concept for a sector-specific SPV for the development and sustainability of ocean economy SMMEs in KZN province.
- To propose an SPV model for the economic development of KZN province's ocean economy SMMEs.

This study aims to answer the following questions:

- What are the challenges faced by SMMEs within the ocean economy in KZN province?
- What is the impact of economic policies and interventions aimed at the development and sustainability of maritime SMMEs in the KZN province?
- What are the existing economic models aimed at the development and sustainability of maritime SMMEs in KZN province?
- What form of an SPV (model) that can be suitably developed as an economic model for KZN province ocean economy SMMEs?

26.3 Conceptual/Theoretical Background or Framework

This paper explores the current architecture for blue economy and its potential for South African economic development, particularly in the context of the maritime sector impacts and SMME development. It further analyses the opportunities and gaps in existing maritime project financing and policy frameworks at the national, regional and sector levels and offers an alternative framework for the maritime sector in the form of special-purpose vehicle model.

The ocean is globally considered as a new economic frontier generating activities worth USD1.5 trillion and providing 31 million jobs in all the ocean sectors (Economist Intelligence Unit, 2015; [16]). The ocean's value is estimated to reach USD3 trillion and create 40 million jobs, expected to triple by 2050 [16].

The blue ocean economy, an alternative development paradigm, is the most dynamic driver of economic development in the maritime economy and embraces economic growth, environmental and social benefits [1]. The concept conceptualises the oceans as shared development spaces. Within the African context, the continent has 13 million km2 of collective exclusive economic zones (EEZs) and a coastline of over 47,000 km (African Union, 2012). The economic value for maritime-related activities in Africa stands at US$1.5 trillion and is forecast to reach US$3.5 trillion in 2020 (Kenya, 2018; World Bank, 2017).

South Africa launched its own version of ocean economy strategy called Operation Phakisa in 2014 focusing on priority maritime areas such as aquaculture, ocean governance and marine protection services and marine transport and manufacturing. The Operation Phakisa since its inception has secured R28.7 billion in infrastructure investments in ports, aquaculture and marine manufacturing, creating 437,694 jobs in the maritime sector. The South African government has committed over US$3bn for expansionary projects and expected to US$14bn. Undoubtedly, the maritime sector has potential in contributing to the country's economic growth, enterprise development and employment creation . Hence, the maritime sector has the potential to act as a catalyst for economic development.

However, considering the level of investment that will be needed to achieve these objectives, the maritime sector ought to find new and creative ways to finance investments in the ocean economy. New innovative ways and paradigm shift are needed to explore both available and optimal use of public-private partnerships (PPP) for blue economy initiatives for entrepreneurship development and sustainability. Public-private partnership (PPP) can be constructed through a special-purpose vehicle (SPV), and if properly crafted, an SPV can address existing system gaps. An existing literature shows that there exist little or no information on the current existence of any SPV model or structure within the South African ocean economy and specifically in KZN. However, current studies on special-purpose vehicles have been focusing on estate and venture capital financing. The concept has not been explored within the maritime studies and emerging context.

26.4 Methods/Techniques

The study will be conducted in KZN province, within eThekwini Municipality. According to Sekaran and Bougie [22], population entails "the entire collection of individuals, items, cases, things or events of importance under evaluation when conducting a study". Babbie and Mouton ([2]:100) define population as "that group (usually of people) about whom we want to draw conclusions". There are currently 200 SMMEs that are actively participating in ocean economy that are listed on the Moses Kotane Institute database. The population chosen will comprise of all the SMMEs, participating in the ocean economy, that are on the database of the Moses Kotane Institute. In other words, the population of the study will comprise of a total of 200 SMMEs, and the population chosen will comprise entirely of all the SMMEs in the ocean economy.

The sample of the study is defined by Newman [14:240] as "a small set of cases a researcher selects from a large pool and generalises to the population". The sample is part of the defined population, and based on this definition, it follows that the sample of the study will be smaller than the actual population. According to Sekaran and Bougie ([24]:50), a population size of 200 elements requires a sample size of approximately 132 elements. Therefore, in this study, the researcher would therefore target 132 SMME owners/entrepreneurs as sample of all SMMEs in the Moses Kotane Institute that are actively participating in the ocean/blue economy. The inclusion criteria would be SMME owners who have been in business within ocean economy from 1 year onwards and who live in KwaZulu-Natal. The author believes that this portion taken from the population is assumed or considered to be representative of the entire population [23].

This research study adopts pragmatist philosophical approach and mixed-method case study in order to achieve its objectives and uses a mixed-method research design which is rooted in the pragmatism paradigm. The pragmatist paradigm was deemed suitable in this study because it tends to include both qualitative and quantitative research, providing a more comprehensive understanding of the research problem and in-depth understanding of the phenomena as well as various modes of analysis combined with a continuous cycle of abductive reasoning. Mitchell ([13]:108) cites that "pragmatism is an attractive philosophical partner for mixed methods" and provides superior research results [11, 19].

Furthermore, this research study adopts a sequential explanatory mixed-method case study strategy, and in applying the mixed-method methodology, the study will utilise both quantitative and qualitative data sequentially starting with quantitative study and then followed by qualitative study.

26.5 Conclusion

This paper explored the current architecture for blue economy and its potential for South African economic development, particularly in the context of the maritime sector impacts and SMME development. It further analyses the opportunities and gaps in existing maritime project financing and policy frameworks at the national, regional and sector levels and offers an alternative framework for the maritime sector in the form of special-purpose vehicle model.

References

1. V.N. Attri, N. Bohler-Muller (eds), *The Blue Economy Handbook of the Indian Ocean Region*. (Africa Institute of South Africa, Pretoria, 2018).
2. E. Babbie, J. Mouton, *The Practice of Social Research* (Oxford University Press, Cape Town, 2011).
3. H. Bhorat, C. Rooney, F. Steenkamp, *Understanding and Characterizing the Services Sector in South Africa: An overview* (2018). https://doi.org/10.1093/oso/9780198821885.003.0014
4. M.C. Cant, J.A. Wiid, C.M. Sephapo, Key factors influencing pricing strategies for SMEs: Are they important? J. Appl. Bus. Res **32**, 1737 (2016)
5. S. Chikunga. Retrieved from www.transport.gov.za/Home/DeputyMinisterofTransport.aspx (2016)
6. Department of Small Business Development, *Strategic Plan 2015 – 2019* (Republic of South Africa, 2018)
7. O. Fatoki, The causes of the failure of new small and medium enterprises in South Africa. Mediter. J. Soc. Sci. 5, (2014). https://doi.org/10.5901/mjss.2014.v5n20p922
8. J. S. Galawe: Endogenous and exogenous risk factors in the success of South Africa wiredspace.wits.ac.za/.../Jabu%20%20PhD%20Thesis%20Final%2031%20March%2020.. (2017)
9. Gasa, *South Africa Maritime Industry Conference* (SAMIC, Durban, 2014)
10. E. Lampadarios, *Critical Success Factors (CSFs) for Small Medium Enterprises (SMEs): An Empirical Study in the UK Chemical Distribution Industry. PhD Thesis* (LBU, Leeds, 2015)
11. G. Makombe, An expose of the relationship between paradigm, method and design in research. Qual. Rep. 22(12), 3363–3382, (2017). Retrieved from https://nsuworks.nova.edu/tqr/vol22
12. S.B. Mchunu, *Examining the Use of Systems Thinking Approach to School Development: A Case Study of Five Schools in the uMgungundlovu District* (Ukzn thesis, 2015)
13. A. Mitchell, A review of mixed methods, pragmatism and abduction techniques. Electron. J. Bus. Res. Methods 16(3), 103–116, (2018)
14. L.W. Neuman, *Social Research Methods: Qualitative and Quantitative Approaches*, 7th edn. (Pearson Education Inc., Boston, 2011)
15. NPC, *National Development Plan: Vision for 2030*. National Planning Commission, (2011). ISBN: 978-0- 621-40475-3
16. OECD, *The Ocean Economy in 2030* (OECD Publishing, Paris, 2016). https://doi.org/10.1787/9789264251724-en
17. OECD, *Enhancing the Contributions of SMEs in a Global and Digitalised Economy* (Meeting of the OECD Council at Ministerial Level, Paris, 2018)
18. G. Pauli, *The Blue Economy: 10 Years, 100 Innovations, 100 Million Jobs* (N.M., Paradigm Publications, Taos, 2010)

19. S. Rahi, Research design and methods: A systematic review of research paradigms, sampling issues and instruments development. Int. J. Econ. Manag. Sci. **6** (2017). https://doi.org/10.4172/2162-6359.1000403
20. SAMSA, *SA Maritime Industry Profile* (SAMSA, Pretoria, 2017)
21. SBP. http://www.sbp.org.za/uploads/media/SBP_Alert_-Examining_the_challenges_facing_small_businesses_in_SA_01.pdf (2017)
22. U. Sekaran, R. Bougie, *Research Methods for Business: A Skill-Building Approach*, 6th edn. (Wiley, New York, 2013)
23. U. Sekaran, R. Bougie, *Research Methods for Business: A Skill-Building Approach*, 6th edn. (Wiley, Haddington, 2014)
24. U. Sekaran, R. Bougie, *Research Methods for Business: A Skill-Building Approach*, 7th edn. (Wiley, Haddington, 2016)
25. T. Walker, *Reviving the AU's maritime strategy*, Policy Briefs, (8 February 2017). Available at https://issafrica.org/research/policy-brief/reviving-the-ausmaritime-strategy

Chapter 27
Comparing Carbon Footprint of Maritime Transport Routes on North Adriatic-South Africa Trade

Bojan Beškovnik and Elen Twrdy

27.1 Introduction

Sustainable transport and supply chains are one of the key directions of global international trade. Reducing greenhouse gas (GHG) emissions in all transport sectors is a globally set strategy that is implemented differently in the transport sectors. Thus, the produced GHG emissions per performed tonne kilometer vary by the transport industry and by the region where the transport service is organized. The International Maritime Organization (IMO) has set ambitious targets for reducing CO_2 emissions from maritime transport. The goal for international shipping is to reduce carbon footprint by 40% until 2030 and by 70% until 2050, compared to 2008 emissions, and by 50% of annual GHG emissions produced in 2008 [10].

Environmental awareness of all stakeholders along the transport chain remains an essential challenge for the entire transport industry [2]. The question of whether the transport industry does enough to inform decision-makers about the environmental impact of selected transport means appears as an operational issue. Planning and knowledge of the transport route and employed ships are especially important in maritime transport, which is usually used over longer transport distances. These elements directly affect the amount of GHG emissions produced [15]. Mathematical models are being developed for optimal service organization [7], but ship rotation must be understood and known by cargo owners and NVOCC. The further the transport route on the port-port line deviates from the optimal sea route, the more the carbon footprint increases, as well as energy consumption. The survey raises the question of whether the price of transport covers the extension of the transport route and the increase in emissions from transport and whether cargo owners and their

B. Beškovnik (✉) · E. Twrdy
Faculty of Maritime Studies and Transport, University of Ljubljana, Portorož, Slovenia
e-mail: bojan.beskovnik@fpp.uni-lj.si; elen.twrdy@fpp.uni-lj.si

S. Bauk, S. D. Ilčev (eds.), *The 1st International Conference on Maritime Education and Development*, https://doi.org/10.1007/978-3-030-64088-0_27

nominated NVOCC agents are informed of the projected carbon footprint of the container service.

The research of operational implementation, commercial aspect, and produced GHG emissions on the maritime transport route between the North Adriatic Port, Port of Koper, and the South African Port of Durban offers answers to the set research problem. Container shipowners (CL) offer different line services between ports, which use different numbers of transshipment ports and different sizes of container ships. A comparison of service data from the six most important CLs touching north Adriatic (NA) and offering weekly liner service to South Africa highlights the need for more comprehensive information to supply chain stakeholders to enable more efficient decision-making in the planning and implementation of green supply chains (GSC). From a scientific and applied point of view, the study deepens the need for a different approach to the evaluation of maritime container services, where it is necessary to put the energy efficiency (EE) and carbon footprint of individual port-to-port service in the foreground.

27.2 Research Background

The objective of reducing pollution in maritime transport is receiving increasing attention. Two separate approaches are most common, namely, the reduction of ship-source pollution during the implementation of primary transport (port-to-port) and the reduction of GHG emissions before and in the port area. Winnes et al. [17] emphasize that stricter policies and regulations on alternative fuels and ship forms are needed to reduce GHG emissions from ships in ports. Zis et al. [19] add that it is also necessary to regulate lower ship speeds in front of and inside ports and to provide a cold ironing procedure. With an onshore powering and ship-port interface, up to 4% reduction in CO_2 emissions from ship operation can be achieved [11]. Reducing the speed of ships is also an important element in reducing GHG emissions, but there are adverse effects on extending transport time, such as higher costs in supply chains [14]. In order to achieve the optimization of the implementation of maritime services, the IMO [9] adopted the Ship Energy Efficiency Management Plan (SEEMP) to use optimal sailing speed, higher EE, and safety. By reducing the sailing speeds of ships, up to 60% reduction in GHG emissions can be achieved; while using larger ships, the carbon footprint can be reduced by up to 30% [11].

Decisions on sailing speed are primarily in the domain of shipowners. Lindstad et al. [13] point out that their operational decisions are highly related to the functioning of the market. When and where a higher degree of demand is present, CLs have no interest in slowing down their ships, as they want to increase revenue, even if fuel costs are high. In times of low demand, however, by lowering the sailing speed, they lower the cost of motor fuel and reduce the available space on ships [16]. As a result, they reduce GHG emissions and increase the EE of transport by moved TEU. Cheaitou and Cariou [3] also point out the discrepancy between the economic and environmental approaches to the implementation of maritime transport. Shipowners can make a significant impact on GHG emissions by making proper operational decisions about vessel schedule and service rotation [6, 15]. The latter is particularly important in the container industry, as container ships generate up to 2.5 times the carbon footprint of general cargo ships [4].

Container liner shipping is characterized by predefined services, hub terminals and ship size, and a fairly closed model that changes only at longer intervals [15]. Such an approach makes it easier to compare different services on a particular port-port transport axis, as the data is available and valid for several weeks. In this way, cargo owners or the NVOCCs can set a decision model for choosing a container service based on price, transit time (TT), or GHG emissions. The emphasis on service selection in terms of lower emissions ensures the organization of GSC as part of green supply chain management (GSCM) [18], where green maritime supply chains pursue higher EE and lower GHG emissions from transport and manufacturing processes [12].

27.3 Research Approach

Data Obtaining Method and Selection

The survey includes data from six global CLs, which have the most important market share of container throughput in the Port of Koper and offer regular eastbound and westbound container services from the Adriatic: CMA-CGM, COSCO, Evergreen, Hapag, Maersk, and MSC. These CLs also offer weekly line connections from NA to South Africa. They have the option of touching the Port of Durban, and some also offer service to Port Elizabeth, Cape Town, and Johannesburg. The Port of Durban was chosen as a study case POD because all CLs involved in the research call it.

CLs and their agents were contacted directly with a request to present a comprehensive operational and commercial offer of maritime transport on the CFR Incoterms parity on the POL Koper-POD Durban route. CLs or their agents submitted electronic offers containing data about the sea freight charge, maritime surcharges, and costs in the Port of Koper, port-port TT, and planned transshipment ports, valid in June 2020. Data on the ship size employed on a single connection POL-hub-POD were not provided. The data were separately checked and analyzed from the available databases about world ship fleet characteristics. The TT offered were double-checked with the published official times on the vessel sailing schedules. Sea freight prices were also checked with the NVOCC agents in Durban, to compare prices and the scope of data offered. Data for each CL are presented by randomly numbered services to ensure the confidentiality of data, following the given request of CL or their agents (*Service 1* S1, *Service 2* S2, *Service 3* S3, *Service 4* S4, *Service 5* S5, and *Service 6* S6).

Applied Tools and Setting Criteria

The analysis of service EE data, carbon footprint, and SO_2 and NO_x emissions was performed by using the EcoTransIT environmental calculator. The tool basis uses the emission calculation methodology that complies with the requirements of the EN 16258 standard and the GHG Protocol and is managed by the independent scientific institutes IFEU, INFRAS, and Fraunhofer IML. Several logistics companies already apply the calculator for internal use and external communication [8].

Ship capacity information was used from available databases and accordingly selected from a set of available ship categories in the calculator (e.g., CC Suez max). For the ship occupancy level, a value of 80% is used for all ships (regardless of size and connection) because the said level usually allows operational profit to CLs. Moreover, it is the value that CLs achieve on the route between Asia and Europe [5]. A 20% reduction in travel speed was used for the value of slow steaming. It is the average selected value, as CLs opt for slow steaming up to 40% of the ship's estimated operating speed. Emission and EE values are calculated per 20′ container (1 TEU) with an average weight of 10 tonnes.

27.4 Results and Discussion

Operational and Commercial Data

Collected data show significant diversity in the implementation of maritime container services among CL. One of the six CLs (S2) offers a westbound rotation through the Mediterranean and the Atlantic Ocean for Durban. Other CLs have established eastbound services. S4 and S5 use a hub point in the Mediterranean when connecting service transiting Suez. Furthermore, hub terminals in the Middle East and Asia are used at eastbound services. The two CLs offer the same port rotation of ships on Koper to Durban trade, but with the differences in transport prices.

CLs use container vessels of different sizes, with no connection using the largest container vessels (ULCVs) with a carrying capacity of more than 18,000 TEU. Four services use smaller feeder container ships with a carrying capacity of up to 3,000 TEU at least on 1 transport leg, while on all analyzed services, a ship with a carrying capacity of up to 4,250 TEU is employed. These ships are older and less energy-efficient per container or tonne of goods transported and result in higher GHG emissions per container compared to ULCV [1]; thus, actual services are not optimally organized from the GHG and EE perspective.

The shortest TT is provided by S1 (38 days), while the longest TT is as much as 58% longer and amounts to 60 days (Table 27.1). A larger number of transshipment ports do not represent a longer TT. S3 and S5 use a connection from Jebel Ali to Durban with ship rotations calling Indian ports and Colombo on the way to Durban,

Table 27.1 Container service data about transit time, port rotation, and vessel capacity

Service	Transit time	Service rotation	Employed vessels capacity(TEU)
Service 1	38	Koper-Salalah-Port Louis-Durban	14,000; 2,450; 8,400
Service 2	44	Koper-Gioia Tauro-Las Palmas-Durban	2,400; 9,400; 11,500
Service 3	60	Koper-Piraeus-Jebel Ali-Durban	1,740; 8,000; 2,800
Service 4	44	Koper-Port Klang-Durban	10,000; 4,200
Service 5	55	Koper-Malta-Jebel Ali-Durban	2,250; 9,570; 4,250
Service 6	45	Koper-Port Klang-Durban	10,000; 4,200

which significantly increases the total TT and the transport distance. On average, CLs offer 48 days TT on maritime transport route Koper-Durban.

Data about sea prices also differ significantly. The CFR price of Koper-Durban maritime transport for a 20′ container varies from USD 1,340 per 20′ (S5) to USD 2000 for the same box (S4), including local surcharges in POL Koper. The difference for 40′ box is not so big, as CLs offer prices from USD 2,160 (on S3 and S5) to USD 2,660 (on S1). The price does not reflect the connection with a shorter or longer TT, as the shortest service (S1) is among the cheapest for 20′ container and, at the same time, the most expensive for 40′ one. The longest service (S3), together with the S5, is the cheapest when transporting 40′ box but is 15% more expensive when transporting 20′ one, compared with S1. The prices obtained in POL Koper differ between 5 and max 10% compared to the prices obtained by the NVOCC in Durban, but the ratios between CL are maintained.

GHG Footprint and Energy Efficiency

The analysis of GHG emissions caused by export maritime container services from the NA to South Africa shows large variations between services, which are expected given the data in Table 27.1. CLs use different container ships by their capacity and different connections between hub ports. The greenest service (S2) produces 1.04 tonnes of CO_2 (Fig. 27.1). This service is also the most energy-efficient with 3,778 kWh consumed for carrying 20′ box, even though two transshipment ports are used in the service. The largest carbon footprint is caused by S3, which stands at 1.83 tonnes and is 76% higher than the carbon footprint on S2, which is not the longest on the sea transport route. It is about 800 NM shorter than S4 and S6.

The second greenest is S1, which emits 1.18 tonnes of CO_2 or 13% more than S2. The remaining three services are less green in terms of carbon footprint, compared

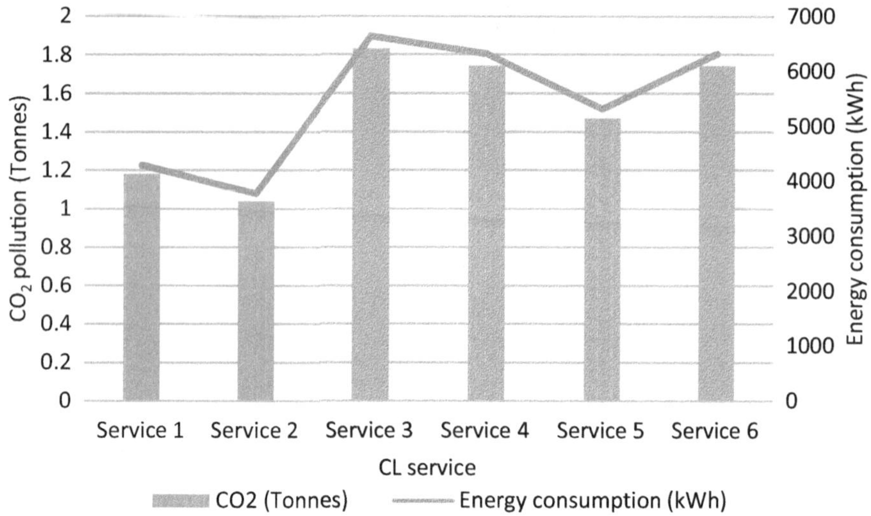

Fig. 27.1 CO_2 pollution and energy consumption by six ocean services from POL Koper to POD Durban. (Source: own research with the use of EcoTransIT)

to S2 (S5 + 41%, S4, and S6 + 67%). These services also consume more energy, as they use 42 to 71% more energy to transport a 20′ container.

The ratio is similar for NO_x and SO_2 emissions. According to performed calculations, S2 generates 24 kg of NO_x emissions and 16 kg of SO_2 emissions. S1 follows it with 17% higher NO_x and SO_2 emissions, while S5 generates 42% more NO_x and SO_2 emissions. The remaining three services achieve significantly higher emissions, which are 70 and 75% higher, respectively. S3, as the least environmentally friendly overseas transport link, generates as much as 42 kg of NO_x and 27 kg of SO_2 emissions (Fig. 27.2).

Discussion

The results of the study highlight the diversity of export maritime container services from the NA to South Africa. All three elements, from the operational to the commercial and environmental aspects, define services differently. Taking into account only the price aspect, cargo owners and NVOCC choose the S5 service, which results in a 41% higher carbon footprint and uses 40% more energy for transport than the greenest S2 service. Another contrast is the price of a 40′ container on the S1. The price is 23% higher than the lowest price on the S5, but the transport service produces a 28% lower carbon footprint compared to the S5.

The ranking of shipowners in terms of total TT, transport cost, carbon footprint, and the energy consumption is based on the total score achieved, separately for transporting 20′ and 40′ containers (best score ranked 1, worst score ranked 6). By

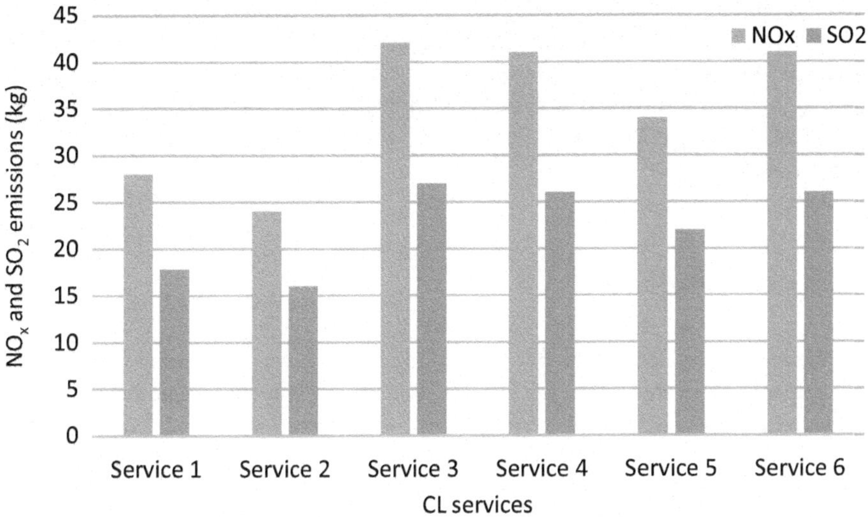

Fig. 27.2 SO_2 and NO_x emissions produced by six analyzed ocean services from POL Koper to POD Durban. (Source: own research with the use of EcoTransIT)

Table 27.2 Container service data about transit time, service rotation, and vessel capacity

Service	TT rank	20′ price rank	40′ price rank	CO_2 emiss. rank	Energy consump. rank	Total 20′/ 40’
Service 1	1	2	6	2	2	7/11
Service 2	2	3	4	1	1	7/8
Service 3	6	5	1	6	6	23/19
Service 4	3	6	5	4–5	4–5	18/17
Service 5	5	1	1	3	3	12/12
Service 6	4	4	2	4–5	4–5	17/15

considering all four elements, S1 and S2 can be best rated when transporting a 20′ container and S2 when transporting a 40′ container, even if they are not the most affordable (Table 27.2).

The analysis also concludes that the offers received by the CLs, shipping agents, and NVOCCs did not contain data on expected GHG emissions and EE of the service. Thus, cargo owners or their forwarding agents cannot decide on sustainably oriented maritime transport on the NA to South Africa trade. The results of the study point out that it is essential to inform stakeholders, who design and decide on the implementation of maritime transport, at least about the carbon footprint and EE of offered maritime transport. Only based on the offered price and TT, they cannot plan GSC design. A proper GSC elaboration is especially important for advanced technology manufacturing companies that promote low-carbon production and the operation of their products and pursue the greenest possible operation of their supply chains.

27.5 Conclusion

Guidelines for the development of green transport are very well defined at the strategic level and in transport policies. The transition to an environmentally oriented implementation of transport chains is still left to carriers and stakeholders who decide on the implementation of the transport process. It is acceptable and necessary for the transport industry to have different transport providers offering different maritime services and more complex intermodal transport. It is up to cargo owners and NVOCC agents to properly evaluate an individual transport service and make decisions based on their criteria or elaborated models in establishing GSC.

The study confirms that the price of maritime transport on the NA-South Africa trade does not reflect the level of GHG emissions from transport and energy efficiency, according to transported TEU. Besides, cargo owners and their nominated NVOCC agents are not informed of the container service's expected carbon footprint. They have a choice of several weekly container services that differ from each other. There are differences at the operational and commercial level, which also has a significant impact on the ecological aspect of the implementation of maritime transport. Thus, a more environmentally acceptable decision when choosing a marine service requires a preliminary analysis and comparison, which is more difficult to elaborate for the owners of the goods.

The results highlight the need for a changed approach to stakeholder information in supply chains, insofar as the implementation level seeks to move more actively to massive low-carbon transport and more efficient operation of the GSCM. Only fully informed and aware transport decision-makers can make sustainable solutions, such as NVOCCs, which typically select maritime services.

References

1. I. Ančić, N. Vladimir, D.S. Cho, Determining environmental pollution from ships using index of energy efficiency and environmental eligibility (I4E). Mar. Policy **95**, 1–7 (2018)
2. M. Björklund, Influence from the business environment on environmental purchasing — Drivers and hinders of purchasing green transportation services. J. Purch. Supply Manag. **17**(1), 11–22 (2011)
3. A. Cheaitou, P. Cariou, Greening of maritime transportation: A multi-objective optimization approach. Ann. Oper. Res. **273**, 501–525 (2019)

4. J. Corbett, H. Wang, J. Winebrake, The effectiveness and costs of speed reductions on emissions from international shipping. Transp. Res. D **14**(539–598), 593 (2009)
5. DSF – Danish Ship Finance: Shipping market review. https://www.shipfinance.dk/media/2000/shipping-market-review-december-2019.pdf (2019)
6. M.A. Dulebenets, Green vessel scheduling in liner shipping: Modeling carbon dioxide emission costs in sea and at ports of call. Int. J. Transport. Sci. Technol. **7**(1), 26–44 (2018)
7. M.A. Dulebenets, M.M. Golias, S. Mishra, The green vessel schedule design problem: Consideration of emissions constraints. Energy Syst. **8**, 761–783 (2017)
8. EcoTransIT., https://www.ecotransit.org/about.en.html (2020)
9. IMO – 2016 Guidelines for the development of a ship energy efficiency management plan (SEEMP) http://www.imo.org/en/KnowledgeCentre/IndexofIMOResolutions
10. IMO – UN body adopts climate change strategy for shipping, http://www.imo.org/en/MediaCentre/PressBriefings/Pages/06GHGinitialstrategy.aspx (2018)
11. ITF – Decarbonising Maritime Transport: Pathways to zero-carbon shipping by 2035, https://www.itf-oecd.org (2018)
12. M.F.A. Jasmi, Y. Fernando, Drivers of maritime green supply chain management. Sustain. Cities Soc. **43**, 366–383 (2018)
13. H. Lindstad, B.E. Asbjørnslett, A.H. Strømman, Reductions in greenhouse gas emissions and cost by shipping at lower speeds. Energy Policy **39**(6), 3456–3464 (2011)
14. H.N. Psaraftis, C.A. Kontovas, Balancing the economic and environmental performance of maritime transport. Transport. Res. Part D15 **15**(8), 458–462 (2010)
15. X. Qi, D.P. Song, Minimizing fuel emissions by optimizing vessel schedules in liner shipping with uncertain port times. Transport. Res. Part E: Logist. Transport. Rev. **48**(4), 863–880 (2012)
16. A. Wiesmann, Slow steaming – A viable long-term option? Wärtsilä Technical J. **2**, 49–55 (2010)
17. H. Winnes, L. Styhre, E. Fridell, Reducing GHG emissions from ships in port areas. Res. Transp. Bus. Manag. **17**, 73–82 (2015)
18. C.C. Yang, An analysis of institutional pressures, green supply chain management, and green performance in the container shipping context. Transp. Res. Part D: Transp. Environ. **61**, 246–260 (2018)
19. T. Zis, R. North, P. Angeloudis, et al., Evaluation of cold ironing and speed reduction policies to reduce ship emissions near and at ports. Marit. Econ. Logist. **16**, 371–398 (2014)

Chapter 28
South Adriatic Connectivity Governance as One of the SAGOV Project Concerns

Nexhat Kapidani, Anja Belojević, Ardian Haçkaj, Đorđije Otašević, Edesa Metaj, and Ernesa Kardović

28.1 Introduction

The South Adriatic (SA) area is located strategically between Eastern Europe and the Mediterranean Sea (Fig. 28.1.).

The European Union (EU) is investing a lot of money to boost cross-border cooperation (CBC) between different regions of the EU and non-EU regions or countries. The aim of cross-border cooperation is to "reinforce the effectiveness of cohesion policy by encouraging exchange of experience between regions on thematic objectives and urban development, including urban-rural linkages, to improve implementation of territorial cooperation programmes and actions as well as promoting analysis of development trends in the area of territorial cohesion through studies, data collection and other measures" [1]. Investing is done through different EU-funded programmes, such as the so-called Interreg programmes.

N. Kapidani
Administration for Maritime Safety and port Management, Bar, Montenegro
e-mail: nexhat.kapidani@pomorstvo.me

A. Belojević · Đ. Otašević · E. Kardović
Ministry of Transport and Maritime Affairs, Podgorica, Montenegro
e-mail: anja.belojevic@msp.gov.me; djordje.otasevic@msp.gov.me; edesa.metaj@infrastruktura.gov.al

A. Haçkaj (✉)
Cooperation & Development Institute, Tirana, Albania
e-mail: ardian.hackaj@cdinstitute.eu

E. Metaj
Ministry of Infrastructure and Energy, Tirana, Albania
e-mail: ernesa.kardovic@msp.gov.me

S. Bauk, S. D. Ilčev (eds.), *The 1st International Conference on Maritime Education and Development*, https://doi.org/10.1007/978-3-030-64088-0_28

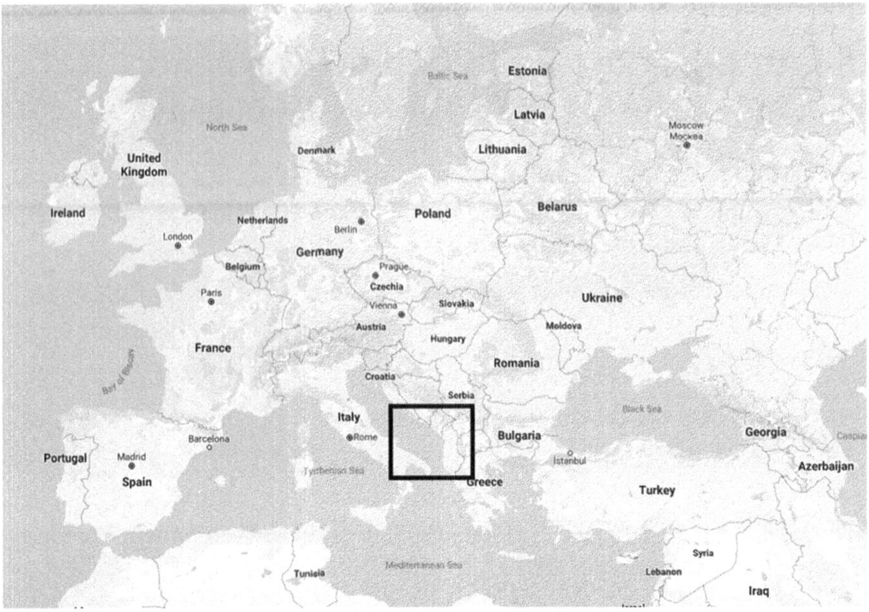

Fig. 28.1 Location of South Adriatic in Europe – black rectangle. (Source: self-edited Google map)

The Interreg IPA II CBC Programme Italy-Albania-Montenegro promotes cooperation between regional and local actors from territories in three different states, namely, Italy (EU Member State), Albania, and Montenegro (Candidate Member States) [2]. This programme encompasses countries and regions on both sides of the Adriatic Sea and includes:

- Italian Provinces of Foggia, Bari, Brindisi, Lecce, Barletta-Andria-Trani (BAT), and Taranto (Apulia Region) as well as Isernia and Campobasso (Molise Region).
- Albania (whole country).
- Montenegro (whole country)[3] (Fig. 28.2).

The programme area has 66.365 km^2 with more than 7.8 million citizens and focuses on four different axes [4]:

1. Strengthening the cross-border cooperation and competitiveness of small and medium enterprises.
2. Smart management of natural and cultural heritage for the exploitation of cross-border sustainable tourism and territorial attractiveness.
3. Environment protection, risk management, and low-carbon strategy.
4. Increasing cross-border accessibility, promoting sustainable transport service and facilities, and improving public infrastructures.

In this paper, we will focus on axe four and how SAGOV project financed by this programme addresses connectivity governance in the SA area.

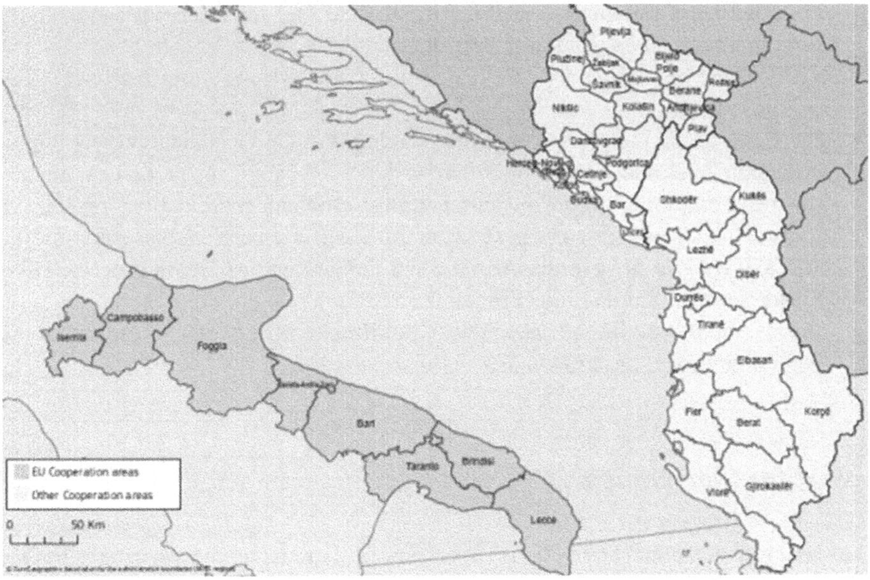

Fig. 28.2 Countries and regions participating in Interreg CBC Albania-Italy-Montenegro. (Source: [3])

28.2 Current Situation

SA is the scene of many strategic connectivity initiatives and projects. Within Western Balkans 6 (WB6) framework, European Commission (EC) and six prime ministers from WB6 (Albania, Bosnia and Herzegovina, Montenegro, Kosovo, North Macedonia, and Serbia) agreed in April 2015 on the indicative extension of Trans-European Transport Network (TEN-T) core network in WB6, so covering the eastern part of SA [5]. The Western Balkans Comprehensive Network is strategically located concerning the European transport system. It constitutes a physical transport corridor that enables the continuity of different parts of the TEN-T, providing connections from the Central European countries to the Black Sea and further beyond to Asia [6]. In June 2015, the transport infrastructure-related ministries of the WB6 and the EC indicatively identified the main transport axes that will be connected to the existing TEN-T core network corridors, and they have identified the maps of the comprehensive and core network, thus extending the TEN-T to WB6 and, respectively, to Albania and Montenegro. Connectivity reform measures (otherwise known as "soft measures") implementing EU standards shall be implemented in parallel during this process [3, 6].

The common challenges that SA area is facing regarding the development of transport connectivity measures are [3]:

- Fragmentation on two sides of the Adriatic Sea. This situation impacts the pattern of spatial interactions between them. The current situation makes visible a

definitive need for improving the intraregional and interregional connectivity through a better organisation of corridors.
- National and regional transport systems are characterised by low interoperability and intermodality.
- There is a detachment between policymaking level and technical/project implementation level: the VTMIS pilot that we have selected is the perfect example.
- The need for joint vision, better planning, efficient coordination, responsive fund-raising, synergies amongst different connectivity platforms such as the EUSAIR (EU Strategy for the Adriatic and Ionian Region), Berlin Process, TEN-T, and National Single Project Pipeline (NSPPs).
- Improvement of skills and capacities of public and private transport actors, being them institutional or individuals.

Need for Cooperation

SA area is a vital connectivity nod crisscrossed with transport infrastructure. To face the need for a shared vision, better coordination, and technical harmonisation, several cooperation structures have emerged, such as the yearly meetings of the National Investment Committees (NIC), different project financing groups like WBIF, or EU-supported CONNECTA project.

However, there is still a need to bring partners together like policymakers, project managers, technical specialists, and other stakeholders such as nongovernmental organisations (NGO) at a cross-border context [3].

At the moment, there is no multi-actor and multi-level coordination platform. The strategic prioritisation of the connectivity projects falls under a strictly governmental competence in respective countries. They define their priority strategic projects and put them in their sector plans and then in the National Single Project Pipeline (NSPP). But they must be part of the comprehensive and core network and fulfil criteria that include CBC [3].

Puglia Region is a special strategic partner in the maritime transport in South Adriatic mainly through maritime routes. The CB cooperation in connectivity is in line with the Berlin Process. Also, according to the EUSAIR Annual Progress Report of 2015 of Pillar 2 "Connecting the Region" covering transport networks, EUSAIR structures already decided to promote a synergic interaction with the "Berlin Process" and to work on the accomplishment of connectivity harmonisation. The importance of transport in bilateral CB relations can be illustrated by the number of bilateral agreements signed amongst partner countries. For example, out of 104 bilateral agreements between Albania and Italy, 15 are in transport. This number goes to 21% (or 10 out of 48) in the case of Albania-Montenegro CBC [3]. By design, the connectivity projects cannot obtain funding from the EU, or other IFI, if they miss the cross-border element. Hence the project partners will draw the benefits from the SAGOV cross-border cooperation.

28.3 SAGOV Project Relevance and Approach

SA area is very dense in cross-border cooperation (CBC) and bi-multilateral initiatives in transport, energy, and other fields. The CBC needs, however, follow-up in concrete actions, as SA suffers from [3]:

- High territorial fragmentation of transport infrastructure that constrains the potential for integrated regional development and accessibility.
- Low interoperability.
- Lack of agreement amongst countries on the priorities of development of existing transport infrastructure and future investments.
- The need for a shared vision and political commitment of states to the connectivity agenda.
- Changing priorities for the beneficiaries of the project.
- Low level of communication amongst stakeholders.
- Local ownership, compliance, and post-investment sustainability.
- Influence of agreements with other stakeholders (China,etc.)

At the policy level, SAGOV will provide the example of an integrated governance approach regarding connectivity initiatives in SA, for higher effectiveness and efficiency of policymaking regarding the strategic connectivity projects. It will be achieved through [3]:

- Coordination: transnational, cross-section (focused on transport but transferable in other sectors), multi-actor, and multi-level (local, regional, national, EU).
- Involvement: project will provide tools and procedures that may be used by all stakeholders in connectivity-policymaking cycle. At the project level, project's approach will be:
  - Multiplatform, multi-actor, and multi-level by bringing together national policymakers.
  - Designed for systemic impact by piggybacking a real-life initiative and through the embedment of its outputs in the policy cycle.
- Build-up of the pilot in the function of real needs and problems coming from the connectivity actors in the SA and based on the partner government policy agenda.
- Based on evidence-based policymaking by using research and analysis in connectivity, territorial development, and cooperation.
- Involvement of NGO at national and regional scale.

The project goes beyond problem-solving, towards the identification of best cases that really work in the complicated context of the SA. It aims to build a working connectivity governance model that is sustainable, applicable, and replicable in other areas. It offers organic interlinking of policymaking in the SA with existing EU cohesion and territorial programmes. The activities foresee the adoption of acquis communautaire and inclusion of accession negotiations in the planning and implementation of goals and objectives. The project provides the perfect ground to further promote this synergic interaction within regional cooperation in transport, to further connect the region, to sustain the rhythm of reforms engaged in Montenegro and Albania, and to speed WB6 accession in the EU. SA can become a laboratory of innovative and pioneering mechanisms of connectivity governance [3].

Project Objectives, Expected Result, and Main Outputs

The main objective of the project is to promote a governance platform for transport connectivity initiatives linking the SA. It contributes to the Programme SO 1, through better understanding, tailor-made support, coordination, and promotion of best cases in connectivity. Specific objectives of the project are to [3]:

- Map and analyse current transport connectivity situation in SA: initiatives, actors, coordination mechanisms, and platforms.
- Support the implementation of one pilot for maritime connectivity initiative that will be part of EUSAIR priority actions for the three countries/regions.
- Establish a connectivity governance platform and a sustainable coordination dynamics completed with proper tools and resources.

The main expected result is to improve the coordination amongst main actors in their endeavour in connectivity initiatives that contribute to sustainable CB connections. The main specific results of the project are [3]:

- At policymaking level, to obtain better understanding and knowledge of the.
- context and factors impacting the connectivity initiatives in the SA, focused on coordination,
- At the project level, to develop and implement joint pilot action and test the feasibility of new coordination mechanisms, tools, and services that could improve the strategic connectivity projects.
- At the institutional level, to enable national, regional, and local stakeholders to exchange knowledge and work together on full project cycle for connectivity initiatives.

The project will offer a shared common vision for joint prioritisation of needs, joint preparation, financing, implementation, and coordinated lobbying at political action at regional and EU level. The final aim is to organise the existing transport corridors better and improve the planning of new potential ones. Main outputs of the project are [3]:

- Transport connectivity e-platform with the so-called semaphore system.
- Trilateral MoU/Agreement with aid for furthering maritime coordination in SA.
- Maturation of connectivity projects for financing.

The main project results will be [3]:

- Policymaking: a comprehensive view of transport connectivity initiatives and of success factors.
- Pilot connectivity project: ToR developed for VTMIS in Albania and sharing experiences on VTMIS between Montenegro and Albania Maritime Competent Authorities.

The VTMIS is instrumental in increasing safety at sea, reducing the number of accidents at sea, and preventing pollution caused by vessels. The VTMIS also tackles maritime security aspects like illegal immigration, smuggling, etc. Project results will contribute to a better knowledge of existing connectivity initiatives (TEN-T, EUSAIR, Berlin Process, IPA CBC, and national planning through NSPP) and better coordination of all partners and stakeholders during connectivity project cycle [3].

28.4 Conclusions

The project aims to increase the efficiency of existing connectivity coordination mechanisms at policy and project level and to pioneer new ones. It will create the conditions for the application of an integrated connectivity governance approach that considers all platforms like EUSAIR, Berlin Process, TEN-T, bilateral agreements, and national prioritisation tools. It will be completed with a multi-actor scene (policymakers, IFI, business, NGO, research bodies, etc.) to allow for political commitment, efficiency, local ownership, and sustainability. The project will deliver following outputs: a comprehensive SA Connectivity Initiative Map, a pilot on transport connectivity and toolbox, a new approach in connectivity based on higher stakeholder interaction and coordination, and a critical mass of transferable connectivity know-how and skills. All stakeholders mentioned above will profit from outputs. The partnership includes central and regional bodies, NGO, and research institutions having a stake to more efficient connectivity. SA is considered as one area; the problems we will tackle are real and concrete. The new approach targets the context where connectivity projects are rooted, through higher involvement of local actors, shared vision, and interaction [3].

Project tackles connectivity that has significant impacts on the category of incoming FDI (foreign direct investment). "Multinational companies (MNC) that primarily seek access to natural resources care more about preferential access to and the land they wish to explore. MNCs that are market-seeking tend to go for bigger and richer markets. Efficiency-seeking FDI, which is most prevalent in the non-MS and WB countries, values more the quality of labor, good infrastructure, policies that facilitate trade, and lower production costs" [7]. FDI is crucial for the development of countries like Albania and Montenegro.

Acknowledgements This work was supported by the Interreg IPA CBC Italy-Albania-Montenegro Programme (SAGOV project), co-funded by the European Union through the Instrument for Pre-accession Assistance (IPA II), Interreg Albania-Italy Montenegro Programme, and the countries of the programme.

References

1. Regulation (EU) No 1299/2013 of the European Parliament and of the Council of 17 December 2013 on specific provisions for the support from the European Regional Development Fund to the European territorial cooperation goal (OJ L 347, 20.12.2013, p. 239)
2. Strategic Environmental Assessment of Italy – Albania – Montenegro Cross-Border Cooperation Programme, Managing Authority – Regione Puglia
3. Application form for SAGOV project, Interreg IPA CBC Italy-Albania-Montenegro Programme, 1st Call for proposals for standard projects
4. Five good reasons for strengthening maritime cross-border cooperation 2020+, Discussion paper, Interreg IPA CBC Italy-Albania-Montenegro Contribution to the Debate on the Future Programming Period (18 April 2018)
5. Final Declaration by the Chair of the Vienna Western Balkans Summit 2015 (27 August 2015)
6. TA to Connectivity in the Western Balkans, EuropeAid/137850/IH/SER/MULTI, Strategic framework for implementation of ITS on TEN-T Core/comprehensive network on the WB6, Final Report (10 December 2018)
7. IMF Working paper, Foreign Direct Investment in New Member States of the EU and Western Balkans: Taking Stock and Assessing Prospects, La-Bhus Fah Jirasavetakul and Jesmin Rahman

Part VII
Safety and Security in Digital Realm

Chapter 29
Cybersecurity IoT Architecture: One Proposed Solution for the Security Risks and Threats

Nedeljko Šikanjić, Zoran Ž. Avramović, and Dražen Marinković

29.1 Introduction

The Fourth Industrial Revolution (4IR) is the era of making a new step on the existing industries and digital technologies that are being revolutionized by introducing and synergizing, among others, artificial intelligence, machine learning, and cloud computing. The Internet of Things (IoT) is taking one of the main roles in the 4IR, by taking advantage of widely spread fast Internet access. The specific IoT architecture, as being assembled from various heterogeneous environments, requires a systematic approach, in order to provide a foundation of system security in general. By entering the device, the attackers can access the wider portion of network by impersonation and steal data from central servers or even send false instructions to other IoT devices. In this paper, we have made a research in order to get results on how to identify and prevent cybersecurity and other types of attacks on the IoT, what kind of methods and types of risks exists, and how to use tools and algorithms that will perform assessment of IoT security in place. Additional goal of this paper is to provide a valuable resource, for practitioners and researchers, of the current methodologies, standards, and architectures being implemented in the IoT cybersecurity field.

N. Šikanjić (✉) · Z. Ž. Avramović · D. Marinković
Pan-European University APEIRON, Banja Luka, Bosnia and Herzegovina
e-mail: nedeljko.s.sikanjic@apeiron-edu.eu; zoran.z.avramovic@apeiron-edu.eu; drazen.m.marinkovic@apeiron-edu.eu

S. Bauk, S. D. Ilčev (eds.), *The 1st International Conference on Maritime Education and Development*, https://doi.org/10.1007/978-3-030-64088-0_29

29.2 IoT Security Frameworks

NIST IoT Security Framework

The National Institute of Standards and Technology (NIST) is an organization that is part of the US Department of Commerce. It was formed in 1901, and its main role or function is to recommend and give frameworks, independent of vendors or technology, on how to implement certain standard within the technology or industry.

As for IoT technology, the NIST has been watching and proposing, based on current occurrences and future prediction, what should be implemented when it comes to IoT implementation. The document NIST.800.183 [1] has defined five primitives within the Network of Things (NoT) which can be translated into the IoT. These five primitives consist of six elements that can influence the security and risk of the IoT (Table 29.1).

Before we elaborate more on the primitives, it is important to know that in addition to primitives, elements that are influencing security and reliability of the IoT system also exist. These elements of the IoT framework are:

- Environment.
- Costs.
- Geographic location.
- Owner.
- Device_ID.
- Snapshot.

When we take a closer look at these elements, we can see that environment helps us determine where the primitives are located or what surrounds them. This way we can assume that for sensors or aggregators in heating systems, for example, needs to assume hardening or physical aspect need to be taken into account. Cost element influences on the design aspect and drives the decision of what security type we could apply. Geographic location is another element that could influence a primitive, for example, RFID location, or to use wireless RF technology used in WAN [2] and on how to comprehend the data we might get or send from sensor or external utility. Owners affect a primitive as in our IoT network, and we might have different or multiple owners of certain primitive, where we need to comprehend what this

Table 29.1 NIST primitive classification as main components of an IoT system

Primitive	Type
Sensor	Electronic utility
Aggregator	Software implementation (data transformation from raw to aggregated)
Communication channel	Medium by which data is transmitted
External utility (eUtility)	Software or hardware product (can be third party solution)
Decision trigger	Conditional expression that triggers an action

might bring in the sense of security. Device_ID is unique identifier of a primitive that helps in recognizing the primitive within our IoT solution. This can be used also for logging of certain actions and identifying the primitive within the network. Snapshot elements contain time stamp of certain action that is being recorded, modified, or deleted. Also, we should have a global time settings so that these data should be measurable and trackable.

OWASP IoT Security Framework

The Open Web Application Security Project (OWASP) [3] is a nonprofit organization that has a large group of scientist and companies working on various projects in order to improve security of software and the Internet in general. One of their projects relates exactly on that, that is, IoT security. Here, it is specific that the OWASP tries to simplify the framework in order to have a more generic approach, which would be suitable for manufactures, users, auditors, and security professionals (Table 29.2).

OWASP framework by making simplification of their proposed model easily removes exposure of IoT device and by implementing gateways and cloud storages builds a secure layer where possible attackers and hackers can't reach to secured IoT device.

Also, their framework identifies ten security weaknesses and risks [2] that should be avoided when planning and implementing an IoT solution:

- Week, none, and hardcoded passwords.
- Insecure network services, for example, during testing, some ports or network services are left open and unsecured.
- Insecure web, API, cloud, and mobile interfaces.
- Usage of third party libraries, firmware, or software that is outdated or unreliable.
- Privacy protection that is not secured properly or stored extensively with no actual need.
- Data flow and storage not being encrypted.
- Default settings of device or system not properly secured.
- With poor physical security, attacker can gain physical access to it, or device can have hardening issues.

Table 29.2 OWASP component classification of an IoT system

Components	Type
Edge	Actual IoT device
Gateway	Device that bridges communication and implement aggregation between cloud platform and edge device
Cloud platform	Storage that contains and collects data for analysis
Mobile device	Device that connects to cloud platform (end users)

29.3 Proposed Solution

As we could see from the existing IoT security frameworks, there is a general set of things that we should consider when implementing an IoT solution. However, we have made a research, based on the tested IoT network, on what model or solution would be more detailed that would also propose some additional insight when creating an IoT solution, based on risks and threats that might occur.

So, the first elements that would make a starting point would be:

- Pay attention to default settings.
- Lock-out implementation.
- In whole, IoT system (all devices, cloud, mobile) has strong authentication.
- Physical protection of IoT devices including network protection (firewalls).
- Have additional intrusion detection/prevention system that will mitigate the possible attacks.
- Encrypt network traffic.
- Implement virtual private tunnels and allow IP addresses list that device is allowed to communicate with.
- Keep open ports that are only required, and watch status of them; in the case of attack, the legit users will be locked/blocked out from the IoT system.

We can act actively or proactively in the sense of IoT security. In the early stages of planning, designing, and developing the IoT solution, we can identify potential risks and threats. Here, we can introduce SecureDevOps (secure development operations). This process integrates best practices and methodologies in terms of security, in standard development and deployment.

In this perspective, we can implement a combination of threat modeling. Threat modeling is a model where we do not wait on finished solution that is deployed in production, but we react proactively, by analyzing the processes and trying to find a possible security breach and potential threats. This would require collaboration between security expert and development and engineering team, starting as early in design phase and continuing till the deployment phase. As all the IT systems, including IoT system, will have upgrades and updates, this will be a continuous process (Fig. 29.1), where this approach and further security auditing and tasks required to monitor security standard would be followed trough. Here, it is important to mention that this kind of implementation, with team participating in various phases of IoT development, is scalable and dependable on various factors that might influence the IoT system. These factors can be external and internal, like what are the physical locations of devices, what software will be used, what type of authentication, etc. The benefit of this approach is that we can have a focus on the entire system or solution.

With threat modeling, we can identify what risk might occur, how to mitigate these threats, and what is the final validation of these mitigations and implementations. Depending on roles with the team in charge, there are possibilities to delegate certain responsibilities to certain team members. This comes in hand when

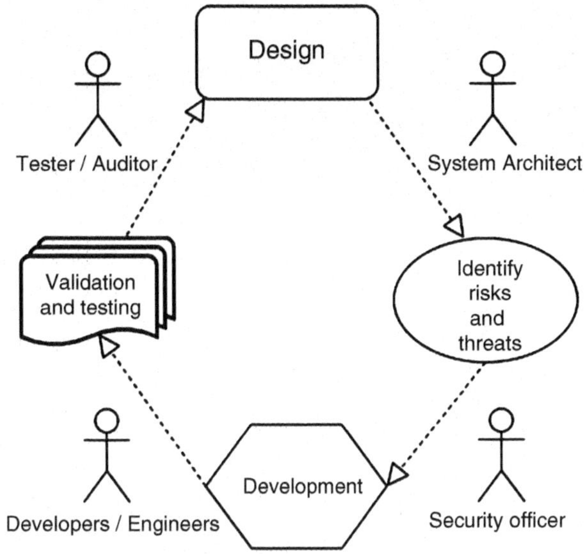

Fig. 29.1 Implementation of threat modeling in IoT security model

implementing SecureDevOps, where we can shorten the path of certain actions, without endangering the process of identifying risks and validation of security. This way, we use the power of automation, especially when having incremental development in place, such as updating or upgrading certain part or whole of the IoT system.

Another major improvement would be to include other technologies in the IoT security such as Big Data. Big Data with its power, beyond storing large volume of data collected from large number of IoT devices, is capable of performing trend analysis and monitoring and processing security events. Also performing a machine learning modeling and AI on large set of data would help us predict behavior and forecast new solutions. This way we can quickly and accurately track and discover potential risk issues that might occur. When combined these powerful technologies and proven methods, we can have an IoT security model (Fig. 29.2), which is generic enough to be followed in any possible IoT solution in various industries. This approach follows the financial aspect of the solution, as costs will be way less, when possible risks and threats are discovered in the design phase, saving on the downtime of the system and on possible loss of other resources, including data loss and financial losses.

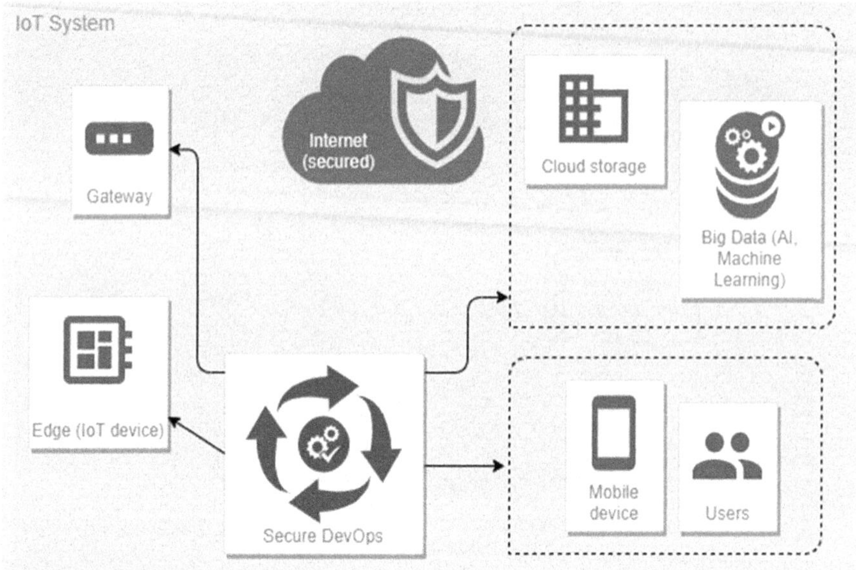

Fig. 29.2 Proposed solution diagram for IoT security model

29.4 Conclusion

We have seen in this science papers the latest recommendations in securing the IoT solutions and devices. Also, we have made our own proposal of the model or framework and on how to design an IoT solution with lowering or completely removing the risks and threats that might occur within the IoT network. As IoT covers various technologies, from wireless sensor network like the MANET and the ZigBee [4], to cloud platforms, big data storages, and AI/machine learning algorithms [5], it creates a challenge to embed security from its design perspective and implementation. We comprehend that IoT industry will continue to grow and that only in 2020 it has reached a 212 billion US dollars [6], and it is forecasted that by 2025, the IoT market will reach 1.5 trillion US dollars. With this in mind, we also need to keep focus, not only on development of the IoT devices and solutions but also on making IoT more secure and reliable, as this influences economy and, depending on industry, human lives.

References

1. Networks of Things, NIST Special Publication 800–183, https://nvlpubs.nist.gov/nistpubs/SpecialPublications/NIST.SP.800-183.pdf
2. Z.Ž. Avramović, D.M. Marinković, I.T. Lastrić, Digitalization of Railways – Ict Approach to the Development of Automation. JITA – J. Inform. Technol. Appl., PanEuropien University

APEIRON, Banja Luka, Republika Srpska, Bosna i Hercegovina JITA **9**(1), 17–23 (2019). https://doi.org/10.7251/JIT1901017A. (UDC: 656.2:[621.313:681.5), ISSN 2232-9625 (print), ISSN 2233-0194 (online)
3. The Open Web Application Security Project (OWASP) Internet of Things (IoT) Project, https://wiki.owasp.org/index.php/OWASP_Internet_of_Things_Project
4. S. Bauk, D.G. Gonzalez, A. Schmeink, Ž. Zoran, Avramović: MANET Vs. ZigBee: Some Simulation Experiments At The Seaport Environment. JITA – J. Inform. Technol. Appl., PanEuropien University APEIRON, Banja Luka, Republika Srpska, Bosna i Hercegovina JITA **6**(2), 63–72 (2016). https://doi.org/10.7251/JIT1602063B. (UDC: 539.163:504.03]:001.892), ISSN 2232-9625 (print), ISSN 2233-0194 (online)
5. Q. Zhang, L.T. Yang, Z. Chen, P. Li, F. Bu, An adaptive dropout deep computation model for industrial IoT big data learning with crowdsourcing to cloud computing. IEEE Transac. Indus. Inform. **15**(4), 2330–2337 (2019)
6. Forecast end-user spending on IoT solutions worldwide from 2017 to 2025, https://www.statista.com/statistics/976313/global-iot-market-size/

Chapter 30
Enhancing Scientific Research Opportunities and E-Learning by Integrating Large Medical ISs

Igor Dugonjić, Mihajlo Travar, Zoran Ž. Avramović, and Gordan Bajić

30.1 Introduction

Information and communication technologies (ICT) have the potential to improve and enhance the quality, efficiency, and safety of health services [1]. Therefore, these technologies are also applied in the healthcare system. Their introduction has opened a whole range of services that may be oriented toward medical service users, medical service providers, hospital administration, technical staff, and medical devices. Some of these are EHR (electronic health record), PHR (patient health record), EMR (electronic medical record), telehealth/telemedicine, HIS (hospital information system), RIS (radiology information system), PACS (picture archiving and communication system), CMMS (computerized maintenance management system), and plenitude of other ICT systems being applied in healthcare. In addition to these relatively independent ICT systems, medical devices contain embedded software packages that support the proper functioning of hardware. They can be visible or hidden from the end user and perform complex tasks based on pre-built algorithms such as analysis of information obtained from external sensors and automatic control functions in life support devices or used for alarming and initiating appropriate procedures. They use hardware to communicate with other physically separate and remote devices of lower, same, or higher level or, in turn, diverse ISs. A very important software application in medicine and other areas is the GUI(graphical

I. Dugonjić (✉) · Z. Ž. Avramović · G. Bajić
Pan-European University APEIRON, Banja Luka, Bosnia and Herzegovina
e-mail: alphaimaging@blic.net; zoran.z.avramovic@apeiron-edu.eu; gordan.z.bajic@apeiron-edu.eu

M. Travar
Regulatory Commission for Energy of RS, Trebinje, Bosnia and Herzegovina
e-mail: mihajlo.travar@reers.ba

S. Bauk, S. D. Ilčev (eds.), *The 1st International Conference on Maritime Education and Development*, https://doi.org/10.1007/978-3-030-64088-0_30

user interface), and the user-friendliness and reduction of the number of user errors depend on the method of implementation [2]. The basic purpose of these systems is patients and enhancing the quality of healthcare [3]. As this system is not closed or isolated, it is necessary to enable interaction with other auxiliary or external systems. Auxiliary systems include, for example, CMMS, while the education system is an example of an external system that relies directly on medical ISs. Medical information systems store and process a large amount of heterogeneous data. These include medical, demographic, economic, organizational, legal, and other data.

30.2 ICT Systems in Medicine

If we look at central monitoring in intensive care, we can see that it communicates with the monitors of individual patients. Monitors are connected to the patient via sensors that convert various biomedical signals (raw signals) of the patient into electrical signals suitable for further processing [4]. All relevant biomedical parameters of the patient are displayed on the patient monitor and central monitor via a graphical interface. In case of discrepancy between these parameters and the set limits, alarms of the appropriate level are activated.

On the other hand, PACS contains multimedia medical data like images obtained from a wide range of heterogeneous DICOM (Digital Imaging and Communications in Medicine) modalities (CT, MRI, analyzers, dermoscopes, etc.). Besides medical data, PACS also contains metadata. Although smaller in size, metadata are a very sensitive set of data because they contain confidential personal information about patients. In one health institution, PACS is connected to other information systems such as RIS, HIS, bookkeeping information systems, procurement, and warehouses. Although seemingly unrelated, very important data can be extracted from the information system intended for the procurement and storage of medical equipment and consumables of a health institution to which a certain geographical region gravitates. The requirement is that the system works properly and that as many parameters as possible that affect the work of the institution and the warehouse as one department are taken into account. These parameters can be the quality of the information system, the level of standardization of the procedures being carried out, the financial status of the institution and society, political influences, etc. Thus, for instance, using AI (artificial intelligence) and analysis of the dynamics of the consumption of certain medical supplies, data can be obtained that can predict the dangers and give time to prevent them. Radiology has always been closely connected with training; today, there are far greater opportunities for this than before. Radiological medical images can be classified and grouped into several categories:

- Patients they refer to.
- Modality of obtaining.
- Type of examination.
- Diagnosis and the like.

When the images are sorted by the patient, the focus is on the patient and the history of the disease. Sorting by modality and by the type of examination is important for the organizational part and workflows in the radiology department. In terms of training, all ways of grouping medical images are important, and PACS systems can filter these data by various parameters. A significant obstacle here is the confidentiality of personal information. Access to private information must be strictly regulated, and the first significant act regulating the confidentiality of medical data is the 1996 HIPAA (Health Insurance Portability and Accountability Act) [5]. To use medical data for educational purposes, they need to be anonymized. It is often necessary to perform this procedure on a whole series of data by grouping medical images by one patient in order to follow the history of the disease and by removing the personal information in a way that the identity cannot be disclosed even by using sophisticated procedures. Personal information could be used by insurance companies, banks, malicious individuals, etc.

30.3 Integration and Sharing of Medical Information

There is a relatively high degree of integration of the said ICT systems in every individual healthcare institution, while the integration among different institutions is relatively in its early stages. This task is facing significant safety; interest; organizational, economic, and legal; and other challenges. On the other hand, scientific research communities may be a very important stakeholder in this process. The reason is the possibility of instant access to a huge amount of relevant, quasi-real-time field data.

As epidemics of Ebola, MERS, and other diseases, as well as the COVID-19 pandemic, threaten the health of nations, ICT combined with social distance and other physical protection measures can be a very powerful weapon against these threats. The use of ICT in healthcare has many aspects, from very complex and advanced technologies to simple technical forms in the form of broadcasts. An example of ICT tools against infectious (but also non-communicable) diseases can be the term "going viral," which is the term most commonly used to describe fast-spreading and vast-spreading online content [6]. There is also an example of the use of a social media platform for cell phones in the fight against coronavirus: namely, using the incorporated GIS (geographic information system) and legally personalized mobile devices, the locations of these devices are marked in people diagnosed with COVID-19 on the virtual map. When an uninfected person is near an infected person, an alarm goes off, and a map appears on the display of the mobile device. Another form of using mobile phones is the timely provision of useful information and censorship of misinformation. The downside of this use of ICT is the violation of privacy.

30.4 Artificial Intelligence

State-of-the-art devices, increased workload of radiologists, and new imaging procedures make the amount of produced and stored medical data on PACS systems increase progressively [3]. On the one hand, this can be a difficulty as it is necessary to provide procedures for storing and managing these data, while on the other hand, this is a great potential in terms of available learning and research material. The obstacle is the centralization of medical ISs in a single institution. In order to enable interoperability, optimal use of data, and quality training in healthcare, it is necessary to change the way of managing centralized sources of medical data and enable integration with external ISs. The integration could use cloud technology and should be able to manipulate a large set of data (Big Data). Although there are numerous algorithms for managing such data, AI (artificial intelligence) is currently the most suitable tool to serve the purpose [3]. Big Data is a large and complex set of data that cannot be managed and processed with traditional database management tools. In this case, relational databases are replaced by the so-called NoSQL databases, which are perfect for storing large amounts of data in distributed systems and the cloud [7].

Many challenges can affect the integration and sharing of medical data for healthcare, e-learning, and research purposes, and we can divide them into technical and organizational human-related. The main technical (including safety) challenges are availability, confidentiality, access control, data ownership, privacy, and authentication [8]. To ensure availability, it is necessary to ensure a systematic and standardized connection between heterogeneous local ISs and the cloud. Cloud should also offer a service that incorporates compatible medical image viewers to avoid losing window and level DICOM functionality in, for instance, format conversions. It is also necessary to parse DICOM files in order to separate the metadata from the medical image [9].

Besides the benefits, the introduction of new technologies brings new challenges like organizational human-related. In the implementation of new ICT solutions for healthcare purposes and training, we distinguish the following: preparing physicians for online educated patients; raising awareness of the benefits of ICT for healthcare technologies; motivating students and technicians to use ICT for information, learning, and development; and, ultimately, changing the medical training methods [10].

Integration should not be limited to large systems and clouds; fog computing should also be taken into consideration. Unlike cloud technology, where data and applications are processed in the cloud, which is a time-consuming task when working with Big Data like medical imaging series, the concept of fog computing and the IoT (Internet of Thing) significantly reduces data size [11]. Ready, prepared, and usable information is sent over the network, while local biomedical signals are not sent to the cloud.

The integration should support DICOM (Digital Imaging and Communications in Medicine) and HL7 (Health Level 7) standards. It should also be in line with IHE recommendations, especially the XDS profile. To achieve data protection, Oliveira proposed an XDS-π (private) profile that combines the two concepts of XDS-I and public cloud [12].

The COVID-19 pandemic has led to a sudden increase in volume of sharing medical data, online notes, deep learning, and forming large repositories. Never before has so much communication activity been recorded between participants in this process; however, there have also been problems related to infrastructure, legal issues, and issues related to standardization [13]. Restrictions imposed by the GDPR (General Data Protection Regulation) have also been observed. The GDPR has strict data protection rules that limit the collection, use, and exchange of medical data and hence the ability to jointly combat the spread of disease [14].

Radiomics, a branch of AI-based radiology, is a method of extracting large amounts of features from medical images using data-characterization algorithms and deep learning. It usually involves several serial steps, starting with image capture, region segmentation, feature extraction, predictive modeling, and model validation (Fig. 30.1). Regarding the use of AI in the fight against COVID-19, opinions are divided; on the one hand, it is believed that CT (computed tomography) of the lungs when using these tools based on a well-balanced and well-controlled data set can answer the question of whether it is COVID-19 or not [14] [15]. On the other hand, we know that SIRM (Italian Society of Medical and Interventional Radiology) does not support the use of CT and AI screening tools for COVID-19, although it

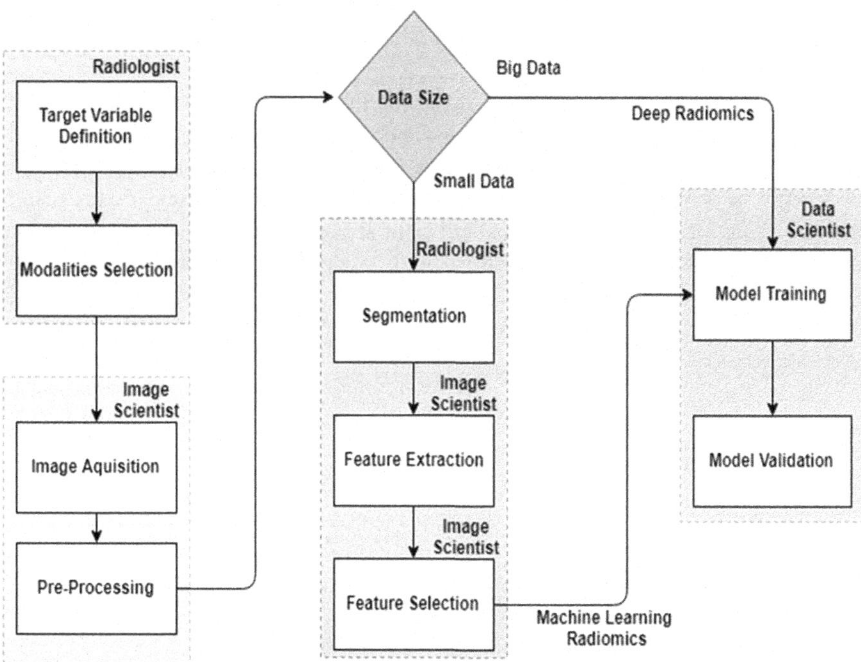

Fig. 30.1 Radiomics workflow of data process and users

provides recommendations for research into the use of AI as a predictive and prognostic DSS (decision support system) [16]. The fact is that radiologists really can't get convincing cases of using AI and deep learning if they don't have the necessary data and infrastructure [17].

For the purposes of e-learning, Ranschaert favors pseudo-anonymization over complete anonymization, where each patient is assigned a code associated with medical images. This method is practically simpler and faster and in line with GDPR [14]. The issue of e-learning in this area is very important and takes several stakeholders into account: students, medical staff, IT staff, etc. The areas of interest of these groups overlap in part, but there are also important specifics.

Medical students are primarily interested in accessing repositories with processed and sorted medical case studies. Medical staff is also interested in this content, but also in training to work with new IT technologies such as AI. There is an initiative from EuSoMII (European Society of Medical Imaging Informatics) to introduce training in the application of AI in radiology as a standard throughout the training of radiologists [13]. It is certainly important to mention that two groups of researchers are directly interested in AI-related research, in medicine and in informatics, each in its own field. As traditional PACS is not sufficiently flexible for scientific research [18] and, on the other hand, it is very suitable for routine operations with its robust, secure, and simple interface for working with imaging data in clinical practice, integration and model for sharing this data should avoid the shortcomings of traditional PACS that occur in the research process. These shortcomings relate to data protection, legislation, loss, duplication and corruption (occupancy) of data, high overhead costs, organizing and indexing files as a volume, increasing the complexity of image studies, etc.

By sharing medical resources and after the integration process carried out to a certain level, the possibilities of new services occur – such as telemedicine, knowledge management, and more optimal workload of professional staff. Accordingly, healthcare at a higher level can be achieved with already existing medical staff, using bidirectional communication and data transmission in places without the physical presence of specialized medical staff. It is primarily in remote and inaccessible areas, islands, overseas ships, etc., with the help of ICT infrastructure and the use of online services of specialist doctors located in another geographical location.

30.5 Conclusion

There are many dedicated ISs that find their application directly or are largely based on healthcare. Some of these systems store medical data accompanied by personal data of patients (PACS, EHR, KIS, etc.). When all these systems are viewed globally in a geographical region, they are a significant resource that can be exploited through adequate integration using ICT. This paper emphasizes the potentials of this resource such as e-learning, health and technical training, fight against epidemics and pandemics, overall improvement of healthcare through new telemedicine services, and significant expansion of opportunities for scientific research in both medicine and ICT. The paper points out great importance of AI in combination with Big Data. This is still an underdeveloped area, and future research should focus in that direction. The paper also points out certain challenges – technical and nontechnical.

References

1. S.G. Sampada, R.V. Kulkarni, Role of information Technology in Health Care, in *Proceedings of the 4th National Conference, INDIACom-2010 Computing for Nation Development*, (Bharati Vidyapeeth's Institute of Computer Applications and Management, New Delhi, 2010)
2. Y. Zhang, P. Masci, P. Jones, H. Thimbleby, User Interface software errors in medical devices: A study of US device recall data. Biomed. Instrument. Technol. **53**, 182–194 (2019). https://doi.org/10.2345/0899-8205-53.3.182
3. E. Ranschaert, S. Morozov, P. Algra, *Artificial Intelligence in Medical Imaging Opportunities, Applications and Risks: Opportunities, Applications and Risks* (2019). https://doi.org/10.1007/978-3-319-94878-2
4. D. Boskovic, O. Lepara, *Biomedicinski signali i sistemi* (ISBN 978-9958-629-68-6, 2018), p. 14
5. HIPAA, Federal Register, Vol. 68, No. 34, Rules and Regulations (February 20, 2003)
6. P. Abbott, S. Barbosa, Using information technology and social mobilization to combat disease. Acta Paul. Enfermeria **28**, 1 (2015)
7. R. Vrbić, Data mining and cloud computing. JITA – J. Inform. Technol. Appl. (Banja Luka) – APEIRON **4** (2012). https://doi.org/10.7251/JIT1202075V
8. M. Anuja, C. Jeyamala, A survey on security issues and solutions for storage and exchange of medical images in cloud, in *International Journal of Emerging Trends in Electrical and Electronics, 11- 6*, (2015), pp. 27–32
9. A.W. Kamauu, S.L. DuVall, R.J. Robison, A.P. Liimatta, R.H. Wiggins 3rd, D.E. Avrin, Informatics in radiology (infoRAD): Vendor-neutral case input into a server-based digital teaching file system. Radiographics: a review publication of the Radiological Society of North America, Inc **26**(6), 1877–1885 (2006). https://doi.org/10.1148/rg.266065707
10. T. Karsenti, B. Charlin, Information and communication technologies (ICT) in medical education and practice: The major challenges. Int. J. Technol. High. Educ. **5**, 68–81 (2009)
11. M. Maksimović, Implementation of fog computing in IoT-based healthcare system. JITA – J. Inform. Technol. Appl. (Banja Luka) - APEIRON **7**, 100–107 (2017). https://doi.org/10.7251/JIT1702100M
12. L. Ribeiro, C. Costa, J. Oliveira, Curr. Trends Arch. Transmission Med. Images (2011). https://doi.org/10.5772/27959
13. P. Ward: Pandemic leads to surge of interest in informatics. https://www.auntminnieeurope.com/index.aspx?sec=log&itemID=618901 (2020)
14. P. Ward: COVID-19 crisis shows limits of GDPR, Ranschaert warns. https://www.auntminnieeurope.com/index.aspx?sec=log&itemID=618746 (2020)
15. X. Xie, Z. Zhong, W. Zhao, C. Zheng, F. Wang, J. Liu, Chest ct for typical 2019-ncov pneumonia: Relationship to negative rt-pcr testing. Radiology **0**(0), 200343., pMID: 32049601 (2020). https://doi.org/10.1148/radiol.2020200343

16. SIRM, SIRM issues guidance on CT, AI use in pandemic. https://www.auntminnieeurope.com/index.aspx?sec=log&itemID=618722 (April, 2020)
17. M. Bassett: The Reality of Deep Learning/Artificial Intelligence in Radiology: They Will Redefine the Specialty. In rsna.org. Available from: https://rsna2017.rsna.org/dailybulletin/index.cfm?pg=17mon05 (2017)
18. S.J. Doran, J. d'Arcy, D.J. Collins, R. Andriantsimiavona, M. Orton, D.M. Koh, M.O. Leach, Informatics in radiology: development of a research PACS for analysis of functional imaging data in clinical research and clinical trials. Radiographics: a review publication of the Radiological Society of North America, Inc **32**(7), 2135–2150 (2012). https://doi.org/10.1148/rg.327115138

Chapter 31
E-Learning When Pandemic Endangers Human Population: Case Studies of Bosnia and Herzegovina and Austria

Nedeljko Šikanjić, Zoran Ž. Avramović, and Nedim I. Smailović

31.1 Introduction

We have provided here two case studies: one is in the country of Bosnia and Herzegovina, and another case study is in Austria, which is a member of the European Union. The goal of these case studies is to provide comparison of two countries' implementation of educational system when the society is being affected by a serious, unexpected, and dangerous situation requiring immediate action in order to provide an undisturbed process in education.

Bosnia and Herzegovina has been relatively a new country in terms of being recognized from the United Nations, came out of the civil war after 30 years, and still despite of some obstacles in political and organizational life has managed to find a way to provide their students and pupils a well-organized e-learning or distance learning system, which has provided a great feedback.

Another case study country is Austria, a really well-organized and stable country and a member of the European Union that has found a good way on how to rise up to the challenge of introducing distance learning system in the extraordinary time.

The goal of this paper is to provide results of this research and contribute to the educational society in providing a resource of successful implementation in two different countries, showing the weaknesses and strengths in planning and the implementation of e-learning systems.

N. Šikanjić (✉) · Z. Ž. Avramović · N. I. Smailović
Pan-European University APEIRON, Banja Luka, Bosnia and Herzegovina
e-mail: nedeljko.s.sikanjic@apeiron-edu.eu; zoran.z.avramovic@apeiron-edu.eu; nedim.i.smailovic@apeiron-edu.eu

S. Bauk, S. D. Ilčev (eds.), *The 1st International Conference on Maritime Education and Development*, https://doi.org/10.1007/978-3-030-64088-0_31

31.2 Case Study Austria

When the COVID-19 pandemic has reached Austria, the whole society was unprepared for this sort of pandemic attack on human population. As first it wasn't expected, like in the rest of the world, how it spreads and how it affects people's health, the Austria government was first to implement "shutting down" of institutions, companies, and every other aspect of possible human contact that could prevent the spread of the virus and with this approach tried to save as many lives as possible.

This has influenced the schools as well, as there was no feasible solution at that moment on how the schooling and education will continue. But before going more in detail about technology being used, we have provided a number of pupils/students in Austria (Table 31.1).

We must mention that there is difference between Austria and Bosnia and Herzegovina in the schooling system, as the former's elementary school last only from first to fourth grade and then it is separated into two secondary school levels called lower and higher degree of secondary school, while the latter's lasts from first grade to eighth grade, and the secondary school has 3 or 4 years in total, depending on the type of the school (i.e., gymnasium, technical school, etc.).

After the pandemic breakout and the few days of no school activities, the approach was to introduce the e-learning in schools and universities. There was certain freedom in choosing the technologies from educational institutions, as they would then know how their staff and pupils will adjust to the technologies. In most cases, there was a combination of different technologies, for example, web mash-up [1], but also extending to mobile technology, as this could give more creativity in fulfilling the daily school activities (Table 31.2).

The list of technologies was added "on the fly," or it was adjusted in short period starting with e-learning processes. As in Austria, as mentioned previously, children age from 6 to 10 (one to four classes) attend elementary schools, and as these children at these age are pretty familiar with mobile devices, it was decided to use the

Table 31.1 Number of pupils of elementary/secondary schools in Austria

Country (Burgenland)	Number of pupils in elementary schools	Number of pupils in secondary schools
Burgenland	10.588	33.199
Kärnten	20.741	67.082
Niederösterreich	65.062	202.319
Oberösterreich	61.211	175.832
Salzburg	21.487	67.107
Steiermark	44.574	156.803
Tirol	28.963	96.883
Vorarlberg	16.824	45.895
Wien	72.942	279.876
Total	**342.392**	**1.124.996**

Table 31.2 Technologies chosen in elementary schools for e-learning processes

Technology	Type
Anton App	Mobile application
Skype	Mobile/desktop application
Signal App	Mobile application
Email	Email clients

Table 31.3 Technologies chosen in secondary/gymnasiums schools for e-learning processes

Components	Type
Office 365/Microsoft Teams School	Cloud software for collaboration and communication for schools
WebEx Cisco	Mobile/desktop communication application
Email	Email client

Anton App [2], co-funded by the European Union, which covers the daily practices in teaching and exercises. This application has a separated teacher and pupils' accounts, where each teacher can monitor the progress of its pupil. The organization of the app itself is the same as in the classes, so it perfectly represents that certain school activity that is being represented in the online form.

For the communication aspect, the technology being used is Skype and Signal App, where it was used for virtual classes and all the pupils and teachers can see each other, have a conversation about school subjects and activities, and also fill in the social gap of at least seeing each other and keeping the team spirit. For paperwork and lessons, when required, it was sent by the email clients (Gmail, Outlook, etc.) depending on the parent and teachers' setting. This way, it bridged the gap with needed exercises or lessons; however, there could be a need for these kinds of documents to be printed out, and this could cause an issue to parents who do not possess printer in-house.

In the secondary schools, of course, the technologies used in e-learning processes are also adjusted to the pupil's knowledge and known usage of software solutions and devices (Table 31.3).

Microsoft teams for education was chosen as it has a structure easy to implement, for example, processes are in classes, where teacher has a direct communication channel with students and also there are collaboration channels for exchanging documents with teacher and students. This way everything is centralized, and there is history of events, including storage for documents. There is option for students to use software like Word for text processing with no cost for students, and everything, including email client, is already loaded in the browser and on one place. This way, teachers and students have all the necessary activities in grasp, which help knowledge to be transferred and assessed.

For communication between teachers and parents, depending on the teacher, there is option to use WebEx, in most cases, where videoconferences are being held and information is being exchanged and communicated.

For the written communication between parents and teachers, email client is being used, depending on their own preferences [3].

In Austria, projects in e-learning have been successfully implemented before, like the FMS and the Academy for New Media in the Transfer of Knowledge (ANMKT) (University of Graz), "Developing an e-learning module for the educational needs" (2011–2012) [2], and the educational society had a quite good experience with them.

31.3 Case Study Bosnia and Herzegovina

In Bosnia and Herzegovina, the pandemic was unexpected in the form as it has appeared; however, especially after seeing how the situation started in Italy, the government, especially in the Republic of Srpska, has responded immediately and implemented "shutting down" of public life, companies, and every other form of contact between people. Companies were instructed, where possible, to start with remote work or home office. This was analyzed and estimated to be feasible and implemented for educational schools and universities. In Bosnia and Herzegovina, as is the case in Austria, elementary schools and secondary schools were using a combination of technologies accommodated to the age of pupils and their knowledge of certain technologies. To acknowledge better how many pupils/students were required to accommodate the distance learning, here we have provided the official statistics to better illustrate the learning process of the Federation of Bosnia and Herzegovina [5] and the Republic of Srpska [4] (Tables 31.4 and 31.5).

Schools implemented various technologies based on the knowledge of pupils in certain classes. In elementary schools, Microsoft Teams for collaboration were used, and depending on teachers, Viber Groups was also used, as Viber is a popular

Table 31.4 Number of pupils of elementary/elementary schools in the Federation of Bosnia and Herzegovina

Kanton	Number of pupils in elementary schools	Number of pupils in secondary schools
Una-Sana	20.873	8.746
Posavina	2.134	1.094
Tuzla	39.267	15.867
Zenica-Doboj	31.223	11.606
Bosnian-Podrinje	1.886	903
Central Bosnia	22.328	8.464
Herzegovina-Neretva	17.549	7.260
West Herzegovina	7.841	3.570
Sarajevo Canton	36.957	16.111
Canton 10	4.411	1.967
Total	**184.469**	**75.588**

Table 31.5 Number of pupils of elementary/elementary schools in the Republic of Srpska

Republic of Srpska	Number of pupils in elementary schools	Number of pupils in secondary schools
Total	**90.995**	**39.831**

Table 31.6 Technologies chosen in elementary schools for e-learning processes

Technology	Type
Microsoft Teams	Mobile/desktop application
Viber/WebEx/Skype	Mobile application
Email	Email clients

Table 31.7 Technologies chosen in secondary/gymnasiums schools for e-learning processes

Technology	Type
Moodle	Mobile/desktop application
Viber/WebEx/Skype	Mobile application
Email	Email clients

communication application in Bosnia and Herzegovina, as well as WebEx. Also, the remote teaching over public television broadcast was implemented, where the daily schedule was published on the Ministry of Education web portal. Schools have communicated on their web sites what the teacher has presented on official YouTube channel lessons, including content of lesson being presented and eventual exercises that followed certain lessons (Table 31.6).

Secondary schools have mostly implemented e-learning platforms such as Moodle, where all the lessons and test exercises have been accessed by the previously authorized pupils/students. The identity management [6] was also a focus in these e-learning systems, where it was controlled, organized, and followed up which student was and when logged into the e-learning system, so that tracking of participation could be followed as it would be in normal school time (Table 31.7).

As structure of Bosnia and Herzegovina is of two main entities or republics, it could vary the use of the abovementioned technologies, depending on the official authorities' recommendation and help that was provided from these governments. However, the knowledge sharing was intensive including best practices that have been proven helpful [7].

31.4 Conclusion

We have seen in this science papers that both countries, Austria and Bosnia and Herzegovina, even though surprised with the pandemic wave that has brought lockdown to societies as a whole, have made a terrific effort and great job in providing and organizing education to the pupils and students, without or with little loss of the quality of provided teaching and knowledge. It has also shown in this time, where human population was endangered across the entire planet, education was not stopped nor paused. Teachers together with their pupils and students have managed to use cloud technologies and Internet applications in the most effective way, which usually in large IT companies would require significant resources of IT experts to make everything work with no issues. This comes to prove that no matter how economically strong or less strong a certain country is, as stated for these two countries, cloud and Internet technologies will overcome this factor and will provide a global bound to human population, with a focus of making our societies generate knowledge that will benefit all of us.

References

1. Schülerinnen und Schüler, Studierende – Statistik Austria Bundesanstalt Statistik, https://www.statistik.at/web_de/statistiken/menschen_und_gesellschaft/bevoelkerung/volkszaehlungen_registerzaehlungen_abgestimmte_erwerbsstatistik/schuelerinnen_und_schueler_studierende/index.html
2. S. Bauk, M. Kopp, A. Zoran, A case study on introducing E-learning into seafarers' education. JITA – J. Inform. tecnol. Appl., Pan-European University APEIRON, Banja Luka, Republika Srpska, Bosna i Hercegovina JITA **3**(1), 34–42 (2013). https://doi.org/10.7251/JIT1301034B. (37.018.43:004.738.5), ISSN 2232-9625 (print), ISSN 2233-0194 (online)
3. F. Mödritscher, G. Neumann, *Victor Manuel García-Barrios, Fridolin Wild* (Web Application Mashup Approach for eLearning) http://www.moedritscher.com/papers/paper_moedritscher_et_al_xomashup_2008.pdf
4. Das Projekt „ANTON – Lernplattform für die Schule" wird kofinanziert durch den Europäischen Fonds für regionale Entwicklung (EFRE), https://www.schule.at/startseite/detail/anton-kostenlose-lern-app.html
5. Federation Ministry of Education – Information of number of pupils / students in 2018/2019, http://fmon.gov.ba/Upload/Dokumenti/3882ddc8-b66e-43d4-a98b9890176421a6_Informacija%20o%20upisu%20ucenika%20skolska%202018_19.pdf
6. Word Bank Report, Review of service efficiency in pre-university education, http://documents.worldbank.org/curated/en/121621571233909890/pdf/Bosnia-and-Herzegovina-Review-of-Efficiency-of-Services-in-Pre-University-Education-Phase-I-Stocktaking.pdf
7. G. Radić, The usage of information technology in the implementation of the bologna principle of the student mobility. JITA – J. Inform. Technol. Appl., Pan-European University APEIRON, Banja Luka, Republika Srpska, Bosna i Hercegovina JITA **1**(1), 24–32 (2011) (UDC 37:331.55(4)), ISSN 2233-0194 (online)

Chapter 32
E-Learning in Secondary Schools in the Republic of Srpska and Improvement Suggestions in COVID-19 Setting

Slavojka Lazić, Tijana Talić, and Dražen Marinković

32.1 Introduction

The pandemic of the new coronavirus has prevented the normal course of the educational process in primary and secondary schools around the world, as well as in BiH and the Republic of Srpska. Globally, over 1.2 billion students from 186 countries attended classes from home. Modern electronic communication that is fast and efficient has come to the fore [1].

The Ministry of Education and Culture of the Republic of Srpska made a decision, according to which, from March 17, 2020, online classes for primary schools were organized through the public television service RTRS. On March 23, 2020, the Ministry of Education and Culture made a decision that obliged all high schools to organize distance learning according to their abilities [2]. Unlike some countries in the region, the Republic of Srpska did not have the technical ability to organize distance learning in high schools through television, so high schools had to "overnight" to manage and organize online classes for their students.

32.2 Organization of Distance Learning in Secondary Schools of the Republic of Srpska

Since secondary schools had a very short time to organize and send a distance learning plan to the Ministry of Education and Culture, most schools were initially organized by creating Viber groups with students in which teachers sent teaching units

S. Lazić (✉) · T. Talić · D. Marinković
Pan-European University APEIRON, Banja Luka, Bosnia and Herzegovina
e-mail: tijana.z.talic@apeiron-edu.eu; drazen.m.marinkovic@apeiron-edu.eu

S. Bauk, S. D. Ilčev (eds.), *The 1st International Conference on Maritime Education and Development*, https://doi.org/10.1007/978-3-030-64088-0_32

and assignments and students sent feedback to professors by mail [3]. This may not have been the best method, but in any case, it was useful and enabled the continuation of the teaching process.

In our country, which belongs to the relatively underdeveloped countries, this type of teaching was something completely new for both students and professors. Bearing in mind that a certain number of students and teachers were not computer literate, at the very beginning of the introduction of distance learning, there were certain technical problems: some students did not know how to send mail, how to use more complex functions in Word, etc. With the help of teachers and classmates, these problems were overcome very quickly.

After this not so simple start, teachers started using better tools and platforms for distance learning such as Google Classroom, cloud folders, Microsoft Teams, Zoom, and others. This way of working made online learning much easier for the students because they had all the teaching contents in one place and were able to communicate more easily with the teachers who were available to them 24 h a day.

32.3 Survey

At the end of the 2019–2020 school year, after 3 months of distance learning, we conducted a survey that included students from two secondary schools in Banja Luka (Republic of Srpska), one private and one public school. The aim of the research was to examine the opinion of students about the implementation and quality of distance learning in their school, the objectivity of assessment, and the problems they encountered. The students answered six questions.

Results and Analysis

The following diagrams show the comparative opinions of students from the two schools in which the research was conducted. Based on the results of the research, we can conclude that students from both private and public schools are satisfied with the distance learning in their school, given that over 90% of students from both schools gave an affirmative answer to this question (Fig. 32.1).

The second and third questions refer to the way in which students better master the material and the form of teaching that gives them a better education. 53.13% of private school students and 33.33% of public school students said that the combination of traditional and online teaching gives the best results in mastering the material (Fig. 32.2).

However, as we can see in Fig. 32.3, the largest percentage of students (81,25% of private school and 94,74% of public school) believe that they acquire a better quality of education through traditional forms of education. From this, we can conclude that students still give preference to the traditional over online learning.

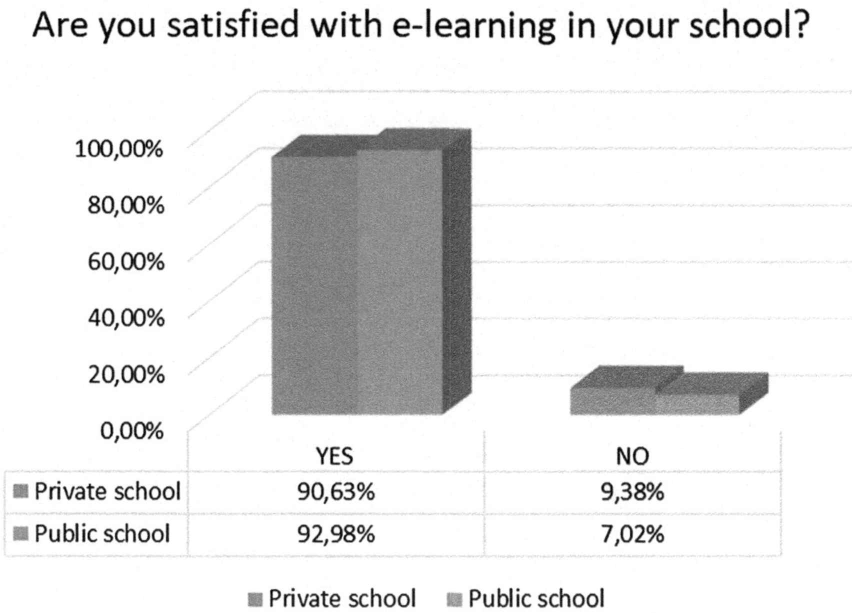

Fig. 32.1 Students' opinions on distance learning in their school

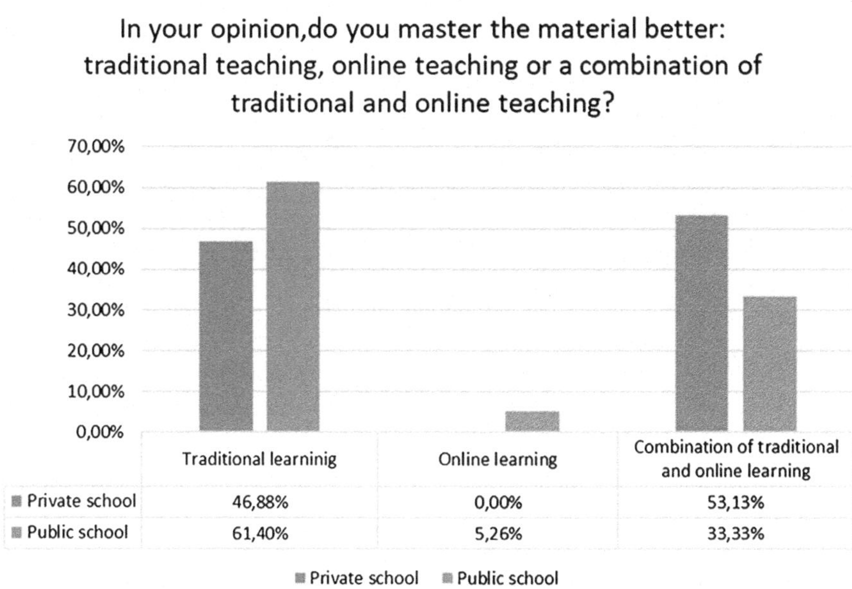

Fig. 32.2 The way students master the material better

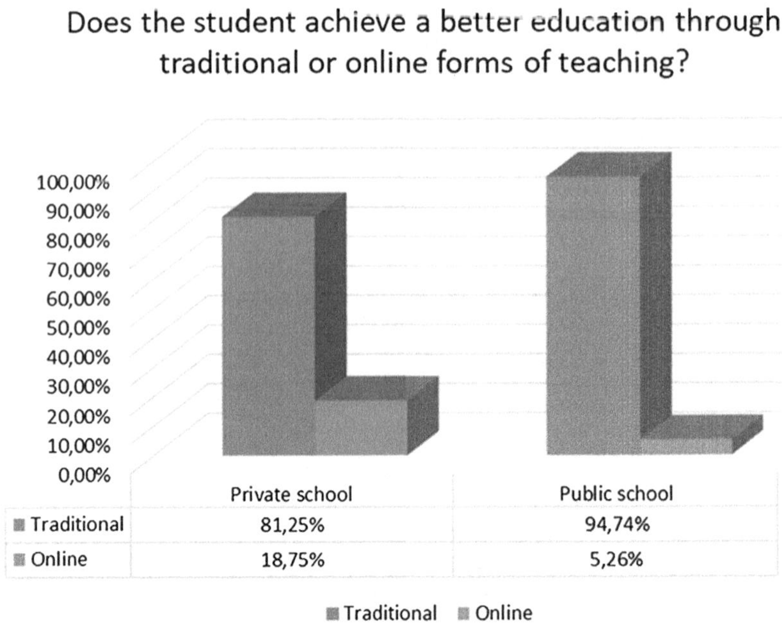

Fig. 32.3 The form of teaching through which students acquire a better education

From Fig. 32.4, we can conclude that the students were satisfied with the cooperation with teachers, which was very important for the students. The teachers were supportive and gave them a motive to overcome all difficulties during online classes.

The issue of objectivity of knowledge evaluation and general evaluation of students in the distance learning system requires the most discussion among professors. As we can see in Figs. 32.5, 84.38% of private school students and 91.23% of public school students stated that the grades they received during distance learning were objective. However, the issue of objective assessment for teachers is one of the major problems in the distance learning system, which will be discussed later in this paper.

Figure 32.6 shows that 90.63% of private school students and 87.72% of public school students, i.e., 88.76% of the total surveyed students, stated that they had no problems during distance learning, while 11.24% of students stated that they had difficulties.

Of that number, 6.74% stated that it was difficult for them to master the material and understand the tasks without the presence of a professor, while 4.49% stated that the slow Internet was a problem.

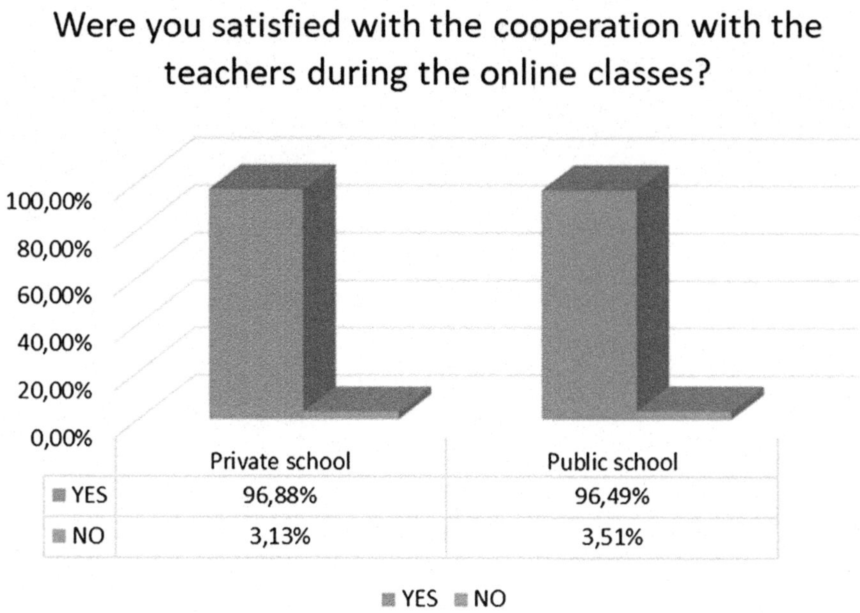

Fig. 32.4 The cooperation with the teachers

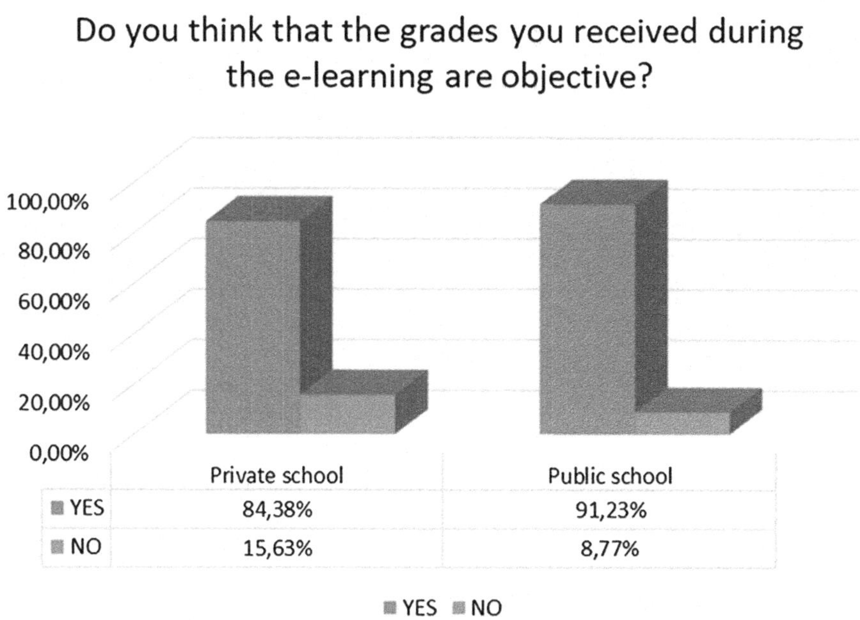

Fig. 32.5 Objectivity of knowledge evaluation

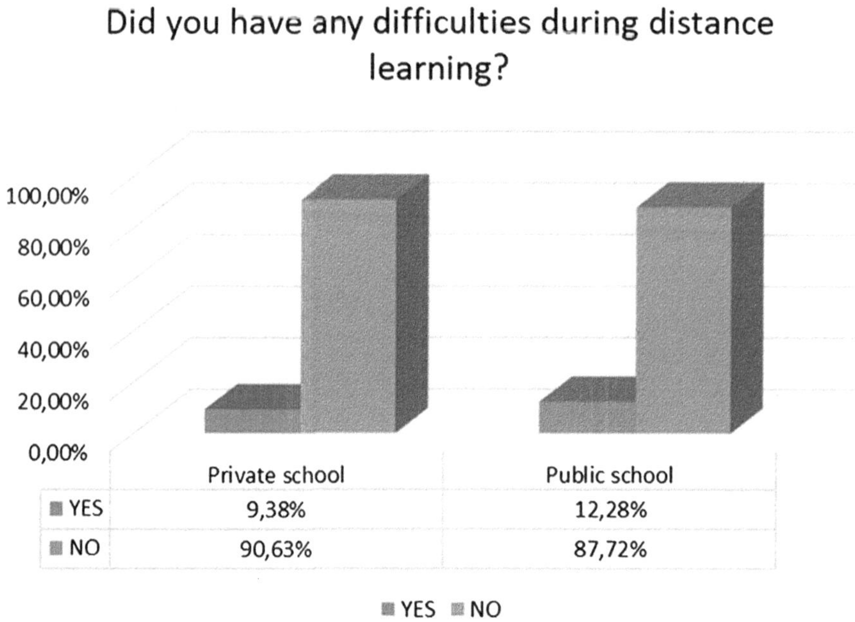

Fig. 32.6 Difficulties during distance learning

32.4 Advantages and Disadvantages of Distance Learning

Distance learning has many advantages over traditional forms of teaching. Modern technologies offer teachers opportunities to create diverse, interesting teaching content, quizzes, and games that make teaching interesting to students. In addition to being interesting, this type of teaching allows students to review the teaching content that the teacher has placed on the platform at any time. Also, by using modern technologies, both students and teachers acquire new skills that will surely be useful to them in business and other challenges of the twenty-first century. On the other hand, according to the opinion of students shown in Sect. 32.3.1, we can conclude that the classroom and communication between teachers and students cannot be completely replaced by modern technologies, online platforms, and other forms of electronic communication.

As part of our research, we asked students and professors what problems they encountered during the online teaching process. In the results of the research, it has already been stated that the students went through the online teaching process without any major difficulties and only a few students stated that the problem was the weak Internet and the difficulty of mastering the material. For teachers, one of the major disadvantages of this type of teaching is the objectivity of student assessment and the real knowledge that students acquire in this way. Although today's modern distance learning platforms allow teachers to create tests and set time limits for creating tasks, the problem of independence of solving these tests remains questionable. This topic will certainly be much discussed in all the countries of the Western Balkans. In addition, teachers of practical classes in vocational technical schools believe that practical classes held online cannot replace those in school offices, companies, workshops, and hospitals.

In the process of distance learning, attention should be paid to the protection and authenticity of data exchanged between teachers and students. One way to authenticate is to use a digital signature in the exchange of messages sent electronically.

32.5 Conclusion

The previous 3 months have certainly been a great challenge for the entire education system of the Republic of Srpska. There have been difficulties that all of us who work in this sphere have encountered, but we can say that we are satisfied because the school year has ended successfully and most importantly the students have not lost the school year.

From the conducted research, we can conclude that students are satisfied with online teaching, although improvement should also be considered. COVID-19 has definitely brought changes in the education system of the Republic of Srpska and opened a new chapter in the use of modern technologies in education.

We hope that students and teachers will welcome the beginning of the next school year in the classrooms. We also believe that teachers will make the best possible use of the IT knowledge they acquired during the COVID-19 pandemic and improve the teaching process in the Republic of Srpska by combining traditional and modern online learning methods.

References

1. T. Talić, G. Radić, Z.Ž. Avramović, Digital signature and organization of decentralized authentication in business environment. JITA– Journal of Information Technology and Applications. Banja Luka, PanEuropien University APEIRON, Banja Luka, Republika Srpska, Bosna i Hercegovina, JITA 9(2019) 1:24–28, (UDC: 316.77:004.773]:658.14), (https://doi.org/10.7251/JIT1901024T), Volume 9, Number 1, Banja Luka (June 2019)
2. Republički pedagoški zavod Republike Srpske (n.d.), www.rpz-rs.org
3. Government of Republic of Srpska, Ministry of Education and Culture (n.d.), www.vladars.net/eng/vlada/ministries/MEC

Chapter 33
The Possibility of Using Distance Learning During the Emergency

Igor Grujić, Vladimir Domazet, and Zoran Ž. Avramović

33.1 Introduction

At the Pan-European University Apeiron in Banja Luka (Republic of Srpska, Bosnia and Herzegovina), classes are held at seven faculties, Faculty of Information Technology, Faculty of Business Economics, Faculty of Law, Faculty of Health Care, Faculty of Sports, Faculty of Transport, and Faculty of Philology, with 20 different courses. Classes are held for full-time students and part-time students. The largest number of students attends classical courses, and a smaller number of students attend distance learning classes. Distance learning is organized as online teaching. Distance learning program attended by students living in other parts of Bosnia and Herzegovina, neighboring countries (Serbia, Croatia), and other European countries or students enrolled on campuses in other cities in Bosnia and Herzegovina is similar to published research conducted in the United States [1]. Part-time study is attended by students who are already working and are generally older than 30 years. Regular classical classes are attended by students aged between 19 and 30, who live in Banja Luka or have moved to the city during their studies. Advantages and disadvantages of distance learning in relation to classical forms of teaching are a frequent topic of research and analysis of a large number of researchers in the world [1–3]. Researchers are particularly interested in examining the quality of distance learning and analyzing quality indicator [4–8].

I. Grujić (✉)
Nova Banka, Banja Luka, Bosnia and Herzegovina
e-mail: igor.r.grujic@apeiron-edu.eu

V. Domazet · Z. Ž. Avramović
Pan-European University Apeiron, Banja Luka, Bosnia and Herzegovina
e-mail: vladimir.n.domazet@apeiron-edu.eu; zoran.z.avramovic@apeiron-edu.eu

S. Bauk, S. D. Ilčev (eds.), *The 1st International Conference on Maritime Education and Development*, https://doi.org/10.1007/978-3-030-64088-0_33

With the outbreak of the COVID-19 pandemic, in order to protect the health of students and professors and prevent the spread of the disease, the competent authorities in the Republic of Srpska [9] prescribed certain measures that universities should follow. One of the measures refers to the interruption of classical classes and the obligation of the university to provide students with adequate conditions for learning and acquiring the necessary knowledge and skills. The Ministry of Scientific and Technological Development, Higher Education, and Information Society of the Republic of Srpska proposed online teaching, without suggesting specific forms of lectures and exercises. Professors provided various forms of teaching using available video and Internet technology (MS Teams, LMS, Skype, Viber). The Pan-European University Apeiron decided to use the existing resources and materials that are already prepared for distance learning. In this paper, the authors set a goal to make an analysis of the efficiency of the teaching process in the emergency situation caused by the COVID-19 pandemic.

33.2 Materials and Methods

During the examination of the efficiency of the teaching process in emergency conditions, three subjects from the second year of study were selected at all seven faculties within the Pan-European University Apeiron. The selection was made on the basis of a random sample: two compulsory courses and one elective course (Table 33.1). The analysis was conducted for two periods: winter semester (October–February) and summer semester (March–April). During the winter semester, classical teaching was organized and during the summer semester online teaching, while students were given learning materials prepared for distance learning.

The following parameters were observed: date of student registration in the e-classroom (shown by months of first connection and date of registration), type of learning materials mostly used by students, time spent reviewing online materials, and number of students by number of courses (e-classrooms) which they approached in the observed period. Data were obtained from the official documentation of the university, and the results were presented as a share (percentage) of the number of students who used e-classrooms in relation to the total number of students who enrolled in the lessons for observed teaching subject.

33.3 Results and Discussion

The results obtained by research are shown in three tables (Tables 33.2, 33.3 and 33.4). Classes at the Pan-European University Apeiron, Banja Luka, are organized in cycles, so that classes in one or two teaching subjects are organized in one cycle. As can be seen from Table 33.2, this way of organizing teaching has an impact on the first approach and registration in the e-classroom. In principle, students

Table 33.1 Overview of teaching subjects covered by the research

Faculty		Teaching subject	
Name	Abbreviated	Name	Abbreviated
Faculty of Information Technologies	FIT	Protection of computer and business systems	FIT 1
		Information systems design	FIT 2
		Business intelligence	FIT 3
Faculty of Business Economics	FPE	Macroeconomics	FPE 1
		Microeconomics	FPE 2
		Securities	FPE 3
Faculty of law	FPN	Criminal procedure law	FPN 1
		Civil litigation law	FPN 2
		Criminal law general part	FPN 3
Faculty of Nursing	FZN	Health care in pediatrics	FZN 1
		Surgical patient care	FZN 2
		Care of neuropsychiatric patients	FZN 3
Faculty of Philological Sciences	FFN	Contemporary pedagogy	FFN 1
		Russian language	FFN 2
		Old Slavic languages	FFN 3
Faculty of Sports Sciences	FSN	Sports medicine	FSN 1
		Fitness and shaping exercises	FSN 2
		Diagnostics in sports	FSN 3
Faculty of Transportation	SF	Passenger transport	SF 1
		Motor vehicles	SF 2
		Traffic regulation and management	SF3

registered in the e-classroom in the month when the classical classes were organized (e.g., FIT 2 just 7.9% of the total number of students, who took the course, joined in November and 13.6% in December). At the end of the cycle, the exam was organized, when these students were probably given a grade. In the months in which the exam period was organized (January and April), the access of new students is visible (3.0% and 2.3%). In other months, this number is negligibly small. After registering a student in the e-classroom, the software does not record subsequent approaches, so it is assumed that students have accessed the e-classroom several times, until they successfully pass the exam (28.7%). A similar analysis can be conducted for other subjects. Furthermore, from Table 33.2, it can be seen that the largest number of registered students was for the subject FSN 2 (100%) and the lowest for the subject SF 3 (6.7%). If the percentage of students registered in one of the 21 e-classrooms is classified in 4 categories, it looks like this: <10.0% for 6 teaching subjects, 10.0%–20.0% for 8 teaching subjects, 20.0%–50.0% for 6, and > 50.0% for 1 teaching subject.

The teaching process is based on interactive technology, owned by the university, and is used for distance learning. Between the learning materials deposited and available in e-classrooms, students most often used PDF documents and video

Table 33.2 Review of the students' first approach to the e-classroom

		Month							Total
Faculty	Subject	Octo.	Nov.	Dec.	Jan.	Feb.	Mar.	Apr.	
FIT	FIT 1	0,0%	1,8%	1,8%	3,6%	0,9%	1,8%	0,0%	10,0%
	FIT 2	1,1%	7,9%	13,6%	3,0%	0,4%	0,4%	2,3%	28,7%
	FIT 3	5,5%	2,2%	0,0%	1,1%	1,1%	1,1%	1,1%	12,1%
FPE	FPE 1	3,8%	2,9%	0,0%	1,0%	0,0%	4,8%	2,9%	15,2%
	FPE 2	1,1%	3,4%	2,3%	1,1%	0,6%	0,0%	0,0%	8,6%
	FPE 3	1,0%	6,7%	3,8%	0,0%	1,0%	1,9%	0,0%	14,3%
FPN	FPN 1	0,0%	0,0%	0,0%	1,1%	1,7%	2,8%	2,2%	7,7%
	FPN 2	0,0%	3,5%	1,2%	1,2%	1,2%	1,2%	2,3%	10,5%
	FPN 3	3,4%	3,4%	1,1%	1,1%	1,1%	3,4%	4,5%	18,2%
FZN	FZN 1	2,3%	4,5%	6,8%	6,8%	13,6%	6,8%	0,0%	40,9%
	FZN 2	0,0%	13,6%	15,9%	0,0%	2,3%	2,3%	0,0%	34,1%
	FZN 3	2,3%	9,1%	4,5%	2,3%	0,0%	0,0%	0,0%	18,2%
FFN	FFN 1	3,8%	0,0%	3,8%	11,5%	3,8%	0,0%	0,0%	23,1%
	FFN 2	0,0%	0,0%	0,0%	0,0%	4,3%	4,3%	0,0%	8,7%
	FFN 3	4,2%	0,0%	0,0%	0,0%	4,2%	0,0%	0,0%	8,3%
FSN	FSN 1	0,0%	1,1%	2,2%	1,1%	2,2%	1,1%	22,0%	29,7%
	FSN 2	0,0%	16,7%	16,7%	33,3%	0,0%	0,0%	33,3%	100,0%
	FSN 3	0,0%	0,0%	11,8%	17,6%	0,0%	5,9%	11,8%	47,1%
SF	SF 1	0,0%	0,0%	4,4%	4,4%	2,2%	2,2%	0,0%	13,3%
	SF 2	2,2%	0,0%	0,0%	2,2%	0,0%	4,4%	0,0%	8,9%
	SF3	0,0%	2,2%	4,4%	0,0%	0,0%	0,0%	0,0%	6,7%

material, while HTML documents were used very rarely or not at all (Table 33.3). This table shows that the total time spent in the e-classroom was the longest in the teaching subjects FIT 2 (105:35 hours), FPN 2 (89:57 hours), FSN 1 (47:25 hours), and FZN 2 (41:11 hours). This data indicated that more materials available for listed teaching subjects were more interesting for students or were prepared in a more acceptable way. It must not be forgotten that the materials available in e-classrooms are intended for distance learning students. Results of the research indicated that university management and faculty deans should pay attention to these data and in collaboration with professors adapt teaching materials and other teaching tools to students in classical studies. Of course, this requires serious preparation of the material and more time, which was not possible to provide in the situation that arose during the state of emergency caused by to the COVID-19 pandemic.

Regarding the analysis of teaching material type that students used, it can be concluded that easily downloading documents dominates. It seems that the students of classical studies used them by transferring and recording to personal computers and used them on the same way as printed textbooks. Classical students used very little HTML documents (Table 33.3), which confirmed the previous statement. In order to improve the quality of teaching in an emergency situation, the university management should consider this problem seriously. At the beginning, faster

Table 33.3 Time that students spent reviewing available content

Faculty	Subject	Hours in first (winter) semester				Hours in second (summer) semester				Total hours for both semesters
		PDF	HTML	Video	Total	PDF	HTML	Video	Total	
FIT	FIT 1	0:00	1:06	0:57	2:04	0:00	0:12	0:15	0:27	2:31
	FIT 2	37:32	0:00	61:34	99:07	3:01	0:00	3:27	6:28	105:35
	FIT 3	0:00	0:00	1:18	1:18	0:00	0:00	3:15	3:15	4:33
FPE	FPE 1	2:14	2:14	3:38	8:07	0:59	0:00	1:06	2:05	10:12
	FPE 2	10:11	0:00	2:20	12:31	0:09	0:00	0:21	0:30	13:01
	FPE 3	21:08	0:00	2:40	23:49	0:03	0:00	0:03	0:06	23:55
FPN	FPN 1	0:33	0:00	19:05	19:38	24:01	0:00	46:17	70:18	89:57
	FPN 2	0:00	0:00	3:14	3:14	0:00	0:00	0:34	0:34	3:48
	FPN 3	0:57	0:00	3:37	4:35	1:26	0:00	13:09	14:35	19:10
FZN	FZN 1	5:36	0:00	10:54	16:30	3:47	0:00	14:16	18:04	34:34
	FZN 2	0:00	0:00	41:05	41:05	0:06	0:00	0:00	0:06	41:11
	FZN 3	0:00	0:00	1:00	1:00	0:00	0:00	0:09	0:09	1:09
FFN	FFN 1	0:00	0:00	1:32	1:32	0:00	0:00	0:00	0:00	1:32
	FFN 2	0:06	0:00	0:00	0:06	0:09	0:03	0:12	0:24	0:30
	FFN 3	0:00	0:00	15:33	15:33	0:00	0:00	0:00	0:00	15:33
FSN	FSN 1	0:00	0:00	0:24	0:24	0:00	0:00	47:01	47:01	47:25
	FSN 2	3:45	0:00	9:58	13:43	0:06	0:00	0:00	0:06	13:49
	FSN 3	2:46	0:00	0:03	2:49	0:15	0:00	0:00	0:15	3:04
SF	SF 1	0:00	0:00	0:09	0:09	0:00	0:00	0:06	0:06	0:15
	SF 2	0:00	0:00	0:21	0:21	0:00	0:00	0:06	0:06	0:27
	SF3	0:00	0:00	1:03	1:03	0:00	0:00	0:00	0:00	1:03

Table 33.4 Distribution of students (in %) who gained first access to e-class teaching subjects by month

No. of teaching subjects	Oct.	Nov.	Dec.	Jan.	Feb.	Mar.	Apr.	Total
1	4,2%	5,8%	7,5%	7,5%	6,0%	10,6%	8,9%	50,5%
2	2,1%	3,9%	3,4%	4,1%	2,7%	5,8%	3,8%	25,8%
3	1,2%	2,1%	1,8%	1,3%	1,0%	2,3%	2,2%	11,8%
4	0,7%	1,1%	0,6%	0,7%	0,5%	1,4%	1,2%	6,2%
5	0,3%	0,6%	0,4%	0,2%	0,1%	0,4%	0,4%	2,2%
6	0,2%	0,2%	0,1%	0,1%	0,2%	0,4%	0,2%	1,4%
7	0,3%	0,2%	0,1%	0,1%	0,1%	0,2%	0,1%	1,1%
8	0,1%	0,1%	0,0%	0,0%	0,0%	0,1%	0,0%	0,4%
9	0,0%	0,1%	0,1%	0,0%	0,0%	0,0%	0,0%	0,3%
10	0,0%	0,0%	0,0%	0,0%	0,0%	0,0%	0,0%	0,1%
11	0,0%	0,0%	0,1%	0,0%	0,0%	0,0%	0,0%	0,2%
Total	9,2%	14,1%	14,1%	13,9%	10,7%	21,2%	16,8%	100,0%

improvement could be proposed to organize training for professors and administration who prepare educational materials and participate in communication with students. In addition, it is necessary to organize IT and other forms of training for students regarding the use of available teaching materials in an adequate manner. Communication between professors and students was two-way. In addition to e-mails, which were regularly used by teachers and students, outside the usual time of lectures, interactions with students also were held by phone, discussions within groups on social networks (Facebook, Instagram), SMS, email, Viber, Skype, telegrams, WhatsUp, and other forms of communication.

Classical students are registered in a different number of e-classrooms (Table 33.4). 50.5% of students used materials for one teaching subject, 25.8% for two, and 11.8% for three. A very small number of students registered to 7 or more subjects (12 subjects students need to pass in 1 academic year). The largest number of students registered in e-classrooms in March (21.2%) and April (16.8%), when we look at the month registration distribution.

During the emergency situation caused by the COVID-19 pandemic, the Pan-European University Apeiron in Banja Luka reorganized its work in a very short period of time and by combining classical forms of teaching with some elements of distance learning ensured a satisfactory level of teaching quality, and students gained a high level of knowledge.

33.4 Conclusion

It is noticeable that students of different faculties (different scientific fields) used materials intended for distance learning students in different forms and that they first made the first connection and registration in the e-classroom during the cycle planned for teaching a specific subject or in the period before exams. Students most often used documents that are similar to the classic textbooks (PDF and video), and during the preparation of the exam, they kept the learning habits from the classical system. In this case, e-learning is a good addition to classical teaching. In order to increase the use of more forms of e-learning, the university should organize courses for the teaching content development (for professors and administration) and courses for the use of e-materials, Internet, video techniques, and other opportunities for better quality of professor-student interaction (e-mail, telephone, chat rooms, and other available modern tools).

References

1. The National Education Association (NEA), *A survey of traditional and distance learning higher education members* (Washington, DC 20036. www.nea.org/he, 2000), pp. 1–54
2. H. Attah, A review on e-learning technology. J. Comput. Eng. Technol. **2**, Nabu Research Academy, 1–3 (2015)
3. E. Gulay, Existing criteria determining course quality in distance education. The Online J. Qual. Higher Educ. **4**(4), 17–24 (2017)
4. V. Kuleto, S. Pokorni, B. Miranović, Obezbjeđivanje kvaliteta veb materijala za elektronsko učenje u LINKgroup i ITS. Infoteh – Jahorina **12**, 747–752 (2013)
5. A. Ogunleye, Quality assurance and quality indicators in opet and distance education: Context, concerns and challenges. Int. J. Educ. Res. Technol. **4**(2), 49–62 (2013)
6. B. Wachter, M. Kelo, Q.K.H. Lam, P. Effertz, C. Jost, S. Kottowski, *University Quality Indicators: A Critical Assessment, Study* (Directorate general for international policies, European parliament, Brussels, 2015)
7. M. Petković, A. Živaljević, Indikatori kvaliteta kao osnov procene kvaliteta i procesa komparacije visokoobrazovnih institucija, XII skup trendovi razvoja. "Bolonjski proces i primena novog zakona" Kopaonik, 120–123 (2006)
8. B. Lungulov, *Analoza ishoda učenja kao indikatora kvaliteta visokog obrazovanja, doktorska disertacija, Univerzitet u Novom sadu* (Filozofski fakultet, Novi Sad, 2015)
9. Ministry of Scientific and Technological Development, Higher Education and Information Society of Republic of Srpska (Ministarstvo za naučnotehnološki razvoj, visoko obrazovanje i informaciono društvo Republike Srpske), Odluka nadležnih vlasti u Republici Srpskoj o organizaciji nastave u vandrednim uslovima nastalim zbog pandemije Covid-19 br. 1905/616-7-1/20 od 16.03.2020. (2020) (in Serbian)

Chapter 34
E-Education During COVID-19 Pandemic: Apeiron University Practical Experiences

Sinisa Tomić, Dalibor Drljača, and Zoran Ž. Avramović

34.1 Introduction

The COVID-19 pandemic has closed educational establishments around the world and tested the educational world's preparedness for emergency management. At the time of writing, educational institutions are discovering that they will have to finish the current school year with many problems. Many universities were not prepared for this situation. This crisis has not surprised e-education professionals who research and cover these topics in scientific journals. First of all, it surprised the university leadership and the state ministries involved in education.

COVID-19 has put on the test secondary student online systems and has exposed their vulnerability. Everybody had some kind of online solutions, but few universities only were able to provide quality response to the total isolation. Some institutions adapted to online students very quickly because they already had the tools and developed the distance learning (DL) system. This resulted in less disruption for these universities and their students. COVID-19 has forced the educational community to use IT technology causing, besides the combined hybrid way of teaching, the creation of a digital online learning environment that could become the primary way of teaching in the future. These changes also require new generations of lecturers. Analytics will be more represented as it will be used to follow up many data related to the activity and efficiency of students and lecturers. Universities will quickly realize that distance learning will bring many benefits such as reduced costs, automation, uniformity and activity control, energy savings, reducing vis-a-vis interaction needs, etc. It would be very interesting in the future to make a compara-

S. Tomić (✉) · D. Drljača · Z. Ž. Avramović
Pan-European University APEIRON, Banja Luka, Bosnia and Herzegovina
e-mail: sinisa.m.tomic@apeiron-edu.eu; dalibor.p.drljaca@apeiron-edu.eu; zoran.z.avramovic@apeiron-edu.eu

S. Bauk, S. D. Ilčev (eds.), *The 1st International Conference on Maritime Education and Development*, https://doi.org/10.1007/978-3-030-64088-0_34

tive study of the competences of students who acquired knowledge by learning online at the time of the pandemic, with the competences of students attending classical teaching process. In the following lines, we will present some of the research performed at Pan-European University Apeiron, Banja Luka, related to the learning process using their distance learning system – Learning Cubes 4.0. We will then provide an analysis of the state of education in the context of the COVID-19 pandemic, as well as the vision of further distance learning system development.

34.2 The Status of E-Education at Pan-European University Apeiron Before the COVID-19 Pandemic

The survey was mostly conducted using a questionnaire [1] that was created in the form of a Web-based application, and as a platform, Drupal 7, a content management system, was used. Most of the collected responds were realized using a printed form on-site at the premises of Pan-European University Apeiron, Banja Luka, with the presence of a control person, who instructed the students on the importance of the survey and gave an additional dose of seriousness during the filling, as recommended by Goran Milas [2, p. 467]. The study included 294 students, which were mostly students in the second, third, and fourth years of the first cycle academic studies. A particular quality of the research is given by the participation of students of the second and third cycle – masters and doctoral studies. The following chart shows the participation of Pan-European University Apeiron students surveyed by year and type of study (Fig. 34.1).

The survey sought answers to the questions:

- What is the representation and utilization rate of ICT on Pan-European University Apeiron, Banja Luka?

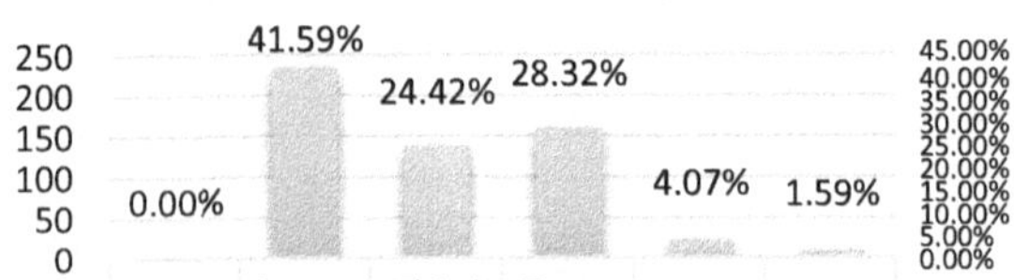

	First year of Bachelor stud...	Second year of Bachelor stud...	Third year of Bachelor stud...	Fourth year of Bachelor stud...	Master studies	Doctoral studies
No of students	0	235	138	160	23	9
%	0.00%	41.59%	24.42%	28.32%	4.07%	1.59%

Fig. 34.1 Year and type of studies

- To what extent are educational institutions prepared to use new ICT technologies in teaching?
- What are the effects of the use of multimedia in e-education on Pan-European University Apeiron taking into account all the specificities of this educational space?
- How much is e-education represented in higher education on Pan-European University Apeiron?

The infrastructures, tools, methods, and concepts of collecting, processing, and publishing multimedia content through e-education system were investigated. The readiness of the teaching staff and students to accept and use new educational concepts based on modern ICTs with the indispensable use of multimedia in e-education is analyzed, and suggestions are given on how to improve the existing e-education on Pan-European University Apeiron.

The research fully or partially answered the following questions:

- To what extent are ICTs represented in e-education in our country?
- What is the willingness of teachers and students to use ICT in e-education?
- What are the most commonly used e-learning models in higher education in Bosnia and Herzegovina, and why?
- What technologies and tools are used in the creation of multimedia content, and what is the quality of the content?
- To what extent does the existing information and communication infrastructure provide the technical prerequisites for quality e-learning delivery?
- In what direction will e-education move in B&H?

This paper does not present all the results and all considerations for the above questions and the results obtained due to its nature and limitations. Therefore, in the following articles, we will only present part of this research.

Survey Claim 1 Teachers and assistants enrich classical teaching in the classroom with educational electronic content (PPT presentations, PDF and Word documents, video, audio, etc.) (Fig. 34.2).

The picture clearly shows that the vast majority of respondents have a positive experience with classroom teaching using multimedia electronic educational materials. It can be seen from the response that the teaching staff in direct teaching is using electronic educational materials to a significant extent. Here are some reasons why: (a) It is easier to teach using multimedia materials already prepared. It is linearly passed through the classes (it is less possible to skip or forget something), and the prepared materials are also used in the next school year in the same form or most often refined to a lesser extent. (b) The teaching institution expects the teaching staff to have a modern and innovative approach to teaching. (c) Students love new technologies and this approach suits them.

Survey Claim 2 I am generally satisfied with the quality and volume of electronic content provided by my institution through the distance learning system (Fig. 34.3).

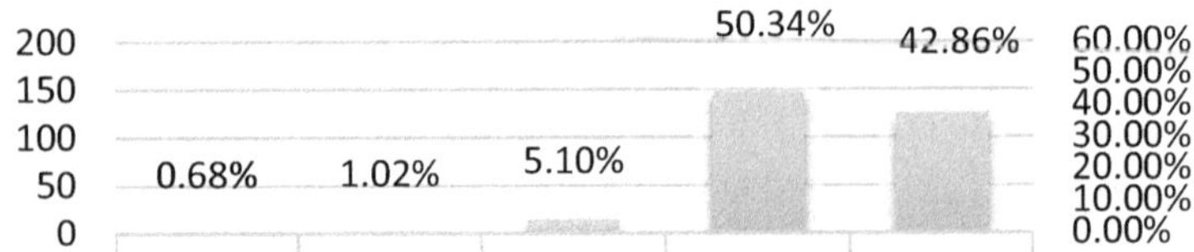

	Completely disagree	Disagree	Undecided	Agree	Completely agree
No of students	2	3	15	148	126
%	0.68%	1.02%	5.10%	50.34%	42.86%

Fig. 34.2 Average rating 4,34

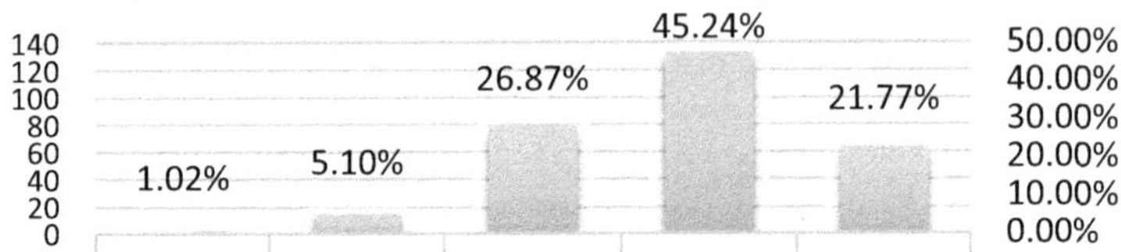

	Completely disagree	Disagree	Undecided	Agree	Completely agree
No of students	3	15	79	133	64
%	1.02%	5.10%	26.87%	45.24%	21.77%

Fig. 34.3 Average rating 3,82

The volume and quality of electronic content in DL systems very often do not depend solely on the teachers in charge. Well-organized and well-prepared educational multimedia material requires the involvement of some other persons, such as video editors, Web developers, graphic artists, speakers, animators, and certainly administrators. A virtual subject often contains materials that are not directly related to the actual lecturer. In distance learning systems, it is often the case that older video material is stored from previous presenters or PPT presentations or scripts from years past. Therefore, in the statement, instead of "... the lecturer provides me ...," it states "... which my Institution provides for me ...." Delivering educational multimedia materials through the distance learning system requires considerable investment in infrastructure, training, preparation, processing and publication of materials, support, etc. In general, the results obtained are encouraging and show that there are educational institutions that support and invest in a distance learning way of teaching and listening to the teaching.

Survey Claim 3 The other electronic resources I find (YouTube video, electronic encyclopedias, electronic libraries, presentations, etc.) that I need to successfully pass the course and prepare for the exam are extremely important to me (Fig. 34.4).

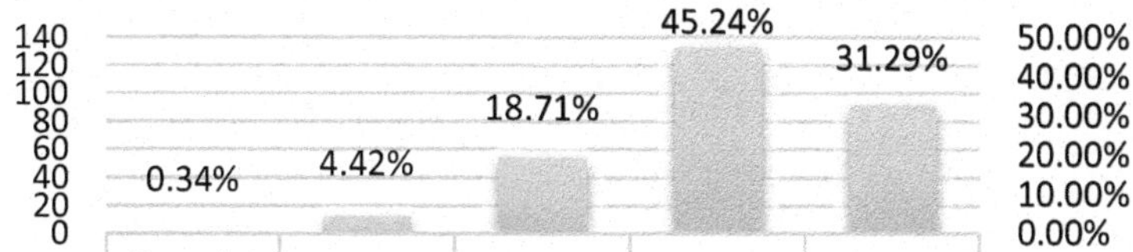

	Completely disagree	Disagree	Undecided	Agree	Completely agree
No of students	1	13	55	133	92
%	0.34%	4.42%	18.71%	45.24%	31.29%

Fig. 34.4 Average rating 4,03

It would be unrealistic to expect the students' research spirit to be limited to classical teaching and predefined electronic multimedia materials. The Internet has globalized knowledge, now available in an incredible variety of forms, with access with a few mouse clicks. For example, the big YouTube has grown from entertainment to the largest educational multimedia video service in the world. Especially today, in a time of rapid change, expansion of knowledge, and new technologies, free access to world resources is getting full sense. Pan-European University Apeiron is aware of the need to acquire knowledge beyond the usual educational forms. That is why it gave its students free access to the most respectable protected databases of electronic books, magazines, journals, etc. Access to protected databases is extremely useful for students, as they can find relevant and up-to-date knowledge that is not published freely on the World Wide Web.

Survey Claim 4 Generally, the institution I study at is ready for the use of new ICTs and invests significant funds in material and technical and logistical support in delivering e-learning (Fig. 34.5).

New multimedia educational concepts require new technologies, considerable financial investments, adequate materials, and technical support, with the main goal of acquiring knowledge in quality and interesting way. The teaching model at Apeiron University fits into a hybrid model of teaching, where e-learning is intertwined with the classical methods of teaching and learning, and this is obviously the model most commonly encountered in B&H. Most respondents found that the distance learning systems they access complement the classic classroom teaching excellence. For example, Pan-European University Apeiron performs screen capture in video format, which is displayed in HD quality combined with associated classroom videos. Thus, for example, the exercises in the higher programming languages – C++ – are performed in a computer room with video projection, where students monitor the performance of the tasks (programming) with the lecturer and then the recorded activities from the lecturer's screen are later thoroughly reviewed and the tasks are taken home (the programming code is seen). Some teaching activities require dominant classical teaching where the role of multimedia via the DL

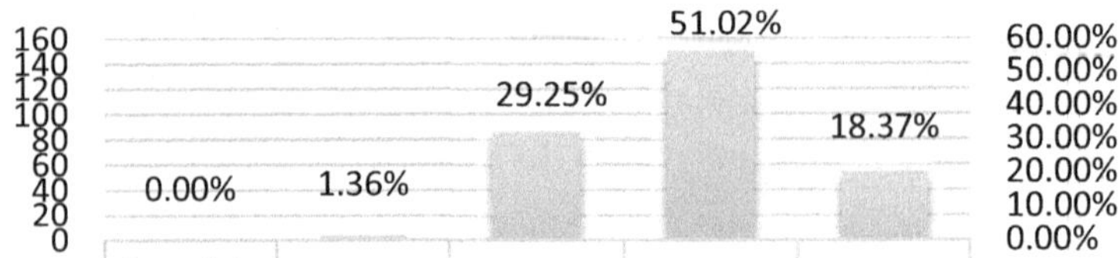

	Completely disagree	Disagree	Undecided	Agree	Completely agree
No of students	0	4	86	150	54
%	0.00%	1.36%	29.25%	51.02%	18.37%

Fig. 34.5 Average rating 3,86

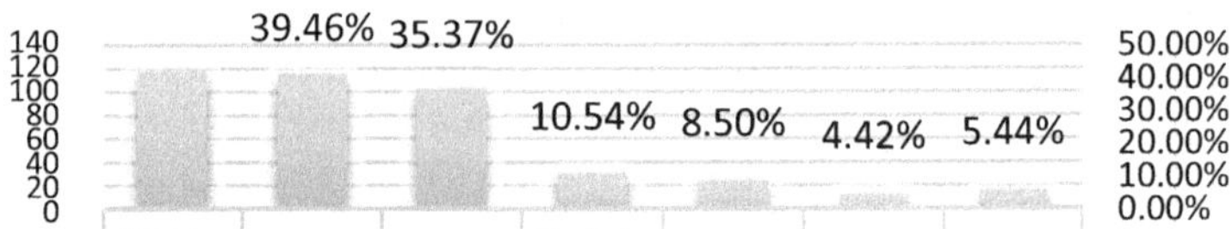

	Video records of the teac...	PPT presentations	PDF files	Scripts and other similar docu...	HTML documents	Animations	Audio files
No of students	118	116	104	31	25	13	16
%	40.14%	39.46%	35.37%	10.54%	8.50%	4.42%	5.44%

Fig. 34.6 Survey results

system is diminished. An example of such activity is the practical fabrication of a denture in the dental laboratory of the faculty.

Survey Claim 5 Which type of multimedia content is most important to you? (Fig. 34.6)

Students were provided with multiple answers for this claim. The results show the affinities of the students toward important multimedia content. Three types of multimedia content stood out by representation: videos of lectures and exercises 40.14%, PPT presentations 39.46%, PDF documents 35.37%. These results can guide Apeiron to which multimedia content should be given special attention and continue to work on their quality. Other types of multimedia content that are less important to the students surveyed should by no means be ignored.

The research sets up one main and three auxiliary hypotheses that have received the following epilogue:

- H.1: By using multimedia in teaching, heterogeneous positive effects of knowledge acquisition and dissemination are achieved. (*Proven*)
- A.H.1: The application of multimedia with the use of modern computer technologies positively influences students' motivation to attend teaching activities. (*Proven*)
- A.H.2: Multimedia in electronic education becomes the standard for quality teaching at Pan-European University Apeiron. (*Proven*)
- A.H.3: The willingness of teachers and students in higher education at Pan-European University Apeiron to adopt and use new educational concepts based on multimedia with the use of modern ICTS is different. (*Not proven*)

Pan-European University Apeiron has built a system where videos from classrooms are synchronized in real time with presentations made by teachers on screen. This simulates the physical presence of students in the classroom. Students who use the system can also view video lectures from past years, so they are more advantageous to those who follow classical teaching in this regard. Access to all required materials, at any time, reduces material costs and gives the flexibility of teaching time. There are around 400 courses taught at Pan-European University, and it is quite clear that it is a huge challenge to translate all of these courses into virtual form. Pan-European University Apeiron has found a great way to translate all subjects into video form, and it involves the automatic recording of all classroom activities via an IP camera, with automatic screen capture of on-screen activities, which combine very useful video material in real time. Later, in simple postproduction, breaks in lectures are eliminated, so only useful video material is published in the distance learning system. About 10,000 hours of television-edited video lectures are currently published in the distance learning system, which is a very respectable knowledge base.

34.3 E-Education in the Context of COVID-19 Pandemic

In response to the spread of the COVID-19 in early 2020, almost all European countries have stopped teaching classrooms and activated e-learning systems at the same time. In the context of the COVID-19 pandemic, students left their student campuses and continued online communication with their universities. Neither state nor university can tell to students how long the pandemic will last and when the universities will reopen their rooms and continue working as usual. Many universities are self-financing from their students' tuition fees, so this situation is a big problem for them as well. The lack of cash inflows can paralyze or destroy such universities. Students have returned to their cities, and some of the cities are in strong isolation. It turns out that many educational institutions in Europe and the world do not have good organization, that their online learning resources are very limited, and that they have not invested enough money or knowledge into the DL systems they now need. In the future, students and their parents are likely to think that their future studies will be at universities that have good online study solutions. We can also expect a transcript of students from institutions that weren't able to provide an adequate online learning environment to institutions able to provide quality online teaching and to complete the academic year. The policy of enrolling educational institutions in marketing campaigns will also change, where their ability to perform online teaching in crises will be enhanced.

The mobility of international students could take up to 5 years to recover from the coronavirus pandemic across the world, as universities continue to grapple with the "extraordinary set of challenges" the crisis has created. This is the view of Simon Marginson [3], Director of the Centre for Global Higher Education, speaking at the first virtual iteration of UUKi's annual conference. Asked if universities should price differently for online programs, he said, "If online is going to become a longer-term substitute for face-to-face learning, as it will in some cases, it needs to be seen as a substantially different product.. and it will need a separate pricing structure. The idea that we charge exactly the same price for any kind of online [product] as we charge for face-to-face has to go" [3].

Teachers will adapt to online teaching, and this is a process that cannot be stopped, but realistically, face-to-face teaching is again the primary form of teaching in the new school year.

A large number of universities had their distance learning systems only as a backup for study, resulting in difficult adaptation to the new situation. Creating a good online student environment requires significant financial, logistical, logistical, and other investments, and this takes considerable time. An interesting statement from the Rector of the University of Belgrade, Ivanka Popovic, says: "The exams at the faculties will have to be postponed until the state of emergency is completed because the University does not have enough experience to provide completely objective conditions for taking the exams online. As soon as the state of emergency is completed, intensive classes in blocks will be organized for those faculties where practical exercises and experimental examinations are required" [4].

Effective e-learning models must be guided by sound pedagogical principles and be flexible in order to adapt to the needs and goals of students. The literature recognizes different teaching and learning strategies – linear and constructivist; some authors advocate teacher-centered and other process- and procedure-based learning, but each model deserves attention and consideration as the choice for selection of the learning model should depend on the goals of the program and the needs for the trainees (students) [5, 6]. E-learning delivery methods, methodology, and even knowledge delivery media vary from state to state and from institution to institution. Again, the old problem arose: How and in which way to conduct final examinations, and how to evaluate students' knowledge? Given the volume of the subject matter and the specific nature of individual teaching areas and subjects, this problem is further complicated. Today, all modern distance learning systems can test knowledge, create tests, lead discussions, measure the advancement of students, etc. The question is: How much these opportunities are used? It is technically no problem to provide basic testing and simple automatic knowledge checks, e.g., selecting offered answers or linking offered terms. Problems arise when descriptive, written answers to questions are required. Automatic scoring is then dropped because the systems do not currently have sufficiently developed artificial intelligence that can intelligently analyze and score such answers (this is being done).

A common problem is that simple automatic forms of assessment are not used fully. The reason for this is a non-systematic approach to solving the problem identified, or there is no clear and firm position of the management that the teaching staff is obliged to create the required number of online tests or quizzes that would be published within the existing LMSs. On the other hand, teachers would have to invest considerable effort in preparing such materials, and when it comes to a large system with a large number of teachers, there are inevitably various problems. Of course, in addition to automatic proficiency testing through DL systems, there is also the possibility of proficiency testing through video chat, e.g., Skype, where the teacher can verbally either examine the students or control their written exam. Videoconferencing has great potential, and it is one of the ways to achieve good communication with students with minimal investment.

Often, educational institutions and their teachers are averse to new solutions because new user environments or new functionalities drive them out of their comfort and already established practices and habits. So the authors (Bauk, Kopp, Avramovic) in their work [7] describe the difficulty of upgrading the Moodle system and raising its functionality to a higher level at the Faculty of Maritime Studies, which is a part of the University of Montenegro: "... Since the program surface of Moodle rather changed with the release of Moodle 2.x FMS decided to stick to the older version. Mainly this is due to two reasons: 1) Teachers and students are used to the look and feel of the 1.9.x versions and it seemed problematical for them to grow accustomed to a new surface especially at an early stage of working with the platform; and/or 2) The installation of Moodle 2.x demands an enhanced technical environment which is not totally available at the FMS at the moment."

Under current laws, a proportion of European countries, including Bosnia and Herzegovina, are not legally allowed to conduct examinations through electronic learning systems. The solution to this specific problem at the time of inability to attend the classical classes due to COVID-19 has come to light, and according to state officials, this problem must be resolved as soon as possible, as the current 2019–2020 academic year needs to be concluded. Laws in this regard must be refined and allow students to take the distance exams because, in a pandemic, this is the only way to end the current school year regularly. The state will have to prescribe standards regarding distance learning systems and online learning in the coming period, as it is its interest and its competence.

A significant part of the funds that could improve the quality of teaching through distance learning systems comes from different funds. Thus, the EU has urgently adopted assistance programs for its member states as well as for non-EU countries in connection with the COVID-19 pandemic. Universities in need of funding for distance learning are eligible to apply and participate in grants.

In the era of the information society, where computer literacy is seen as a part of general literacy, information technologies have been given the function of education, and as such, they have brought some new ways of learning. By modernizing the teaching methods supported by appropriate hardware and software and by using multimedia teaching resources, the main directions of modern education development are defined [8, 9]. Regardless of the student motivation and satisfaction with the teaching forms of today, it is necessary to introduce continuous innovation also at the level of an individual. As this and other studies show, students and teachers are still not enough aware of the possibilities to make their jobs easier and practical with an appropriate form of providing information [10].

Peter Brandt – Head of the "Knowledge Transfer" Department at the German Institute for Adult Education, Leibniz Centre for Lifelong Learning in Bonn [11], points out that it is crucial to note that the corona crisis has caused a clear lack of alternative – digital tools for educational processes need to be used, where these tools no longer compete with face-to-face learning but are a prerequisite for organized learning in a state of emergency. Perhaps in a few years, the 2020 crisis will be identified as the central catalyst for digitization in education.

34.4 Conclusion

The academic community must be able to embrace the development of new IT technologies and outstanding multimedia capabilities in order to provide more dynamic teaching and learning and more efficient use of space, time, and financial resources. The vision for further development of higher education in the world was given by the Minister of Scientific and Technological Development, Higher Education, and Information Society of the Republic of Srpska, and he emphasizes the following: "This emergence of the corona virus situation, in addition to all the negative consequences we are facing and will face, has shown us something important, namely that the teaching process can be innovated in line with modern, technologically supported trends. The assessments by foreign experts that I have been able to see indicate that, concerning higher education, after this pandemic, it will experience a lot of changes and new practices globally, that higher education institutions will increasingly turn to new teaching methods, and that the traditional practice of physical presence continues to become the exclusivity of the most expensive." Knowledge delivery systems, communication, and basic control activities have reached an enviable level.

However, the systems for final assessment are still not sufficiently developed, and we can expect their rapid development soon. In the future, the right solutions lie in intelligent two-way communication [12] between intelligent tutoring systems and students, where the e-learning system contains intelligent methods for analyzing, evaluating user knowledge and skills, as well as controlling e-learning processes, monitoring, and optimization.

References

1. S. Tomić, Efekti i induktivni karakter multimedijalnog koncepta elektronskog obrazovanja u Bosni i Hercegovini, PhD thesis, Sveučilište/Univerzitet Vitez, Vitez (2017)
2. G. Milas, *Istraživačke metode u psihologiji i drugim društvenim znanostima* (SLAP, Zagreb, 2009)
3. www.thepienews.com, 20 Mart 2020. [Online]. Available: https://thepienews.com/news/five-year-recovery-period-predicted-for-global-student-mobility/. Accessed 2 Apr 2020
4. www.blic.rs, blic, 31 Mart 2020. [Online]. Available: https://www.blic.rs/vesti/drustvo/popovic-ispiti-tek-kad-prode-vanredno-stanje/dlj92hy. Accessed 3 Apr 2020
5. S. Tomić, D. Drljača, DDLM – Quality standard for electronic education programs in higher education of Bosnia and Herzegovina. JITA J. Inf. Technol. Appl. **9**(2), 67–79 (2019)
6. C.J. MacDonald, E.J. Stodel, L.G. Farres, K. Breithaupt, M.A. Gabriel., https://elearnmag.acm.org, [Online]. Available: https://elearnmag.acm.org/featured.cfm?aid=609737. Accessed 23 Nov 2019
7. S. Bauk, M. Kopp, Z. Avramović, A case study on introducing e-learning into seafarers' education. JITA J. Inf. Technol. Appl. **3**(1), 34–42 (2013)
8. S. Tomić, D. Drljača, Principles of multimedia learning. Knowl. Int. J. **26**(2), 557–564 (2018)
9. S. Tomić, D. Drljača, I. Grujić, The quality standard for multimedia content in e-education on the example of higher education in BiH. Knowl. Int. J. **34**(1), 263–270 (2019)
10. Ž. Pekić, S. Jovanovski, N. Pekić, The impact of Felder's learning styles index on motivation and adoption of information through e-learning. JITA J. Inf. Technol. Appl. **6**(2), 93–100 (2016)
11. P. Brandt, epale.ec.europa.eu, [Online]. Available: https://epale.ec.europa.eu/hr/blog/corona-und-die-weiterbildung-krise-oder-katalysator. Accessed 3 Apr 2020
12. A. Potode, P. Manjare, E-Learning using artificial intelligence. Int. J. Comput. Sci. Inf. Technol. Res. **3**(1), 78–82. (2015). T.F. Smith, M.S. Waterman, Identification of common molecular subsequences. J. Mol. Biol. **147**, 195–197 (1981)

Chapter 35
MET in Coronavirus Disease (COVID-19) Setting: A Curse or a Blessing in Disguise?

Nhlanhla Rodgers Mtshali

35.1 Introduction

The world is facing a scourge of COVID-19 which has created a sense of insecurity to humankind as there is still no vaccine to combat it at this stage. It is alleged that this virus started in a place called Wuhan (China) in November 2019 and has spread to a number of other countries in the world. Thousands of people have succumbed to the virus, with Italy being the country with the most number of deaths having surpassed China in recent weeks.

Several measures have been put in place by different countries to try and curb the spread of this deadly virus, and these measures include travel bans, lockdown and others to mention just a few. Life as we know it has changed, and people are engaged in various activities aimed at preventing the spread of the virus. Though the measures implemented are mostly seen as being negative, however, there are some positives that may be taken from the spread of the virus, and these will be looked at individually and discussed in the following discussions throughout this paper.

Firstly, I will discuss the negative impact that the virus has caused, and then the positives will be entertained thereafter. It will be noted that the impact both negative and positive will be discussed in the context of South Africa, but they can also be applicable to other countries in the world.

N. R. Mtshali (✉)
Durban University of Technology, Durban, South Africa
e-mail: NhlanhlaM6@dut.ac.za

S. Bauk, S. D. Ilčev (eds.), *The 1st International Conference on Maritime Education and Development*, https://doi.org/10.1007/978-3-030-64088-0_35

35.2 Disadvantages

Poor Families South Africa has high levels of poverty and unemployment, and as such, most people depend on selling different items as vendors in order to obtain some form of income. Considering what has been said, it is evident that the spread of COVID-19 has limited the movements of people from one place to the other due to lockdown which has put into effect as of midnight of Thursday, 27 March 2020. The resulting factor is that the source of income for the poor has been restricted to such an extent that more people will suffer as a result of the virus [1, 2, 3].

Socialisation and Interaction Limited movements of people have also reduced socialisation and interaction amongst individuals in the country. People are afraid to be in contact with others because one cannot tell if someone else has contracted the virus or not by just looking at an individual.

Boredom Most people find themselves indoors with no activities to engage in during the lockdown phase. Boredom is also linked to the point discussed in the paragraph above about socialisation. Many activities such as sports, education, entertainment events, etc. have all been banned, postponed or even cancelled.

Health As people find themselves locked-up in their house, unable to attend the gym and engage in exercises such as road-running, it is envisaged that there might be prevalence of diseases, e.g. hypertension, stroke, diabetes and so forth, which may have been reduced by exercising daily. People are advised to exercise indoors, but the limited space in some of the homes does not allow for effective training.

Housing In South Africa, there is a large number of people without proper housing, some are living in the streets, whilst others are found in cramped spaces such as in hostels or in shelters. The effect of this is that the possibility of contraction of virus by those individuals is increased hugely.

Water The country is still having a shortage of water supply in some areas (e.g. Hammanskraal et al.). Hygiene is very important especially during the spread of COVID-19; however, water shortage makes it difficult to maintain the required standard of hygiene. Households with sufficient clean water are expected to pay more for their water bills as people try hard to maintain a level of cleanliness.

Jobs and Unemployment A number of businesses have been closed during this difficult time of COVID-19, and many of them cannot generate revenue. The impact of this would be evident when many people are being laid-off because businesses won't be able to pay their staff after the virus has been contained or a cure is found.

Economy The economy of South Africa has particularly been unstable in the recent past, and this is reflected through the strengthening of dollar or the weakening of the rand against the dollar. The spread of COVID-19 has exacerbated the situation, and the levels of unemployment and poverty are expected to rise in the not so distant future.

35.3 Advantages

Emissions vs. Climate Change With the fact that most people will not be going to work during COVID-19 spread, it is expected that the emission of toxic gases, such as CO_2 from vehicles, will be reduced. The fact that less emissions than normal will occur serves as a positive to the protection of the environment. Climate change, though at a limited scale, will be reduced, and this is good for the environment.

Technology The rapid spread of COVID-19 has somewhat forced countries to accelerate their efforts in the implementation of the Fourth Industrial Revolution (4IR). Institutions such as companies, universities, schools and other organisations are forced to execute their duties remotely, staff members are required to work from home, and this is all possible with the use of technology.

Family Interaction Family members are seen to be spending more time together as parents and kids find themselves locked-up in their homes. The time spent should be seen as a way to strengthen the bond amongst family members, and not be used as time for children and women abuse.

Transmission of Diseases Individuals and families will be at home; consequently, the transfer of diseases among a large group of people will be reduced drastically during this time of COVID-19 pandemic.

Alcohol and Drug Abuse The lockdown as ordered by the South African Government, if implemented correctly and respected by all citizens, will result in the reduction of alcohol and drug abuse, which might lead to others quitting those hazardous substances all together.

Crime The lockdown will result in the deployment of a large number of the South African Police Services (SAPS) and the South African Defence Force (SANDF) members which must be seen as a plus in the prevention of crime in the country. It should be noted that criminals do not abide by the law in most instances; however, the deployment of these entities will act as a deterrent to those criminal activities.

Cleanliness The eruption of COVID-19 has forced or coerced people into looking after their hygiene by making sure that hands are washed or sanitised frequently than normal, which is supposed to be the case all the time. Strict measures are being put in place in stores and other amenities to safeguard against the spread of COVID-19.

35.4 Conclusion

In conclusion, South Africa is no different from other countries of the world in terms of the spread of COVID-19. The number of infections is seen to be on the rise every day, and it is expected that more people will be tested positive during the lockdown. However, there are positives to take from the spread of virus; in other words, it is not all doom and gloom. Just as the President of South Africa, Honourable Cyril Matamela Ramaphosa, indicated, we need not panic about the situation, but adhere to the regulations as stipulated by the government.

References

1. South African Government, President Cyril Ramaphosa: Measures to combat Coronavirus COVID-19 epidemic, https://www.gov.za/speeches/statement-president-cyril-ramaphosa-measures-combat-covid-19-epidemic-15-mar-2020-0000 (15 Mar 2020)
2. South African Government, President Cyril Ramaphosa: Escalation of measures to combat Coronavirus COVID-19 pandemic, https://www.gov.za/speeches/president-cyril-ramaphosa-escalation-measures-combat-coronavirus-covid-19-pandemic-23-mar (23 Mar 2020)
3. COVID-19 Corona Virus South African Resource Portal, Statement by President Cyril Ramaphosa on Progress in the National Effort to Contain the COVID-19 pandemic, https://sacoronavirus.co.za/2020/07/12/statement-by-president-cyril-ramaphosa-on-progress-in-the-national-effort-to-contain-the-covid-19-pandemic/ (12 July 2020)

Part VIII
Miscellaneous

Chapter 36
Usage of Information Technologies in the Digitalization Process of the Modern Architectural Office

Boris Pauković and Zoran Ž. Avramović

36.1 Introduction

Analyzing today's approach to and organization of business, regardless of the field, industry, whether production, construction or social, we conclude that information and communication technologies have a crucial role and little could be done in a fast, effective, safe and reliable way without those technologies. Of course, in order to be effective, it is necessary to perform precise and detailed planning of the entire information and communication system and adapt it to specific needs and solutions that will enable effective work.

After analyzing the work process of typical work of an architect, which is in itself demanding and complex when it comes to a single project or design and especially in a medium-sized company with more than 100 employees, we can see that the classic "two-dimensional" approach significantly slows work down, reduces competitiveness, and vastly increases the percentage of mistakes that ultimately lead to the loss of project and money or in the worst case miscalculations with far-reaching consequences. Also, the standard approach often leads to incompatible information and misunderstandings in the case of several architectural offices collaborating, from different locations, and working on the same project. The traditional way of transmitting information by telephone or sending drafts by regular mail or e-mail without a verification system is insufficient. It often leads to, as previously mentioned, mistakes, incompatibilities, and inefficiencies.

On the other hand, the introduction of information systems in the entire process, with proper planning and implementation of the appropriate software, allows the neutralization of all the said "weaknesses" and elevates the whole process to an

B. Pauković (✉) · Z. Ž. Avramović
Pan-European University APEIRON, Banja Luka, Bosnia and Herzegovina
e-mail: zoran.z.avramovic@apeiron-edu.eu

S. Bauk, S. D. Ilčev (eds.), *The 1st International Conference on Maritime Education and Development*, https://doi.org/10.1007/978-3-030-64088-0_36

impressive level. Building information modeling (BIM) as a digital approach to collaboration and modeling in architecture and construction represents a great help and progress in solving the said problems. After the proper implementation of basic BIM functions, i.e., the introduction of this type of collaboration when working on projects and educating users on it, we come to another great advantage of this system which is to expand functionality with various plug-ins/add-ons, which enable control of elements such as statics, geometry, stability, integration, physical security with quality control, rendering and three-dimensional representations, and as the last step virtual and augmented reality with animations that allow the entire project to be visually portrayed and presented to investors before implementation and construction. As a result of a successful assembly of these parts into a whole, we get a modern architectural office that can work smoothly and synchronously, regardless of the size of the project, the location of the office, and the speed of the Internet connection.

36.2 Definition of BIM: Advantages and Risks

Advanced and complex collaboration that is an integral part of most architectural projects today would not be successful without the use of BIM. Therefore, BIM can be defined as a collaboration platform that allows for more effective operation and represents a step up from 2D CAD implementation where each part is designed separately. It allows the use of three-dimensional modeling that is more intelligent and dynamic, with project status actualization in real time as well as simultaneously. The entire process integrates all parties by presenting the final product in a three-dimensional display with all necessary information for design and construction. In order to successfully implement BIM, it is necessary to have a combination of technology, hardware, infrastructure, processes, and users themselves. It is important to emphasize that collaboration, in this case, cannot be defined as a classic exchange of information electronically, nor as visualization or online sharing of documents. Collaboration, in this case, represents a systematic communication strategy, protocols, and organization all implemented for better cooperation, coordination, and integration. In order for BIM to be successfully implemented, it is necessary to go through a digital transformation of architectural processes and skills [1].

The team in charge of information and communication technologies plays a key role in the process of digital transformation. The role of the IT team is crucial for the selection of hardware, software installation and implementation, and solutions for creating backups and recovery in case of failure and technical support in the process of implementing BIM solutions. Furthermore, after the implementation, the IT team provides further availability as well as updates and maintains the system.

Some of the risks that the implementation of the BIM collaboration system brings are the unwillingness to accept innovative solutions and the feeling of ineffectiveness among employees (a conclusion reached based on internal conversations with architect colleagues). Many did not feel technologically proficient and computer literate enough to switch to a "nonstandard" new solution. Leaving the design of information and communication systems, hardware selection, creating network infrastructure to insufficiently competent engineers and IT architects presents a significant risk as well. In addition to improper maintenance and failure to regularly update and create backups, in that case, downtime is inevitable, which can partially or entirely stop the entire collaboration and lead to unproductivity, unavailability of solutions, and general dissatisfaction with the introduction of this design method.

36.3 IT Requirements for BIM Implementation

Proper implementation of BIM solutions includes detailed planning of hardware and software, licensing methods, and computer network design. It is crucial to choose the right hardware performance and configuration of the server that will host the project to have the precise design of the computer network that will allow clients to connect to the servers and the configuration of the workstations themselves which should be powerful enough to communicate with servers, process data, and send feedback. Improper initial configuration can lead to congestion in communication and latency, resulting in termination of communication with the server.

Server infrastructure and storage – to ensure smooth operation and reduce the risk of downtime, redundancy needs to be provided. Most of today's BIM solutions are based on Microsoft Windows architecture, so the Microsoft Windows environment is the most reasonable choice. Proper implementation means that the infrastructure consists of several servers, where each is in charge of a particular service. For example, we need one server for licensing purposes; a user database and central management of the entire BIM environment; several servers that will house projects organized by size, criticality, and priority; a server that will be the central file repository; a server for construction software; etc.

Furthermore, the Windows environment requires an active directory, domain controllers, print servers that will allow centralized printing of large format designs, and more.

Based on the previously defined number of servers, virtualization is proposed as the most elegant solution and Microsoft Hyper-V as a reliable option. To achieve redundancy, it is necessary to set up a failover cluster that allows "live migration" of virtual machines between hosts and the proper allocation of resources. Also, configuring central storage has proved to be the best practice, independent of hosts, one which contains SSD and HDD configured in a hybrid pool where the so-called hot data is stored in SSD, which "act" as cache, while the rest represents standard storage. The hybrid configuration significantly reduces costs because the so-called "all-flash" solutions are still quite expensive.

To ensure the smooth and fast flow of information and communication between servers and clients, it is necessary to carefully and correctly design the computer network. Redundancy is also necessary where critical, core switches are duplicated, and in order to allow flexibility and scalability, access switches are connected into stacks. It is also necessary to segment the computer network and configure VLANs to avoid congestion, increase the security of the management segment, and optimize the flow of information during communication between the client and server.

Using the example of Graphisoft Archicad software solution for BIM collaboration, we can see what the recommended hardware and software features which keep the process running with as little interference and downtime as possible are [2].

Hardware specifications for a BIM server:

- Processor: 64-bit processor with four or more cores.
- RAM: minimum 8 GB, for servers with multiple active users and projects optimally 32 GB.
- The minimum amount of free disk space is 10 GB for an active project.
- Bandwidth: in order for communication to be free of congestion and downtime, it is necessary to ensure a high-speed connection. The biggest requirement for high bandwidth is during the first log-in of the user to the project because then the process of downloading the entire project and the library to the local computer in the temporary folder is performed.

In the case of a computer network configurations that do not boast high bandwidth or use VPN, the first download of the project can take several hours or even days, depending on the amount of packages that must be downloaded.

The following table shows the recommended bandwidth rate, depending on the size of the project itself [3] (Table 36.1).

Latency – the higher the latency, the slower the round-trip between server and client. Since communication between the client and the server requires the exchange of a large amount of packages, latency needs to be reduced to a minimum. The higher the latency, the slower the communication between the server and the client, which results in very slow data exchange, long project loading, and generally poor performance of the overall work environment.

Practice shows the following:

- Good performance – low latency 1–2 ms
- Average performance – latency between 10 and 50 ms
- Acceptable but already slow performance – latency between 50 and 100 ms
- Poor performance – 100+ ms latency

In order for the whole process to run smoothly, it is not only necessary to optimize the server infrastructure and computer network. Clients also play an important role, i.e., workstations, where client software installations are located and where the

Table 36.1 Bandwidth depending on the size of the project

Project size	Bandwidth (upload/download)	Speed (upload/download)
100–300 MB	10/10 Mbps	~1 MB per second
300 MB–2 GB	100/100 Mbps	~12 MB per second
2+ GB	1/1 Gbps	~125 MB per second

actual designing is done. It is necessary to take into account their characteristics. If it is a schematic, less demanding type of project, the processing power of the Intel i5 processor is enough, with a minimum of 8 GB of RAM and a recommended graphics card with 4 GB of RAM. By increasing the complexity of the project and by introducing 3D visualization, hardware requirements increase as well, so it is recommended to configure clients with high-end configurations (Intel i9 processors of the latest generation, up to 64 GB RAM and graphics cards with 16 GB of RAM). The use of SSDs is definitely recommended for every client workstation [4].

Almost all BIM software works on the principle of multithreading. Considering that, processors must be multi-core or at least support hyperthreading. The advantages of using such processors are the speed of creating 3D models and rendering, background conversions, and other demanding calculations and operations. We can conclude that when configuring workstations, it is necessary to take into account the type of processor that will suit the complexity of the project and also to match the processor speed and the amount of RAM with a strong enough graphics card.

36.4 Visualization and Virtual and Augmented Reality

Collaboration plays a central role in successful BIM implementation and use. One of the more important possibilities that collaboration provides is the visualization of models. Visualization tools are a great relief and help to architects in all phases of design. "Real-time rendering" technology and virtual and augmented reality are imposed as primary visualization solutions in architectural practice.

With the introduction of BIM technology, visualization and virtual reality have become more accessible and easier to implement. It is important to note that software solutions such as Graphisoft Archicad and Autodesk Revit do not have a native 3D rendering capability and it is necessary to implement additional software solutions, configure render servers (powerful hardware machines), and integrate individuals into the larger whole using plug-in and add-on options.

Rendering is one of the most expensive and hardware-intensive processes, especially when it comes to real-time rendering. Real-time rendering can be defined as an animation that is generated in a very short time interval (usually less than a second). In order for real-time rendering to faithfully represent the 3D model, it is necessary to enable such hardware components that can generate between 24 and 50 fps. Reaching these speeds allows the user to see animated scenes in real time [5].

Just like virtual reality, which has gained great popularity in BIM environments, augmented reality is becoming more and more noticeable. Augmented reality enables the "placement" of virtual models in "our reality." Visualization is most often realized with the help of tablets or a mobile device.

With the help of augmented reality, the model becomes interactive and has the possibility of manipulation. The primary difference between virtual and augmented realities is that virtual reality creates a completely new and independent environment of "reality." In contrast, augmented reality includes virtual elements that are "placed" in real space. The advantage of augmented reality is in the combination of design with the already existing world. From the point of view of implementation and use of information technologies, it is more expensive and technically demanding, but much more effective [6].

No matter what type of visualization is chosen, it is necessary to focus on a careful selection of software (whether it supports complex 3D models, animation, collaboration), tracking systems, sensors, and ease of manipulation of 3D CAD projects.

36.5 The Importance of Backup and Snapshotting

From the very beginning of doing business, archiving and data protection have played a big and important role. Prior to digitalization, the option of rapid data restoration was not considered, and storing was done physically and in "tangible" storage. There weren't many options – to protect copies; data were mostly stored in paper form, archived in special rooms where it was attempted to maintain optimal temperature and humidity, so as not to destroy the archived data and information. With the advent of digitization, such paper forms of archives have been digitalized with the help of scanning and OCR software.

However, with the advancement of technology, new trends have also developed. With the expansion of digitalization, there was a need for data storage in multiple places, fast synchronization, and creating more protections and copies. Alongside that, technologies have advanced at high speed. New possibilities for data protection and storage arose in case of a need for fast recovery.

Today, all organizations and architects, regardless of size, use some kind of data protection. The goal is to return to operational condition in the shortest possible timeframe in the event of a breakdown (hardware failure, software instability, force majeure), which means to continue business as normal with minimal loss of time and data.

The absence of a data protection system would lead to the loss of critical data and information, creating instability in doing business, which further leads to large financial losses that often end in breakdowns, job losses, and overall business collapse.

In order for data restoration to be successful and effective, it is necessary to create a data protection plan and identify and test all steps in the event of a breakdown, in order to return all business activities to normal flow in the fastest and best possible way.

These are basic methods of data protection and their restoration; their use is defined by the very needs of the organization and by the speed of restoration and financial capabilities:

- Full backup
- Incremental backup

- Differential backup
- Synthetic complete backup
- Unlimited incremental backup

This general overview of backup and recovery capabilities is fully applicable in a BIM architectural environment. Most BIM software solutions have sophisticated methods in case of data loss and design errors so that the recovery itself can be performed quite easily, automatically, from the software environment itself.

Most software solutions are configured to have the so-called “Autosave” option. At certain time intervals, it creates a “snapshot,” stores it, and, if necessary, performs a restoration to a predefined “snapshot point.” A snapshot is a reasonably reliable option in a BIM environment. At certain time intervals, it allows you to create a backup and, depending on the architecture and performance, allows a very quick “return” to the previous point. The existence of such an option is quite advantageous, especially in the case when many clients access and make changes to the same project. Therefore, the percentage of errors is higher, and the possibility of a quick return to the previous good state is of great importance [7].

However, manual restoration of data and projects is sometimes required. In order to avoid the problem of not having an alternative solution in case of an impossibility to use snapshots, the most acceptable solution is the use of specialized software for backup and recovery, especially if it is the previously described virtual infrastructure. Based on the previous analysis, some of the existing solutions on the market are selected. Like most application software requirements, the backup/recovery solution must support application-aware processing and VSS snapshoting because there are databases, among other things.

36.6 Conclusion

Digitization of architectural offices and the implementation of the BIM environment are becoming increasingly popular in recent years. The main advantage is the increase in efficiency and quality of work. Survival in the market and strong competition require a restructuring of doing business and the implementation of digitalization.

However, it is important to emphasize that the implementation of a BIM environment in a medium-sized architectural office is a time-consuming and demanding process. Based on all the above, it can be concluded that the implementation depends on many factors, combining more technologies, from computer, network, audio, and video through the transition from classical to digital work and thinking of users, i.e., architects, and their education, and finally maintaining and updating the entire system.

In order for the entire process to be successful and effective, it is necessary to form quality and competent team of experts from all the mentioned fields and perform precise operational and strategic planning of all steps.

References

1. H.R.Z. Mohammad, BIM implementation in architectural practices, Doctoral thesis, Universitat Politecnica de Catalunya, Barcelona
2. Graphisoft Archicad: Hardware and Software Requirements. Accessed 11 May 2020. https://helpcenter.graphisoft.com/knowledgebase/25883/
3. Graphisoft Archicad: Network Requirements. Accessed 11 May 2020. https://helpcenter.graphisoft.com/knowledgebase/43354/
4. Cadalyst: Find the Optimal Hardware Configuration for ArchiCAD. Accessed 10 June 2020. https://blog.cadalyst.com/cadspeed/2011/08/23/find-the-optimal-hardware-configuration-for-archicad
5. C.B. Alan, *Understanding Augmented Reality* (Morgan Kaufmann, Waltham, 2013)
6. Archidaily: 9 Augmented Reality Technologies for Architecture and Construction. Accessed 5 June 2020. https://blog.cadalyst.com/cadspeed/2011/08/23/find-the-optimal-hardware-configuration-for-archicad
7. Asiabim: Data Safety, File Recovery and Troubleshooting. Accessed 11 June 2020. https://asiabim.wordpress.com/2016/10/04/data-safety-file-recovery-and-troubleshooting/

Chapter 37
Control Towers in Supply Chain Management: Standardization and Documentation Preparation

Tijana Talić and Zoran Ž. Avramović

37.1 Introduction

The control tower was originally conceived as a central place where analysts collect data from multiple trading partners. In this way, the management of the entire supply chain and its visibility is enabled. Such a control tower remained closed within the four walls of the company that formed it. With the development of technical capabilities, cloud technology is increasingly used. The control tower becomes a kind of virtual control panel. In this way, it goes beyond the four walls of the company and becomes visible to all members of the supply chain.

According to research by Geodis [1], the vast majority of executives ranked visibility as the third most important challenge to the functioning of the tower. Only 6% of them claimed to have full visibility, while 62% of them claimed to have visibility only with direct trading partners.

To successfully plan and manage a supply network, it is necessary to see the entire supply chain. Nigel Duckworth [2] compares this to playing chess when you can see only part of the chessboard.

Control towers are often built on poorly interconnected legacy systems. This leads to the fact that although we have invested considerable effort and significant funds, we have not obtained the desired financial effect. In the development of control towers, network access gives better results. Network-based control towers have the following advantages: visibility outside the company walls, connecting trading partners and providing real-time visibility, ability to respond to changing business needs or goals, quick engage, and activate online partners. The network approach allows each partner to connect to the control tower only once, which contributes to

T. Talić (✉) · Z. Ž. Avramović
Pan-European University APEIRON, Banja Luka, Bosnia and Herzegovina
e-mail: tijana.z.talic@apeiron-edu.eu; zoran.z.avramovic@apeiron-edu.eu

S. Bauk, S. D. Ilčev (eds.), *The 1st International Conference on Maritime Education and Development*, https://doi.org/10.1007/978-3-030-64088-0_37

simplicity. If a certain economic entity and all its partners are connected to one control tower, then the visibility and management of the supply chain are ensured by proper functioning. At the same time, each of his partners connects only once to the control tower. If now his partners connect to the same tower, then they also gain visibility and the ability to monitor and manage the supply chain. In a network environment, information is collected and exchanged in real time. Connecting direct and indirect partners now introduces a new quality where artificial intelligence and software agents can plan the supply chain far better and more accurately. They notice anomalies on time and make decisions and corrections of the flow faster and more precisely.

A control tower built on a common network platform puts the entire supply network under control, eliminates blind spots, and improves compliance. The system built in this way provides the conditions for joint planning of all involved partners, which gives a fluent flow of the supply chain. We know that such an uninterrupted supply flow significantly reduces the need for stockpiling, which directly and indirectly reduces operating costs.

37.2 Standardization of Trade Documents

Significance and Role of Standardization: The Layout of Trade Documents

Standardization is a mechanism that facilitates and enables the exchange, interchangeability, and use of standard elements in different situations. Manufacturers of complex products can use standard elements and thus make their production cheaper, freeing it from the need to produce standard elements in small quantities. At the same time, it gives them the freedom to choose the manufacturer of standard elements either according to its geographical location, possibilities of delivery, delivery, or the like. At the same time, it is possible to later replace the standard elements with the same standard elements without fear of a malfunction.

Trade is the exchange of goods and services. Trading partners communicate with each other through trade documents. The basic requirement related to these documents is their comprehensibility and readability by all participants. This is where the importance of standardizing the layout of a document as one of the key elements of comprehensibility and readability of a document comes to the fore. At the same place on a series of documents is the same information. This allows for quick reading whether the documents are read by a human or scanned for the purpose of performing translation into electronic form via OCR. In business practice, at least in Bosnia and Herzegovina and the surrounding countries, there is no clearly defined standard or recommendation on what trade documents such as invoices, orders, and other documents related to business communication between business partners should look like. There is standardization of appearance only when it comes to

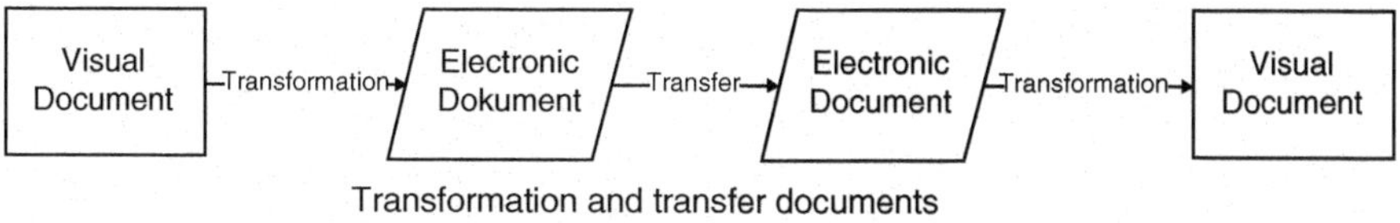

Fig. 37.1 Transformation and transfer documents

documents defined by international conventions, such as the international consignment note CMR. Standardizing the appearance of a document increases readability and speed and reduces the cost of exchanging data between partners.

We can say that the standardization of the layout of a document, in a same way, is a precursor to the electronic exchange of data. Why? In addition to the visual position of which data is located, the standardization of the document layout also defines which data must be located on the document. In this way, the definition of data to be transmitted by electronic communication was performed. This is the first step toward the digitization of commercial documents. Why document layout standardization is also important in control towers whose imperative is electronic data exchange can best be seen from the following diagram (Fig. 37.1):

The document is visually (via appropriate software) created by one of the members of the supply chain within the control tower. It transforms it into an electronic form and sends it to the control tower. The recipient of the document can transfer it electronically from the control tower and perform its visualization. If the standardization of the document layout is followed, the recipient or other participant in the supply chain will visually see an identical document. Visualization can also include the translation of standard document labels. For example, if the sender is from China, the labels on his document can be in Chinese and in B&H in one of the languages spoken in B&H or at will or need in the original language.

UNECE is involved in the development of standards and recommendations for the appearance of trade documents. In its recommendation, Recommendation No. 1: United Nations Layout Key for Trade Documents [3] deals in detail with the layout of data on trade documents. The recommendation for the document layout development process is presented in Fig. 37.2.

Invoice: Layout

The invoicing process is used to exchange invoices between suppliers and to deliver goods or services that have been ordered, delivered, received, consumed, etc. In recent years, a great deal of work has been done on standardization and electronic invoicing. When it comes to the appearance, we consider Recommendation 6: Aligned Invoice Layout Key for International Trade as the first and umbrella document [4] (Fig. 37.3).

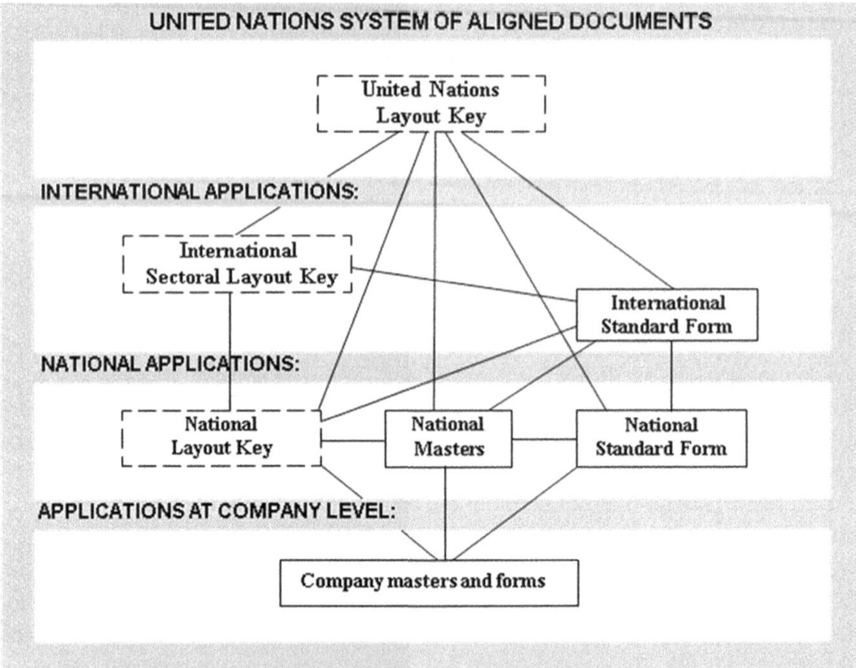

Fig. 37.2 Develop company master document. (Source f2g)

For the needs of the functioning control tower of supply chain, it is necessary to visualize the received or sent invoice. The logical sequence is to choose this recommended layout in communication within the control tower. The original invoice will of course always accompany the goods. Examining the available invoices in both domestic and foreign trade, we noticed that most of them are burdened with image elements (oversized logo) as well as a lot of descriptive and contact information that significantly affect the deviation of the appearance of the invoice from this in the recommendation. For the visibility of the document within the control tower, this layout is the right choice because it contains all the essential elements. When defining the layout of the invoice for actual use, it is necessary to take into account and include all regional and national standards and rules, and they are mainly related to taxation and are visible in the umbrella document. When it comes to international trade with the EU and countries with which there are bilateral agreements, within the space of "free disposal," we define the space for declarations of origin of goods because they must be visible and important in customs procedures and must be specially processed.

We will use this layout whenever we render the received data inside the control tower. We will give recommendations to other participants to use it. After the electronic transfer of the document, the client can use another layout within his system in accordance with his software solutions.

INVOICE LAYOUT KEY

Seller	Invoice date and N°. Other references
Consignee	Buyer (if other than consignee)
	Country of origin of goods
Transport details	Terms of delivery and payment

Shipping marks; Container N°.	N°. and kind of packages; Goods description (in full and/or in code)	Gross weight. kg.	Cube. m3

Specification of commodities (in code and/or in full)	Quantity	Unit price	Amount
FREE DISPOSAL			

	Included above	Not incl. above
Packing		
Freight		
Other costs (Specify)		
Insurance		
Total invoice amount		

Fig. 37.3 Aligned invoice layout key. (Source [4])

Electronic Transmission

Electronic data interchange is not new in the business world. It has been present since the 1960s of the last century and took place on mutually agreed standards of business partners. The development of exchange standards is being developed by the UN/EDIFACT (United Nations/Electronic Data Interchange for Administration, Commerce and Transport). In 1987, the UN/EDIFACT syntax and grammar were approved as ISO standard 9735. Many countries and international organizations support the development of the UN/EDIFACT.

European Article Number Communication (ANCOM®) is a subset of the UN/EDIFACT internationally standardized standard for messages. EANCOM® is based on the GS1 numeric system for the physical identification of products, addresses (companies), and logistics units (GTIN [Global Trade Item Number], GLN [Global Location Number], and SSCC [Serial Shipping Container Code]). It is the most commonly used and internationally unique standard format for electronic data interchange (EDI).

The GS1 system is a set of standards that enable efficient management of global, multi-industrial supply chains through unique identification of products, transportation units, assets, locations, and services. It facilitates e-commerce processes including full tracking of shipments

All the systems listed above are in active use. Many countries even more so enable the electronic exchange of EDI under the auspices of their tax administrations. Active use of these systems facilitates our proper selection of standards for use in our control tower.

The UN/EDIFACT is a complete well-developed and well-documented system. It is publicly available. It is the source of all other standards. It is clear then why we choose to have the UN/EDIFACT fully implemented in our electronic exchange control tower, not only for invoices but also for all procedures in the supply chain. Cross Industry Invoicing (CII) is a complete standard that covers almost all aspects of invoicing, including elements for displaying tax impost. Based on it, the European CORE INVOICE data model was made. This model is based on EU standards EN 16931-1.

Our modules need to include a GS1 system for the unique identification of entities, locations, and, if necessary, goods that are the subject of trade. When it comes to foreign trade, CII in dictionaries has well-defined fields for participants in customs procedures: Customs Export Agent and Customs Import Agent.

The European Union has clearly defined rules for determining the preferential origin of goods. For the purposes of customs procedures, invoices contain elements that affect the amount of customs duties. These are invoice declarations either given by an approved exporter or up to a certain amount by any exporter for any goods. Unfortunately, specific support for customs clearance is outside the scope of norm EN 16931-1. We can overcome the problem by introducing a special document "statement" and stating its reference in the invoice. This would not compromise the integrity of the e-invoice. But that is not a good practice. These declarations must be specially signed either by the exporter or an approved exporter. But in any case, it is good to give a signal to the customs broker to prepare customs clearance as if there is a correct statement and at the time of arrival of the goods to make a final check. There are defined protocols for the transfer of information on the status of authorized importers, but they are at the level of customs administrations. This indicates that work should be intensified on defining international standards for the international exchange of invoices, i.e., adding special support for customs procedures.

In modern electronic business, the key issue in any exchange of electronic messages is the issue of security: verification of the identity of the sender and the authenticity of the transmitted messages [5]. In the control tower, we must implement a special software agent that deals with the verification of digital signatures and the authenticity of the transmitted data.

E-CMR

The CMR Convention was drawn up in 1956 by the United Nations with the aim of providing a single legal framework for national and international road transport. Most European countries have ratified the convention, as have several countries outside Europe. The CMR document not only contains accurate cargo data but is also an agreement between the three parties involved in the transport (consignor, carrier, and consignee). The CMR regulates liability, damages, and compensation. The CMR automatically applies to transport contracts for international road transport, where the place of receipt and place of delivery are in the signatory country.

In February 2008, an additional e-protocol was added to the CMR convention, which entered into force in June 2011. The e-protocol provides the legal framework and standards for the use of electronic means to record the CMR consignment note [6]. So far, the use of e-CMR has been ratified by 26 countries.In this paper, we do not intend to analyze the existing standards and protocols in wide application and officially verified by a number of countries.

This introductory research work was initiated by the freight forwarding company Ciambella export-import doo, which is based in a country that is not a signatory to the e-CMR protocol. Therefore, we will only discuss here the obligation to include this protocol in the future control tower and the guidelines for its creation. The CMR contains three different signatories as defined by the Geneva Convention. These are the sender, the carrier, and the recipient [6]:

- The customer of the transport service gives consignment instructions to the transport service provider.
- E-CMR is proof of the existence of a contract between the consignor or consignee and the carrier who has a legally binding condition provided for by the CMR Convention.
- E-CMR must allow recording of the receipt of the goods by the consignor to the carrier, including reporting of any difference in the received goods with instructions on the consignment instructions.

- E-CMR must allow the recording of proof of delivery of the goods by the carrier to the consignee and report any damage or difference in the delivered goods in relation to the shipment instructions.

We see that these are all functions that are needed in the functioning of the control tower. So the implementation of e-CMR in the control tower is necessary. Building a control tower so that it is also an e-CMR service provider is very demanding and coupled with many necessary certificates from the official authorities. It is desirable that the control tower, based on the authorization of the previously mentioned three e-CMR signatories, be set up as a kind of intermediary between the e-CMR service provider and the e-CMR signatory. Such support could also be built into the client software of the control tower users. In any case, it is necessary to develop and implement agents for receiving, translating, sending, visualizing, and recording e-CMR in control tower.

37.3 Conclusion

In this paper, we have analyzed the situation in the standards for electronic transmission of documents with an emphasis on invoices and CMR, which we consider a very important function. We have shown that there are internationally accepted standards by which this can be done. We have also shown that there are certain ambiguities in the transfer of invoices when it comes to foreign trade between the EU and countries that have preferential agreements as well as with CEFTA countries. Let this paper be a definite incentive to remove these ambiguities call for their faster definition in e-protocols.

References

1. GEODIS 2017 Supply Chain Worldwide Survey. https://geodis.com/fr/en/newsroom/press-releases/geodis-unveils-its-2017-supply-chain-worldwide-survey
2. N. Duckworth, *Elitism for All: How Network Control Towers Create Value Throughout the Supply Chain.* https://www.sdcexec.com/software-technology/blog/21012580/one-network-enterprises-elitism-for-all-how-network-control-towers-create-value-throughout-the-supply-chain 2020
3. UNECE, *United Nations Layout Key for Trade Documents, Guidelines for Applications* (United Nations, Geneva/New York, 2002)
4. UNECE, Aligned Invoice Layout Key. Recommendation no. 6, revision 1983. UNECE (1975)
5. T. Talić, G. Radić, Z.Ž. Avramović, Digital signature and organization of decentralized authentication in business environment. JITA J. Inf. Technol. Appl. Banja Luka, Pan-European University APEIRON, Banja Luka, Republika Srpska, Bosna i Hercegovina, JITA **9**(2019) 1:24-28, (UDC: 316.77:004.773]:658.14), (https://doi.org/10.7251/JIT1901024T), Volume 9, Number 1, Banja Luka, June 2019
6. UN/CEFACT, e-CMR Business requirements specification. United Nations Economic Commission For Europe – UN/CEFACT (02 2018)

Chapter 38
Space Technology and the Management of Water Infrastructure in South Africa

Matlou Lesley Mokgobu and Roger B. Mason

38.1 Introduction

The issues surrounding water infrastructure are always challenging to local authorities and water boards. These issues are, among others, burst pipes, unlocated leaks, vandalism, blocked pipes, and aging water infrastructure. As a result, the challenges experienced are street inaccessibility, limited or no access to homes, traffic jams, unbarricaded deep trenches, leaking water pipes, water and electricity supply interruptions, vandalism, and theft. To this end, some of these challenges may be addressed via the space technologies of monitoring and remote sensing.

The problem of lack of knowledge about the management of water infrastructure may be prevalent. The introduction of space technology may provide success to some of the challenges being faced by South African (SA) local authorities. The involvement of South Africa in space research started as far back as 1820, with the establishment of the South African Astronomical Observatory in Cape Town [1]. The country today is leading African space research and is benefitting tremendously from the investments. Africa's expansion into space activities has been realised by collaborating with organisations like National Aeronautics and Space Administration (NASA) in conferences and at workshops in Africa to share knowledge [2].

Space technology refers to spacecraft, satellites, space stations, back-up infrastructure, tools, and processes established and utilised by the aerospace environment in spaceflight, satellites, or space exploration [3]. This literature review will focus on the application of space technology and the challenges of managing water infrastructure in South Africa. The literature will explore not only the internal issues of organisations managing water infrastructure but also local, national, and international

M. L. Mokgobu (✉) · R. B. Mason
Durban University of Technology, Durban, South Africa
e-mail: rogerm@dut.ac.za

S. Bauk, S. D. Ilčev (eds.), *The 1st International Conference on Maritime Education and Development*, https://doi.org/10.1007/978-3-030-64088-0_38

applications of space technology regarding water infrastructure. For example, space technology could help in the detection of the location of buried infrastructure. Kganyago and Mhangara [4] noted efforts to address data gaps and sustainable development programmes by most African countries progressively embracing Earth observation (EO) and geospatial technologies. The following African countries have established their own space agencies with increasing investment: Nigeria, Egypt, Algeria, Kenya, South Africa, and Gabon. They are the twenty-first-century players in the space industry with their own EO satellites. Parliamentary Monitoring Group [5] recalls the meeting held on 14 September 2014, where the South African Council for Space Affairs (SACSA) outlined the importance of space technology to South Africa (SA). Satellite communication; satellite-aided position, timing, and navigation; and Earth observation were mentioned as the three pillars of space technology. The benefits of space technology to society and the importance of policy formulation to regulate the industry were discussed. Parliamentary Monitoring Group [5] further details the services offered by the three pillars of space technology, namely, food security, water security, urban management, disaster management, nature preservation, land use, drought, climate change, desertification, treaty conformance, human rights exposure, and environmental destructions.

The South African National Space Agency (SANSA) is an entity of the Department of Science and Technology (DST) and derives its mandate from South African National Space Agency Act, 2008 [Act No. 36 of 2008], and National Space Strategy [4]. Since it is a worldwide occurrence that the space sector is undergoing swift change, global players have to adapt to the changes, especially new entrants into the sector. The South African National Space Agency [6] envisions four vital fundamentals that motivate space industry innovation, namely, the national security and scientific goals, downstream space application growth, human space exploration, and Fourth Industrial Revolution (4IR).

38.2 Aims and Objectives

Aim of the Study

The aim of this paper is to provide a background to possible ways space science can assist with the management of water infrastructures in a developing country and thus form the basis for more detailed empirical research into such applications.

Objectives of the Study

- Identify the application of space technology to the management of water infrastructure.

- Determine barriers to African countries becoming business players in the space industry.

38.3 Methodology

Research Design

This study applied secondary research to explore the application of space technology in South Africa and in Africa and globally. Secondary research was further used to establish barriers to the African continent's full participation in the space industry.

Research Approach

Various searches were conducted for relevant literature on space technology in e-text books, e-journals, and e-newspapers, using a number of search themes that were established in advance. These themes were helpful in guiding the search. The themes used in the search for literature were space technology, managing Earth from space, and managing water infrastructure from space. These themes produced a number of words in guiding the search process. The words were space, space technology, managing Earth, Earth observation (EO), water infrastructure, managing water infrastructure from space, SANSA, ISS, remote sensing, and Fourth Industrial Revolution (4IR).

Data Extraction

Data was extracted from Google.Com and Google Scholar (https://scholar.google.com) search engines. However, other search engines such as Emerald and ProQuest were also visited but did not produce the required results. The reference lists of other authors were also helpful in forming part of the search for the latest literature.

Analysis

A review process aimed at generating an assessment of the research topic in comparison with the existing literature was followed. The review was created with the assistance of the themes and keywords as the guiding principles of the review process.

Interpretation

Much of the literature that the researcher found covered the technical aspects of the topic. Literature related to the management aspects of water infrastructure were very limited, and so it was necessary to be very selective and to extract the limited management successes and shortfalls included in the technical literature and apply it to this study. Furthermore, management principles were applied to the extant literature to find a way to develop a meaningful literature review.

38.4 Managing Water Infrastructure

Management of water infrastructure requires a combination of organised managerial expertise and technical expertise. Management of infrastructure in South African (SA) cities has become problematic due to utilities' lack of skills, budget reductions, and poor employee retention strategies. Innovation is one of the gaps identified by various municipalities regarding water infrastructure. Some SA cities have been identified as examples of cities with innovative water infrastructure management [7].

Space technology is one of the innovations in water infrastructure management that could help African cities. This technology can be a valuable asset considering theft and vandalism experienced by many water infrastructures. Earth observation can enable satellites to locate, view, and monitor these assets from space at any given distance and time for theft and vandalism prevention. Ojogba et al. [8] noted that satellites are infrastructures experiencing swift growth and that can support economic improvement.

Space technology could close this gap, experienced over many years, of challenges with water infrastructure. Observing the activities of the infrastructure requires advanced monitoring which is possible with the help of space technology. The following are phases of remote-sensing and geospatial technologies as outlined by Kurnaz and Rustamov [9]: detection, preparedness, prevention, protection, and response. Alsdorf, Rodríguez, and Lettenmaier [10] explain that many satellites are used for global observation of Earth from space.

Dindar, Kaewunruen, and Osman [11] highlight the remote sensing (RS) for monitoring a railway system from space using satellites. This technology from space records observes and identifies objects from extremely far away without having direct contact. The sun strikes the object on Earth, and the satellite's optical sensors detect and receive solar reflections, and information gets transmitted to the receiver. Photographs are then taken by camera from space and sent to the Earth by means of satellite dishes. Many remote-sensing images (RSI) can be generated from this system. This technology is a perfect fit for monitoring the installed water infrastructure location against vandalism, theft, etc. Magidi and Ahmed [12] suggested the use of high-resolution satellite and remotely sensed data for better identification of land patterns and features. There are various fields of research in as far as space technology is concerned [13].

The use of mechanical means such as back-actors, excavators, etc. is costly and often results in broken pipes and electrical cables. Chang [14] proposes that space technology is therefore necessary to save time and locate the civil infra-structures such as buildings, railways, roads, bridges, dikes, dams, quays, and pipelines. Synthetic aperture radar interferometry (InSAR) is a "precise and efficient technique to monitor deformation on Earth with millimetre precision". This technology therefore could help a municipality with monitoring aging pipe deformations, reservoir locations, etc. The use of this technology, however, re-quires an advanced degree of knowledge for its application.

Water supply infrastructure is perceived to be the most steady, "well-developed, and long-lasting infrastructures" – it is robust. Geography of technology and science and technology studies (STS) can be introduced to create a bond between communities and infrastructure, and possible inspiration of minor technology is largely ignored [15].

38.5 Space Technology in Africa and South Africa

The high growth of investments realised gives Africa a good playing field for expansion of space technology use. Africa's expansion will further be realised if all states collaborate in the fields of agriculture, security, telecommunications, teaching and learning, Earth observation, remote sensing, and so on. The in-formation drawn from the satellites' data helps authorities to draw up management plans and to influence policies.

African countries have made outstanding progress in the space programme. There have been 41 African satellites launched, 3 of which have multilateral ownership [16]. The African Space Industry 2019 report outlined the exponential growth of African countries from 1998 to May 2019. The African space industry collected more than 7 billion USD annual income, with a growth projection of 7.3%, expected to exceed 10 billion USD in 2024 [17]. Egypt hosts the African Space Agency with approval granted by African heads of states during the African Union's last summit in January 2019 [6].

Table 38.1 African countries' satellites constructed, purchased, or rented

Satellite	Country or organisation	Year launched
ALSAT1	Algeria	2002
ALSAT2A	Algeria	2010
ALSAT1B	Algeria	2016
ALSAT2B	Algeria	2016
ALSAT1	Algeria	2016
ALCOMSAT-1	Algeria	2017
AngoSat-1	Angola	2017
GhanaSat-1	Ghana	2017
NILESAT 101	Egypt	1998
NILESAT 102	Egypt	2000
EGYPTSAT1	Egypt	2007
NILESAT201	Egypt	2010
EGYPTSAT2	Egypt	2014
EGYPTSAT-A	Egypt	2019
NARSSCube-2	Egypt	2019
NARSSCube-1	Egypt	2019
TIBA-1	Egypt	2019
ETRSS-1	Ethiopia	2019
1KUNS-PF	Kenya	2018
Maroc-TUBSAT	Morocco	2001
MOHAMMED V1-A	Morocco	2017
MOHAMMED V1-B	Morocco	2018
RascomStar-QAF-1	Multilateral	
RascomStar-QAF-1R	Multilateral	
NewDawn	Multilateral	
Nigeriasat-1	Nigeria	2003
NIGCOMSAT1	Nigeria	2007
NigeriaSat-2	Nigeria	2011
NigeriaSat-X	Nigeria	2011
NIGCOMSAT1R	Nigeria	2011
NigeriaEduSAT-1	Nigeria	2017
RwaSat-1	Rwanda	2019
SUNSAT	South Africa	1999
ZACUBE	South Africa	2003
SUMBADILA	South Africa	2009
KONDORE	South Africa	2014
nSight1	South Africa	2017
ZA-AEROSAT	South Africa	2017
ZaCube-2	South Africa	2018
XinaBox ThinSAT	South Africa	2019
SRSS-1	Sudan	2019

Source: [16]

There are 19 African countries forming the African space programme with an escalating number of space technology businesses interested in offering services to the continent. Africa is capable of solving serious problems in agriculture, security, telecommunications, and other fields with assistance of the current space technology. For example, satellites have assisted Mali herdsmen to locate water for cattle. Further, satellite assistance in Africa's TV programmes and satellite internet connection to rural classrooms were realised in Angola and Rwanda [17].

Table 38.1 outlines the country of ownership, names of satellites, and the year each satellite was launched.

38.6 Practical Role of Space Technology

The following practical roles could be performed by space technology in SA to manage water infrastructure, communication, health, safety, security, advanced water filtration and purification system, detection of water levels in reservoirs, monitoring Earth's natural resources such as ground water, water leak detection, underground water pipe detection, and monitoring water quality. Space technology is currently used for various applications in SA such as Earth observation, remote sensing, aviation industry, research and development, weather forecasting, and many more. Various SA universities are part of the space science programme which is forming part of their curricula. South African National Space Agency (SANSA) has started recruiting students in maths and science to join their initiatives in space science.

However, there are potential problems associated with the application of space activities. One of the major drawbacks of space technology is that it is expensive. Another challenge posed by some satellites is their discontinuity in mapping some data [18] – satellites do not have any concern for politics and know no demarcations [19]. The advantage of space technology is that observations can be made of larger areas in a short space of time compared to humans doing it in the same time/space. Giardino et al. [20] confirmed the agility of remote-sensing capability to reproduce spatial and temporal observations, whereas similar in situ observations of surface water quality are impossible at the same space and time dimensions.

Other challenges in the use of space technology are the teaching of outdated curricula by higher learning institutions, poor funding, and insufficient resources for Earth observations, satellite communication, satellite systems, navigation and positioning, and space agencies. Furthermore, there is no sharing of datasets by space agencies due to the restrictive bureaucracy of obtaining EO data in other African states, except for SA [21].

The South African Department of Science and Technology [22] pointed out that space science and technology can assist in research and development for services, as a monitoring tool and as an evaluation of serious land resources, and for better decision-making. South Africa has been purchasing and importing space technology and has started developing its own systems to support local industry requirements. There are three key priorities identified by the government in space exploration, namely, environmental and resource management; health, safety, and security; and innovative and economic development.

A good example of SA having started to develop the local space industry is the Cape Peninsula University of Technology (CPUT) Satellite Programme. The programme is aligned to the National Space Strategy and receives funds from Department of Science and Technology and the National Research Foundation (NRF). The satellite programme has been successful in graduating 60 students in the master's programme and developing, building, and launching a CubeSat. The programme is jointly hosted by the French South African Institute of Technology and the African Space Innovation Centre [23].

Campbell [24] advises of the launch of a non-profit-making company, ZASpace Inc., in Pretoria. Its purpose is to encourage local space industry growth particularly in the area of geospatial technology for South Africa and Africa and for skills development and innovative funding for beginners, small and medium enterprises and microenterprises. During the launch, ZASpace Chief Executive Officer (CEO) Kamal Ramsingh announced that the world's geospatial market was valued at $193 billion in 2013 and had risen to $299 billion in 2017 and was anticipated to hit $500 billion by the year 2020.

Campbell [24] supported Kamal Ramsingh's statement that the drivers for the geospatial sector worldwide are travel and hospitality, disaster management, mapping agencies, banking and financial services and insurance, defence and security, mining and energy, retail and logistics, e-government, utilities, infra-structure, and smart cities. All these drivers are listed according to size from the smallest to the largest. The sector also consists of global satellite navigation systems, geographic information systems and spatial analysis, Earth observation, and three-dimensional scanning.

Campbell [24] further noted the USA, the UK, and the whole Europe as the leading geospatial markets. However, Africa has the fastest growth, with annual compound growth rate of 16.8%, whereas the spatial analysis accounts for 30% of the African market. The continental growth comparison between the years 2018 and 2020 for Africa's geospatial market is 21%, with 11% for Europe and 10% for the USA. ZASpace CEO further noted the shortage of skills in South Africa with an even greater shortage of skills in other African countries. Regarding global geospatial preparedness, South Africa is ranked in position 31.

Campbell [24] highlighted the CEO of SANSA Dr. Val Munsami's announcement that for the local space agency, Africa is the way to go considering its growth. He further reminded that one of the objectives of the National Space Policy was to improve the national space industry. The creation of SANSA from organisations that existed before was difficult because of the shortage of a business model, unpredictable business models, and fragmented business models split among the divisions, each with dissimilar operational needs. Finally, the CEO pointed out that ZASpace was viewed by SANSA as a key role player in the local space industry.

The South African Department of Science and Technology [22] noted that to further support the priorities, there are three programmes that have been developed to make space initiatives a success. These programmes are thematic, functional, and supportive. Thematic programmes are Earth surveillance, navigation, communication, space science, and discoveries. Functional programmes entail aiding technologies, mission improvement, space mission manoeuvre, and space mission application. Support programmes include human capital advancement, infrastructure, and partnering international communities.

Space technology has been used to solve social challenges in the areas of management, environment, usage of natural assets, growing movement of individuals and products, increase in security threats, and the move towards the knowledge economy. Space exploration is going to solve many present and future challenges faced by South Africa. Space exploration has produced many benefits and continues to bear fruit internationally. The costs associated with space activities are extremely high, but the returns associated therewith are worthwhile. Job creation, technological advancement, scientific familiarity, and space derivatives are some of the advantages associated with space technology. It is for that reason that this study investigates space technology as a tool to help manage water infrastructure [22].

Ojoyi [3] maintains that many African countries are limited as far as space initiatives are concerned due to inadequate funding. They depend on external donations, and for that reason, adequate knowledge, infrastructure, tools, and education are unachievable. This technology is necessary for location, imagery, and security for infrastructure under threat of vandalism and other threats in the African continent.

38.7 Linking Water Infrastructure to Space Technology

Space technology can be applied to water infrastructure through the following activities: water quality detection in lakes [18] and in a dam [20]; dam and pipeline location [14]; imagery, location, and security [3]; and security and management [22]. Another good example is the remote sensing of water quality assessment and chlorophyll a (chl-a) with the application of Landsat 8 Operation-al Land Imager (OLI) data. The technology tested the red to near-infrared (NIR-red) bands in the Vaal Dam of SA for the categorisation of chl-a concentrations [18]

Regarding water infrastructure specifically, there is little literature. However, detection of underground water sources is one of the advantages offered by space technology. The Israelis launched a freshwater leak detection and prevention technology from outer space. They have adopted a technology previously used on Mars and Venus for detection of water. The technology was invented after an estimated annual worldwide freshwater loss of 32 billion cubic metres was reported by the World Bank in developing countries. Synthetic-aperture radar (SAR) has been adopted by universities and research institutions for detection of water underground and on other planets. The CEO of Utilis, in a telephone interview, confirmed the technology's usefulness for underground treated water detection in urban settlements [25].

This technology is necessary in the detection of water infrastructure, especially buried pipes below the surface of the Earth. According to Evagorou et al. [26], space technology can achieve optical satellite images for bathymetric data to a depth of 30 metres below sea level. Azambuja [27] notes that space technology saved community lives with the innovative filtration and purification systems developed aboard the space station. The joint partnership of aid organisations and NASA technology demonstrated the effectiveness of space research in response to global difficulties. This resulted in global collaborative efforts between Water Security Corporation and other organisations by organising systems with NASA water processing technology.

Another space technology benefit is remote image sensing of water quality from space by means of Hyperspectral Imager for the Coastal Ocean (HICO). The HICO was invented by the US Naval Research Laboratory for use in coastal ocean water quality assessment. The quality parameters for detection are water clarity, phytoplankton concentrations, light absorption, and distribution of cyanobacteria. HICO's obtained data is used by US Environmental Protection Agency (EPA) researchers to develop a smartphone application for detection of harmful concentrations of pollutants [27]. Greenblat and Anzaldua [28] outline Earth observation functions as support of agricultural production, fisheries management, freshwater management, managing forests, and monitoring detrimental activities.

38.8 How Space Science Technology Can Assist in Resolving the Challenges

The use of space technology can aid managers in issues such as positioning of the infrastructure, remote sensing of underground water availability, and leak detection. This will be a valuable time-saving for local authorities and other organisations. It has benefits for managers and policy-makers in decision-making. To this end, the technology has proven itself to be a good management tool in groundwater exploration, climate change, and flood hazard monitoring [21]. This section of the paper summarises the countries and communities of sub-Saharan Africa, the Kurdish village of Kendala in Iraq, and Morocco, India, Spain, and many other countries that have enjoyed the benefits of space activities.

Space technology has previously been used locally and internationally to address the challenges of water infrastructure and management. The National Aeronautics and Space Administration (NASA) [29] notes that an advanced water filtration and purification system developed for the International Space Station (ISS) is being used in sub-Saharan Africa for managing water resources. A representative (Todd Harrison) from a non-profit organisation, Concern for Kids (CFK), sourced funds from NASA to help Iraq, Malaysia, and Indonesia with aid and disaster relief programmes.

The community of the Kurdish village of Kendala in Iraq drank dirty, fabric-sifted water together with their livestock. They were aided by NASA-developed technology for the provision of clean water so that now they enjoy clean drinking water. Iceland, Luo and Donchyts [30] show a satellite image that, through the use of space technology, detected a 60% drop in water level of the second largest Al Massira Dam in Morocco in 2016. Various drops in dam levels in many parts of the world, such as Indira Sagar Dam in India, Mosul Dam in Iraq, and Buendia Dam in Spain, have been identified with the help of space technology. This technology can be applied in SA cities for monitoring water levels in reservoirs without having to physically travel to the reservoirs for inspections.

Similarly, the assistance offered by NASA engineers to the Kurdish village of Kendala could be sourced for SA cities, organisations, and communities. This is an opportunity for SA cities experiencing the challenge of overworked water purification plants due to high sewage volumes. The technology could help overcome the health risks faced by communities due to malfunctioning water purification systems.

38.9 Conclusion

With regard to space technology, South Africa must dedicate this era of the 4IR to embrace this technology which can bring investment into the country. The country needs to be equal to the task.

Any barriers to execution of this global transformation require a concerted effort to overcome the resistance. New African space initiatives must be undertaken in collaboration with the international community. African states must avail themselves of space technology to become future space champions. African countries must join hands to make sure that resources to educate people in space sciences are directed to the youth. This would involve the introduction of space technology curricula in high schools and all institutions of higher learning in South Africa and the encouraging of researchers to follow a career in space technology with funding for inspiration.

Space technology must be used to benefit South Africa and the local authorities for the management of water infrastructure. South Africa should collaborate with global countries which are already ahead with space affairs to help train learners and to further the application of technology used in the ISS for terrestrial water filtration and purification and other space activities. The South African government must try to contain cost escalations for space technology to accommodate new entrants, and people must be upskilled to operate this highly advanced technology.

The study objectives have been met by exposing some issues which may be hindering the African continent from optimal participation in space affairs. The study identified African countries which are making slow progress in participation in space affairs. Exploitation of space technology in the management of water infrastructure also exposed other avenues for further research. This study has provided a helpful literature of space activities which other researchers and authors can apply for their studies. Managers and water specialists may also apply this literature for better application of managerial functions and to influence policy decisions. Learners in schools and institutions of higher learning can also use this to increase their knowledge about space technology and its applications.

38.10 Recommendations for Future Research

Further empirical exploratory research and empirical quantitative research need to be undertaken to identify how big the potential industry of space-based water monitoring might be. Further research is recommended on:

- Breaking the barriers preventing cooperation and business between African countries in space technology
- Exploring in greater depth the benefits of space science technology to water infrastructure
- Possibilities of Africa's collaboration with developed countries in the space industry
- Africa's challenges for increased participation in space science technology

References

1. F. Ghadaki, An overview of past and future South African space activities. SpaceRef, 26 October 2010. Available http://www.spaceref.com/news/viewnews.html?id=1458. Accessed 7 July 2020
2. O.H. Ligate, Space research in Africa: Challenges and opportunities. Int. J. Sci. Technol. **4**(2), 1–17 (2015) Available https://doi.org/10.4314/stech.v4i2.1. Accessed 22 Apr 2020
3. M.M. Ojoyi, *Harnessing Space Technology for Africa's Growth and Development* (2016). Available: https://saiia.org.za/research/harnessing-into-space-technology-for-africas-growth-and-development/. Accessed 07 May 2019
4. M. Kganyago, P. Mhangara, The role of African emerging space agencies in earth observation capacity building for facilitating the implementation and monitoring of the African development agenda: The case of African earth observation program. ISPRS Int. J. Geo-Inf. **8**(7), 1–22 (2019) Available: https://doi.org/10.3390/ijgi8070292. Accessed 05 Feb 2020
5. Parliamentary Monitoring Group, Space Policy and Regulation: South African Council for Space Affairs briefing. Department of Trade and Industry, Cape Town, South Africa (2014). Available https://pmg.org.za/committee-meeting/17554/. Accessed: 03 Feb 2020
6. South African National Space Agency: South African National Space Agency Annual Performance Plan 2019/20/20, Pretoria: South African National Space Agency (2020). Available https://www.sansa.org.za/wp-content/uploads/2019/09/Annual-Performance-plan-201920.pdf. Accessed 20 May 2020
7. E.S. Mnguni, Water infrastructure asset management: A comparative analysis of three urban water utilities in South Africa. WIT Trans. Ecol. Environ. **217**, 927–938 (2019). Available https://www.witpress.com/Secure/elibrary/papers/SDP18/SDP18078FU1.pdf. Accessed 02 Oct 2019
8. Ojogba, O.S., Yommy, A.S., Christopher, T.T. and Adesina, L.O.: Space technology, an essential infrastructure for rural area development. Am. J. Environ. Resour. Econ. **4**(1), 25–31 (2019) Available https://doi.org/10.11648/j.ajere.20190401.13. Accessed 26 March 2020
9. S. Kurnaz, R.B. Rustamov, Application of space technology in support of security and safety of critical infrastructure, in H.G. Coskun, H.K. Cigizoglu, M.D. Mak-tav Integration of Information for Environmental Security. NATO Science for Peace and Security Series C: Environmental Security (Springer, Dordrecht, 2008), pp. 149–154 Available https://doi.org/10.1007/978-1-4020-6575-0_9. Accessed 07 Dec 2019

10. D.E. Alsdorf, E. Rodríguez, D.P. Lettenmaier, Measuring surface water from space. Rev. Geophys. **45**(2) (2007). Available https://agupubs.onlinelibrary.wiley.com/doi/epdf/10.1029/2006RG000197. Accessed 13 May 2019
11. S. Dindar, S. Kaewunruen, M.H. Osman, October. Review on Feasibility of Using Satellite Imaging for Risk Management of Derailment Related Turnout Component Failures, in IOP Conference Series: Materials Science and Engineering (Vol. 245, No. 4, 042025). IOP Publishing (2017)
12. J. Magidi, F. Ahmed, Assessing urban sprawl using remote sensing and landscape metrics: A case study of City of Tshwane, South Africa (1984–2015). Egypt. J. Remote Sens. Space Sci. **22**(3), 335–346 (2019) Available https://doi.org/10.1016/j.ejrs.2018.07.003. Accessed 26 Mar 2020
13. National Aeronautics and Space Administration, NASA's Journey to Mars: Pioneering next steps in Space Exploration (2015). Available https://www.nasa.gov/sites/default/files/atoms/files/journey-to-mars-next-steps-20151008_508.pdf. Accessed 13 May 2019
14. L. Chang, Monitoring civil infrastructure using satellite radar interferometry. Doctoral, Delft University of Technology (2015). Available https://repository.tudelft.nl/islandora/object/uuid:f4c6a3a2-73a8-4250-a34f-bc67d1e34516. Accessed 08 May 2019
15. K. Furlong, Small technologies, big change: Rethinking infrastructure through STS and geography. Prog. Hum. Geogr. **35**(4), 460–482 (2010) Available https://doi.org/10.1177/0309132510380488. Accessed 14 May 2019
16. Space in Africa: 2019, Might Be The Best Year So Far For The African Space Industry. Space in Africa, 23 December (2019a). Available https://africanews.space/2019-is-the-best-year-so-far-for-the-african-space-industry/ (Accessed 24 April 2020).
17. Space in Africa, African Space Industry Now Generating Over USD 7 Billion Annually. SPACEREF, 10 June (2019b). Available http://spaceref.com/news/viewpr.html?pid=54163. Accessed 23 April 2020
18. O.E. Malahlela, T. Oliphant, L.T. Tsoeleng, P. Mhangara, Mapping chlorophyll-a concentrations in a cyanobacteria- and algae-impacted Vaal Dam using Landsat 8 OLI data. S. Afr. J. Sci. **114**(9–10), 1–9 (2018) Available https://doi.org/10.17159/sajs.2018/4841. Accessed 30 June 2020
19. P. Sekhula, The law governing the acquisition and use of Earth observation data in South Africa: A need for legal harmonisation. S. Afr. J. Geomatics, **2**(2), 139–151 (2013). Available https://www.ajol.info/index.php/sajg/article/viewFile/106987/96895. Accessed 29 Jan 2020
20. Giardino, C., Bresciani, M., Villa, P. and Martinelli, A.: Application of remote sensing in water resource management: The case study of Lake Trasimeno, Italy. Water Resour. Manag. **24**(14), 3885–3899 (2010) Available https://doi.org/10.1007/s11269-010-9639-3. Accessed 02 July 2020
21. T. Woldai, The status of Earth Observation (EO) & Geo-Information Sciences in Africa – Trends and challenges. Geo-spatial Inf. Sci. **23**(1), 107–123 (2020) Available https://doi.org/10.1080/10095020.2020.1730711. Accessed 03 July 2020
22. South Africa, Department of Science and Technology, National Space Strategy. Pretoria: (2018b). Available https://www.sansa.org.za/wp-content/uploads/2018/05/National-Space-Strategy.pdf. Accessed 05 Oct 2019
23. French South African Institute of Technology, About F'SATI (2019). Available http://www.cput.ac.za/blogs/fsati/. Accessed 14 Nov 2019
24. R. Campbell, New entity launched to stimulate SA space sector. Engineering News, 13 December (2019). Available https://m.engineeringnews.co.za/article/new-entity-launched-to-stimulate-and-coordinate-the-south-african-space-sector-2019-12-13. Accessed 15 Dec 2019
25. F. Maccioni, Outer space tech used by Israeli startup to find water leaks on Earth. The Times of IsraeL, 24 February (2019). Available https://www.timesofisrael.com/outer-space-tech-used-by-israeli-startup-to-find-water-leaks-on-earth/. Accessed 07 May 2019
26. E.S. Evagorou, C. Mettas, A. Agapiou, K. Themistocleous, D.G. Hadjimitsis, Bathymetric maps from multi-temporal analysis of Sentinel-2 data: The case study of Limassol, Cyprus.

Adv. Geosci. **45**, 397–407 (2019). Available https://ktisis.cut.ac.cy/handle/10488/13660. Accessed 22 April 2020

27. E. Azambuja, How is space exploration improving life on Earth? Electronic Specifier, 12 October (2017). Available https://www.electronicspecifier.com/industries/aerospace-defence/how-is-space-exploration-improving-life-on-earth. Accessed 21 May 2020
28. J. Greenblatt, A. Anzaldua, How space technology benefits the Earth. The Space Review, 29 July (2019) Available http://www.thespacereview.com/article/3768/1. Accessed 24 Sept 2019
29. National Aeronautics and Space Administration, Advanced NASA Technology Supports Water Purification Efforts Worldwide (2019). Available https://www.nasa.gov/mission_pages/station/research/benefits/water_purification.html. Accessed 17 Nov 2019
30. C. Iceland, T. Luo, G. Donchyts, *It's Not Just Cape Town: 4 Shrinking Reservoirs to Watch*. World Resources Institute (2018). Available https://www.wri.org/blog/2018/04/its-not-just-cape-town-4-shrinking-reservoirs-watch. Accessed 17 Nov 2019

Chapter 39
The Analysis of Global Navigation Satellite Systems (GNSS) and the Numerical Estimation of Multipath Propagation

Bheki Madonsela, Innocent Davidson, and Chrispin Mulangu

39.1 Introduction

The requirements of reliable timing and navigating systems are increasing drastically; hence, today we have the second generation of the GNSS that incorporate the European Galileo and the Chinese BeiDou. These are not stand-alone systems since they are designed and developed to be interoperable with the previous systems: the American Global Positioning System (GPS) and the Russian GLONASS system [1]. The GNSS are radio navigation systems that have space satellites and GNSS receivers that are equipped to use radio frequencies (RF) at the dedicated frequencies. This enables the receiver to estimate the time and the position, and the navigation messages appropriate the ranging code, while the time references are modulated into the wave carrier [2, 3]. Furthermore, the time difference between the transmission and the received signal is crucial also. The GNSS signal received on Earth is relatively weak and prone to jamming and other influences. Time references are modulated into the wave carrier. Furthermore, the time difference between the transmission and the received signal is crucial also. The GNSS signal received on earth is relatively weak and prone to jamming[4] and other influences.

B. Madonsela (✉) · I. Davidson · C. Mulangu
Durban University of Technology, Durban, South Africa
e-mail: 21208384@dut4life.ac.za; innocentD@dut.ac.za; mulangu@mut.ac.za

S. Bauk, S. D. Ilčev (eds.), *The 1st International Conference on Maritime Education and Development*, https://doi.org/10.1007/978-3-030-64088-0_39

39.2 The GNSS Architecture

The basic architecture of the GNSS is made up of three segments: the space, control, and user segment as illustrated in Fig. 39.1. The space segment comprises of the spacecraft satellites that are responsible for emitting the navigation and timing signals [5]. The second one is control segment, which is also known as ground segment. The ground segment is there to ensure the proper and reliable operation of the whole system. The last one is user segment that is equipped with GNSS receiver and antenna to receive the satellite signal for further processing to obtain the positioning, velocity, and time[3].

The space segment generates and transmits the navigation signal that contains the carrier and code phase. Furthermore, it stores and broadcasts the navigational messages that are uploaded from the control segment. The transmission of these data is controlled and governed by the atomic clock within the satellites and is highly stable to ensure reliable operations [1, 3]. For the GNSS user to calculate the positioning and timing, the space segment must have at least four or more satellites available to use simultaneously from the point of user location. Hence, we have the GNSS satellite constellation, and there are four global satellite constellations, the GPS constellation, GLONASS constellation, Galileo constellation, and the BeiDou constellation [3, 6]. In Fig. 39.2, the designs of these satellites are provided, and each satellite has its own operating characteristic.

From Fig. 39.2, we have the GPS IIR-M satellite (top left), the GLONASS-M satellite (top right), the Galileo-IOV satellite (bottom left), and the BeiDou-M satellite (bottom right).

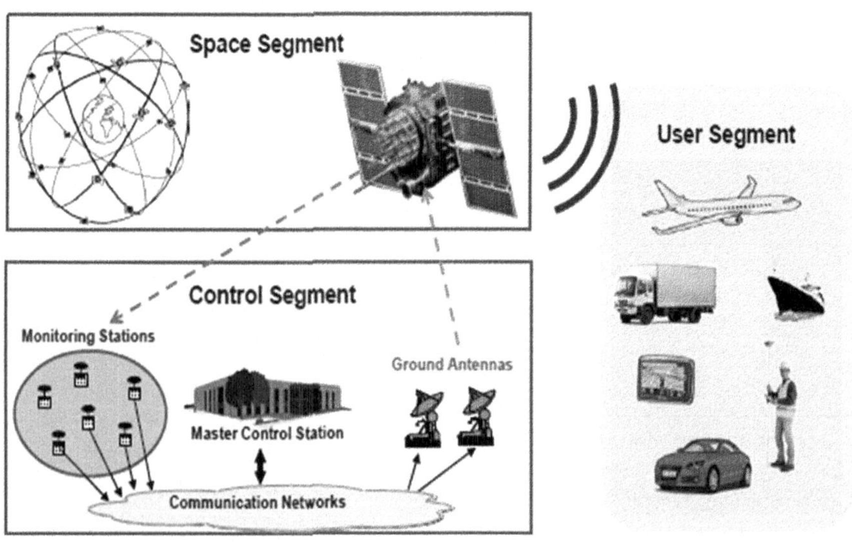

Fig. 39.1 The GNSS architecture [3]

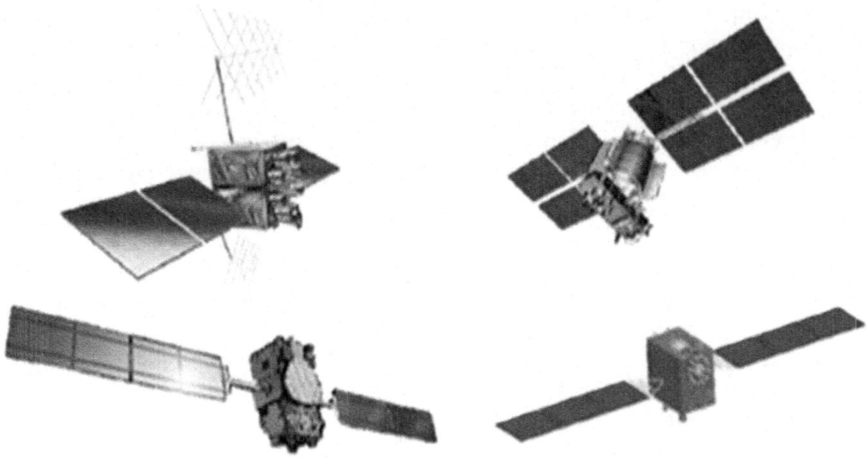

Fig. 39.2 Satellite constellation in the space [3]

Table 39.1 Basic characteristics of GNSS1

Title	GNSS1 – first generation	
GNSS1 network	GLONASS (Russia)	GPS (US)
Quantity of satellites	24	24
Types of satellite cancellations	Leo	Leo
Satellite orbit inclination	64.8°	55°

Table. 39.2 Basic characteristics of GNSS2

Title	(GNSS2) Second GNSS generation	
GNSS2 network	BeiDou (China)	Galileo (Europe)
Quantity of satellites	35	30
Types of satellite cancellations	27 MEO/5 GEO/3 IGSO	30 MEO

The satellite constellations are represented in terms of their characteristics as illustrated in Table 39.1. The characteristics of GLONASS and GPS are almost the same beside the orbit inclination, altitude, frequency, and modulating techniques [7]. The GPS have seen the numerous developments and upgrades for the past few years with the addition of L5 frequency Link (L) at 1176.45 Mhz as in [3] (Table 39.2).

The Galileo and BeiDou satellites are medium Earth orbit (MEO) [8]. However, the Chinese BeiDou has five geosynchronous equatorial orbit (GEO) and three inclined geosynchronous orbit (ISGO) [9]. The full operational of the Galileo is expected in 2020, and the final satellite of the BeiDou constellation was sent to space on 22 June 2020 from the Xichang Satellite Launch Center in China using the Long March 3B rocket. In addition to the global navigation systems, there are two popular regional navigation systems, the Japanese Quasi-Zenith Satellite Systems

(QZSS) and the Indian Radio Navigation Satellite System (IRNSS), that were developed to improve the availability and reliability of the GNSS and satellite communication in areas with high buildings that can cause signal masking [10]. This is considered as the augmentation systems to the Galileo and GPS since it provides signals at the higher elevation angles. The Regional Navigation Satellite Systems (RNSS) are designed to be compatible with both GNSS and Satellite-Based Augmentation System (SBAS) [11]. Alongside the GNSS and RNSS, SBAS are also being implemented to augment the satellite navigation regionally. SBAS is specifically used for the purpose of modifying and improving the precision and integrity of data provided by the global navigation systems. There are four SBAS in space, the European Geostationary Navigation Overlay Service (EGNOS), Wide Area Augmentation System (WAAS), Multifunction Transport Satellite Space-Based Augmentation System (MSAS), and GPS-Aided GEO Augmented Navigation (GAGAN) [12]. These systems are used for different purposes. SBAS uses the supplementary messages from GEO satellite transmission to mitigate the signal attenuation.

The GNSS Frequency Allocation and the GNSS Receiver

The distribution of the frequency band along the satellite constellation is the more complex process because of the numerous users and services that need to fall within the given range [3, 13]. The sharing of the radio spectrum is coordinated by the United Nation agency called the International Telecommunication Union (ITU), and it must incorporate all services that are frequency-dependent such as radio, cell phones, televisions, microwave, radar satellites [14], etc.

The electromagnetic spectrum is divided into two frequency bands, the Radio Navigation Satellite Services (RNSS) and the Aeronautical Radio Navigation Services (ARNS). Furthermore, the ARNS is divided into L bands, the lower band and the upper ban, respectively [15]. The upper band comprises of the GPS L1, Galileo E1, GLONASS G1, and BeiDou B1, and the signals for the lower bands are GPS L5, GLONASS G3, BeiDou B1, and Galileo E5. The GPS L2, Galileo E6, BeiDou B3, and GLONASS G2 were assigned to the ground radar and RNSS. Hence, these signals are more susceptible to the signal interference likened to the ones in upper and lower bands. The dual-frequency band measurements for both L1 and L2 represent the significant advantages of eliminating all ambiguities relating to ionosphere more especially if the measurements of the high precision are required [16]. L5 codes have the highest chipping rate that provide high-performance abilities and improved code measurements compared to L1 code. L2 code is the long code that has better correlation, and such abilities are useful in circumstances where the GPS signal is relatively weak such as indoor applications and forest areas. The recent designs of the GNSS receivers are made up of digital components that are integrated to enable and ease the digital communication between the devices as illustrated in Fig. 39.3. The architecture of the GNSS receiver is very crucial to this paper, since

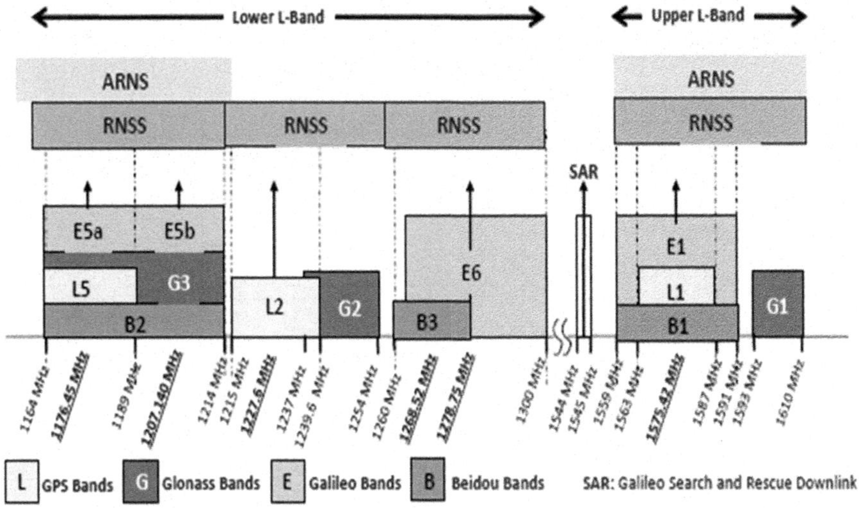

Fig. 39.3 Space frequency allocation [3]

the background of the numerous multipath mitigation techniques is dependent on the design of the GNSS receiver [17]. The multipath is the topic of interest of this paper. The antenna is the first element in the architecture of GNSS receiver and is used to collect the signal from the space. In order to improve the reliability and the signal strength, the antenna must be omnidirectional to collect the signal from all possible directions. The navigation signal collected by the antenna is then transferred to the low-noise amplifier (LNA) to block the noise [18]. The LNA also function as the filter [19] (Fig. 39.4).

The wider bandwidths have more noise power that is passing through the LNA and is statistically expressed as:

$$P_{\text{LNA}} = kT_{\text{Eff}}B \tag{39.1}$$

where the Boltzmann's constant $k = 1.3806 \times 10^{-23}$ KJ^{-1} for free-space analysis and the effective temperature T_{Eff} have the great influence to the power that is passing through LNA. $T_{\text{Eff}} = 513$ K for all GPS receivers and the bandwidth is denoted by B. The LNA itself have the gain between 20 and 40 dB. Adhering to the frequency of L1 C/A, the signal-to-noise ratio (SNR) is computed, and the satellites in view are assumed to be transmitting 478.6 W of the effective isotropic power from the space with the free space attenuation of:

$$\xi_{\text{Free Space}} = \left(\frac{\lambda_C}{4\pi r} \right)^2 \tag{39.2}$$

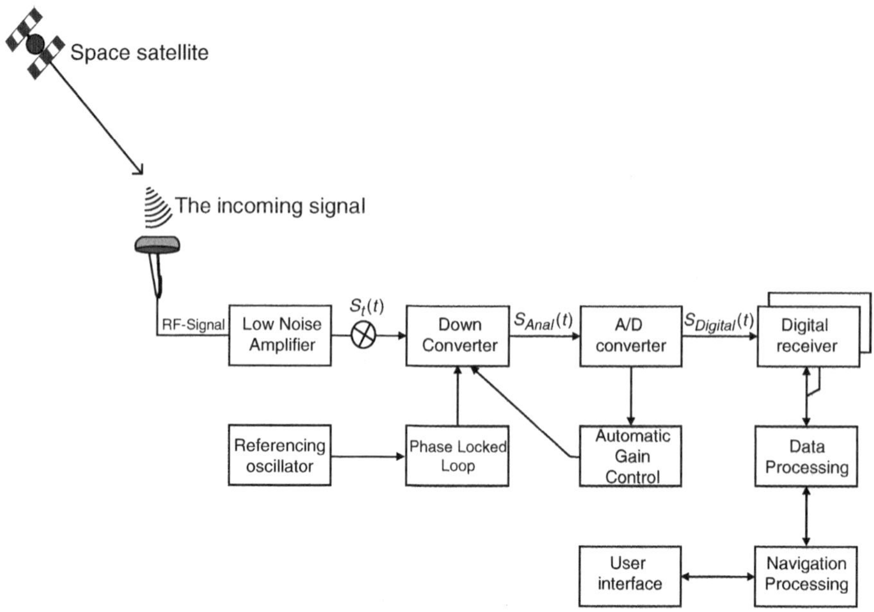

Fig. 39.4 The architecture of GNSS receiver

where the carrier wavelength $\lambda_C = 0.19$ m and the average distance between the transmitter and receiver r is equal to 2×10^7. The IF signal is digitized after LNA and is fed to the numerous digital channels and then to data processing before reaching the user interface.

39.3 The GNSS Multipath

Numerous studies in the past have proved that the multipath meddling normally leads to a positioning inaccuracy for both static and dynamic applications, respectively. Such imprecisions jeopardize the GNSS accuracy and reliability. As a result, the multipath issue is the burning topic to the field of satellite navigation, more especially the designers of the satellite receivers, and the foremost justifying techniques for multipath are grounded on the signal processing algorithm and antenna techniques. The concept of the multipath propagation signal is illustrated in Fig. 39.5.

As stated in the abstract, the GNSS multipath cannot be easily mitigated using the basic differential techniques; hence, the thorough analysis of the multipath is provided in this section.

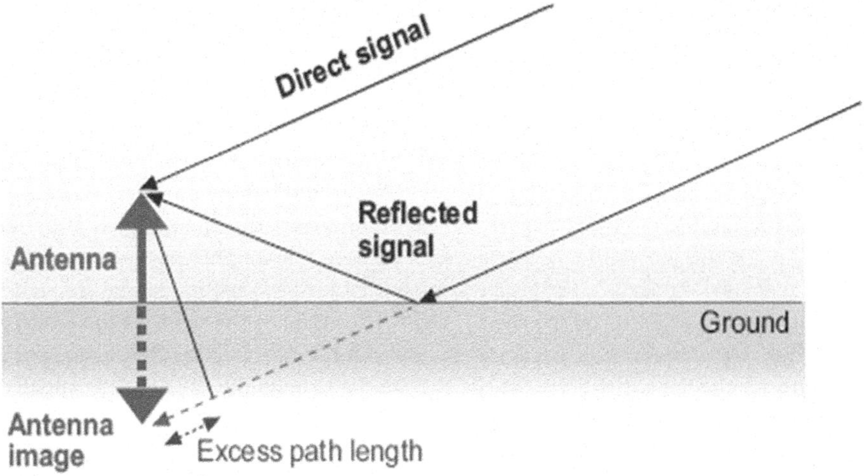

Fig. 39.5 The concept of the multipath propagation [20]

The Receiver and Satellite Multipath

The reflection of the signal does occur near the receiver and sometimes in the transmitting unit itself (the satellite). Henceforth, the satellite multipath is referring to the reflection in the specific sections of the satellite part, and the receiver multipath is the reflection that is anticipated in the receiving antenna [21].

The Diffuse Multipath and the Specular Multipath

The signal reflection can be either diffused or specular depending upon the roughness of the reflecting surfaces. The smooth surfaces normally result in specular multipath, and the rough surfaces result to diffuse reflection [22].

The Hardware-Induced Multipath

This multipath is caused by the improper installation of both the transmitter and receiver. This is very important to this research because the paradigm to mitigate the multipath will be through the modification of the receiver installation and alignment of the receiver antennas.

39.4 The Numerical Algorithms to Estimate the Multipath Along the Receiving Channel

The multipath signals have the serious influence in the direct signal that is emitted by the satellites [21]. Hence, the influence of the multipath is illustrated using the simple signal model that contains the numerous multipath parameters. Before the estimation and the analysis of the multipath influence, the direct signal is anticipated to contain the following parameters, namely: the amplitude of the signal a_s, the signal carrier phase ζ_θ, the propagation duration τ_p, the angular signal frequency $\omega_0=2\pi f$ inclusive of the Doppler shift, the direct signal component for the Doppler shift $\Delta\omega_0$, the frequency of the carrier f_c, and p for the data codes and range. These parameters change in the presence of the multipath signal. From these investigative parameters, the $s_t(t)$ is expressed as follows:

$$S_t(t)=\alpha(t-\tau_p).\cos(w_0 t+\zeta_\theta) \tag{39.3}$$

Equation (39.3) is the general expression to represent the direct signal for any satellite in the space. Considering the GPS systems, the equation is expanded to Equation (39.4) for L1 and 5 for L2:

$$\begin{aligned} S_{\text{GPS},\ L1} &= a_s P_k(t) D_k(t)\sin(w_{L1}t+\delta_{L1}) \\ &\quad + a_s C_k(t) D_k(\text{t})\cos(w_{L1}t+\delta_{L1}) \end{aligned} \tag{39.4}$$

$$S_{\text{GPS},\ L2} = c_s P_k(t) D_k(t)\sin(w_{L2}t+\delta_{L2}) \tag{39.5}$$

From Eqs. (39.4) and (39.5), $C_k(t)$ represents the civilian code also known as course/acquisition (C/A) code, $P_k(t)$ denotes the precision code that used authorized civilians and military, $D_k(t)$ denotes the navigation message, and k denotes index for GPS satellites in view agreeing to [3] . In the presence of the multipath, the expression changes and the multipath propagation signal is:

$$\begin{aligned} S_M(t) &= X\sum_{P=1}^{N}\alpha_p\beta(t-\tau_0-\tau_p)\cos \\ &\quad \left[\omega_0 t+\theta_0+\Delta\Phi_{M.P}+(\Delta\omega_p-\Delta\omega_0)t\right] \end{aligned} \tag{39.6}$$

The multipath signal from Equation 6 contains the following parameters:

- $(A\alpha_p\beta)$ represents the relative amplitude of the multipath, where (α_p) denotes the coefficient of the reflection and further characterizes the signal attenuation along the satellite path and the gain pattern of the antenna.
- The (τ_p) also known as the geometric path delay is used to represent the multipath delay and the time shift between the direct and multipath signal components.

- The $(\Delta\Phi_{M.P})$ is used to represent the phase shift between the direct and multipath signal constituents.
- $(\Delta\omega_p - \Delta\omega_0)$ denotes the Doppler difference between the direct and multipath signals, also called fading frequency.

Due to the combination of the several receiver parameters and signals, the parameters of the multipath signal are used to determine the concept of how the phase observation and the code are influenced by the propagation of multipath signal.

The Numerical Algorithms of the Transfer Function Along the Receiving Channel

The number of the multipath parameters in [equation 39.22] can be used in the modeling of the GNSS received signal where the receiving channel is signified as the linear filter. This is characterized by the transfer function that uses three computing methods. The secondary signal normally travel the longer propagation path compared to the direct signal; this will result to the relative time delay between the secondary and direct signal. Similarly, the cause of the radial velocity differences is due to the moving receiver. The multipath signal is the basic expression of the multipath condition. In the real-life environments where *(M>2)*, the secondary echoes are expected to be very high, the expression is manipulated, and the bit data dependency is detached. The input correlator will then be:

$$S_M(t) = \sum_{m>1}^{M} \alpha_m e^{j2\pi\left(f_{dm}-f_{d0}\right)t} \sum_{k=0}^{L_c-1} P_k(t)\delta_k(t-\tau_m-K\tau_k) + x_k(t) \tag{39.7}$$

where the:

- $L_c - 1$ is the chip of a spreading code.
- α_m is the time delay of the signal replica.
- (δ_k) function represents the transmitted signal pulse filtered by the pre-correlation filter.
- τ_m represents the phase differential carrier.
- τ_k is the satellite phase carrier.
- $x(t)$ is the thermal noise.

It is assumed that the synchronization of (f_{dm}) and (f_{d0}) by the tracking loop carrier is perfect. The impact of the multipath signal in the GNSS receiver can be computed by giving attention to the multipath delay-locked loop (DLL) code delay approximation and the expression of the GNSS power signal broadcasted by the satellite. The impulse baseband response for the wideband model is capable of totaling the L1 and L5 received power signal in the antenna without noise and is given by:

$$S_T(t) = S_{T,DC,L1}(t) + S_{T,DC,L5}(t) + S_{I,PC,L5}(t) \tag{39.8}$$

where

- The functions of $S_{T,\,DC,\,L1}(t)$ and $S_{T,\,DC,\,L5}(t)$ represent the data component for L1 and L5.
- $S_{T,\,PC,\,L1}(t)$ and $S_{T,\,PC,\,L5}(t)$ denote the pilot component for L1 and L5

Signals L1C, L5, E1, and E5a have pilot and data channel with the existence of the numerous delay-locked loop (DLL) architectures, and some only use the data component, while others are using both pilot and signal channels, respectively.

Multipath Parameter Definition and Computation

In this section, we attempt to estimate the multipath signal propagation by developing the numerical expression that will incorporate the parameters that are anticipated to be in the output of the GNSS receiving antenna. Those parameters are α_p, τ_p, $f_{D,\,P,}$and θ_p, where the virtual characteristics of both $\alpha_p$and θ_p depend on the parameters of the receiving antenna and the propagation channel. Hence, these parameters are computed as follows:

Computing the amplitude for L1 frequency link:

$$\alpha_{p.L1} = \frac{E_{L1}(e_{r.p}).I_{L1}(e_{r.p})}{V_o} \tag{39.9}$$

Computing the amplitude for L5 frequency link:

$$\alpha_{p.L1} = \frac{E_{L5}(e_{r.p}).I_{L5}(e_{r.p})}{V_o} \tag{39.10}$$

The phase shift for L1 frequency link:

$$\theta_{P.L1} = E_{L1}(e_{r.p}).L_{L1}(e_{r.p}) \tag{39.11}$$

The phase shift for L5 frequency link:

$$\theta_{P.L5} = E_{L5}(e_{r.p}).L_{L5}(e_{r.p}) \tag{39.12}$$

where:

- $e_{r.\,p}$ is used to represent the direction of the unit vector for the arrival of the multipath signal.

- $E_{L1}(e_{r.\ p})$ and $E_{L5}(e_{r.\ p})$ represent the electric field of the multipath for both L1 and L5 frequency links.
- $L_{L1}(e_{r.\ p})$ and $L_{L5}(e_{r.\ p})$ represent the effectiveness of the vectorial heights for the GNSS receiving antenna for both links L1 and L5, respectively.
- V_o is the open-circuit voltage obtained to adopt the ideal configuration parameters.

The characteristics of the propagation channels are strongly influenced by the parameters of Doppler shift $f_{D,\ P}$ effect and the code delay τ_p. If the 3D scene and the spectrum satellites are assumed to be immobile, the Doppler shift effect for both L1 and L5, respectively, are:

Doppler shift effect for L1 frequency band:

$$f_{D,P,L1} = \frac{f_{L1}}{C}.v_{r,k} \tag{39.13}$$

Doppler shift effect for L5 frequency band:

$$f_{D,P,L5} = \frac{f_{L5}}{C}.v_{r,k} \tag{39.14}$$

where:

- C represents the speed of light in the vacuum.
- v is the vectorial speed of the GNSS receiver.
- $v_{r.\ k}$ is equal to $v.e_{r.\ p}$; therefore, this term represents the effectiveness of the receiver vectorial speed.

Most of the multipath error simulating software for maritime, land, and aeronautical applications are design to compute the required navigational parameters in 3D scene to necessitate the identification of multipath.

Impact of Multipath on GNSS Code Pseudo-range Measurements

In this section, the arithmetical expression to estimate the impact of the multipath signal in the code delay and the pseudo-range measurement is provided. It must be noted that the pseudo-range measurement is anticipated to incorporate all delays along the space, ionosphere, and troposphere [23]. The delay codes of the DLL approximation are converted to be pseudo-range measurement codes as soon as they reach the receiver. Hence, the multipath code delay error is induced as an error of the pseudo-range management, is called raw multipath ranging error, and is expressed as follows:

$$M\varepsilon_{L1C} = C_{data} . \varepsilon_{tracking} \tag{39.15}$$

where:

- $M\varepsilon_{L1C}$ represent the multipath error of the Navstar GPSL1C signal.
- C_{data} is the data waveform.
- $\varepsilon_{tracking}$ is the multipath code tracking error.

There are three forms and expressions of the multipath that fall under the raw code multipath ranging error.

- The L1 code pseudo-range measurement multipath error that is represented by $M\varepsilon_{L1C}$ for GPSL1C in Eq. (39.19) and $M\varepsilon_{E1}$ for GalileoE1 in Eq. (39.20).
- The L5 code pseudo-range measurement multipath error that is represented by $M\varepsilon_{L5}$ for GPSL5 in Eq. (39.19) and $M\varepsilon_{E5a}$ for GalileoE5a in Eq. (39.20).
- The iono free L1 + L5 code pseudo-range measurement multipath error that is represented by $M\varepsilon_{LC1-L5}$ for GPSPL1C and GPSL5 in Eq. (39.19) and $M\varepsilon_{E1-E5a}$ for GalileoE1 and GalileoE5a in Eq. (39.20).

The ionosphere medium of the Galileo and the Navstar GPS is deemed to be dispersive since it permits the different components of frequencies to travel at different speed. Therefore, the dedicated receivers are designed to measure the pseudo-range from the satellite using two different frequencies. The first-order residual of the raw code ionosphere ranging error for both L1 and L5 frequency bands is defined using $\varepsilon_{rc\ Iono,\ L1}$ and $\varepsilon_{rc\ Iono,\ L5}$, respectively. The following expression is developed, where f_{L1} and f_{L5} denote the frequency links for L1 and L5:

$$\varepsilon_{rc\,Iono,L5}\left(f_{L5}\right)^2 = \varepsilon_{rc\,Iono,L1}\left(f_{L1}\right)^2 \tag{39.16}$$

Therefore, GPSL1C/A, GPSL5, GalileoE1, and GalileoE5a pseudo-code measurements are given by the following expressions:

The GPSL1C/A-GPSL5:

$$\beta_{L1C-L5} = f_a\beta_{L1C} + f_b\beta_{L5} \tag{39.17}$$

GalileoE1-GalileoE5a:

$$\beta_{E1-E5a} = f_a\beta_{E1} + f_b\beta_{E5a} \tag{39.18}$$

where:

- β_{L1C-L5} and β_{E1-E5a} denote the pseudo-range code for GPSL1C/A and GPSL5 and GalileoE1 and GalileoE5a, respectively.

- f_a is the frequency regulation from L1 to L5 $f_a = \dfrac{f_{L1}^2}{f_{L1}^2 - f_{L5}^2}$.

- f_b is the frequency regulation between L5 and L_1 $f_a = \frac{f_{L5}^2}{f_{L5}^2 - f_{L1}^2}$.

From Eqs. (39.17 and 39.18), new numerical algorithms for pseudo-range measurement multipath error are developed:

$$M\varepsilon_{L1C-L5} = f_a\ M\varepsilon_{L1C} + f_b\ M\varepsilon_{L5} \tag{39.19}$$

$$M\varepsilon_{E1-E5a} = f_a\ M\varepsilon_{E1} + f_b\ M\varepsilon_{E5a} \tag{39.20}$$

Impact of Multipath on the Code Delay Estimate

The principle of evaluating and understanding the impact of the multipath on the code delay error is to firstly consider the instance where there is no echo signal to affect the original received GNSS signal:

$$\mathrm{AC}_{\mathrm{data}}\left(\tau_{\mathrm{data}} + \frac{\Delta}{2}\right) = \mathrm{AC}_{\mathrm{data}}\left(\tau_{\mathrm{data}} - \frac{\Delta}{2}\right) \tag{39.21}$$

where:

- $AC_{\mathrm{data}}(.)$ is the autocorrelation function for the data waveform, and the stability point is achieved if the following condition is met: $\tau_{\mathrm{data}} = 0$.

Raw Code Multipath Ranging Errors in the Presence of a Single Echo Signal

If the direct signal and the single resonance are delivered to the receiving antenna, the minimal to the utmost tracking error code is represented by the multipath error envelopes. The multipath error envelopes for GPSL1C-GPSL5 and GalileoE1-GalileoE5a adhering to the functions of the amplitude and code delay phase shift of the resonance signals fall under the raw code multipath ranging with the presence of the single echo signal. The minimum and maximum tracking code errors are obtained between $\Delta\theta_1 = 0$ and $\Delta\theta_1 = \pi$ or 180°. The maximum amplitude of the multipath error for GPSL1C-GPSL5 and GalileoE1-GalileoE5a signals depend upon the relative amplitude of the resonance signal. Hence, the maximum error of the GPSL1C-GalileoE1 is considered to be two times the maximum error of the GPSL5-GalileoE5a adhering to the relative amplitude modeling of the echo signal. The multipath error model and representation define the multipath error similar to raw code ranging from using Binary Phase Shift Keying (BPKS) modulated signals in the presence of the single resonance signal. This general equation represents the multipath code tracking error observant to the relative amplitude a, the relative code delay $\Delta\theta_1$, and the relative carrier phase $\Delta\tau_1$ for the resonance signal:

$$M\varepsilon_{\text{model}} = \frac{a^2 + a\cos(\Delta\theta_1)}{1 + a^2 + 2a\cos(\Delta\theta_1)} - c\Delta\tau_1 \tag{39.22}$$

If the satellite is assumed to be stationary 3D scene, the following condition applies to the stationary GNSS receiver:

$$\Delta\theta_1 = -2\pi f_{L5}\Delta\tau_1 + \Delta\Phi_1 \tag{39.23}$$

where:

$\Delta\Phi_1$ represents the phase shift difference between the echo signal and the direct signal.

39.5 Conclusion

The GNSS receivers calculate their location using several signals received from the satellites in view. This paper describes the brief concept of the GNSS satellite constellation, frequency allocation, and architecture of the GNSS receiver to show the impact of the multipath signal to the receiver. Hence, the estimation of the multipath signal was the fundamental topic for this paper. The numerical algorithms for multipath estimation were provided adhering to the receiver-based approach. The GPS signals were used for analysis, and it was noted that the GPS satellites are interoperable with the Galileo satellites and their combination is GPS L5+E5a both using Binary Phase Shift Keying (BPSK (10)) at the same frequency. As we conclude, the numerical algorithms provided are essential for GNSS users to estimate the level of the multipath. Estimating the multipath is the crucial step prior to multipath monitoring and mitigation.

References

1. P.D. Groves, Principles of GNSS, inertial, and multisensor integrated navigation systems, [Book review]. IEEE Aerosp. Electron. Syst. Mag. **30**(2), 26–27 (2015)
2. X. Li, X. Zhang, X. Ren, M. Fritsche, J. Wickert, H. Schuh, Precise positioning with current multi-constellation global navigation satellite systems: GPS, GLONASS, Galileo and BeiDou. Sci. Rep. **5**, 8328 (2015)
3. J. S. Subirana, M. Hernandez-Pajares, J.E.M.J. Zornoza, *GNSS Data Processing: Fundamentals and Algorithms*. European Space Agency (2013)

4. K.D. Wesson, J.N. Gross, T.E. Humphreys, B.L. Evans, GNSS signal authentication via power and distortion monitoring. IEEE Trans. Aerosp. Electron. Syst. **54**(2), 739–754 (2017)
5. B. R. Rao, K. McDonald, W. Kunysz, GPS/GNSS Antennas. Artech House (2013)
6. M.G. Petovello, C. O'Driscoll, G. Lachapelle, D. Borio, H. Murtaza, Architecture and benefits of an advanced GNSS software receiver. J. Glob. Positioning Syst. **7**(2), 156–168 (2008)
7. J. Wang, C. Rizos, M.P. Stewart, A. Leick, GPS and GLONASS integration: modeling and ambiguity resolution issues. GPS Solut. **5**(1), 55–64 (2001)
8. Q. Zhao, C. Wang, J. Guo, X. Liu, Assessment of the contribution of BeiDou GEO, IGSO, and MEO Satellites to PPP in Asia—Pacific Region. Sensors **15**(12), 29970–29983 (2015)
9. J. Tegedor, O. Øvstedal, E. Vigen, Precise orbit determination and point positioning using GPS, Glonass, Galileo and BeiDou (2014)
10. A. Bose et al., Current multi-GNSS Status from India: A Brief Review
11. F. Van Diggelen, *A-gps: Assisted Gps, Gnss, and Sbas*. Artech House (2009).
12. M. Siergiejczyk, A. Rosiński, K. Krzykowska, Reliability assessment of supporting satellite system EGNOS, in *New Results in Dependability and Computer Systems* (Springer, 2013), pp. 353–363
13. D. M. Bevly, S. Cobb, *GNSS for Vehicle Control*. Artech House (2010).
14. C. Wu, C. He, *Interference analysis among modernized GNSS*, in International Conference on Computational Problem-Solving (ICCP), China, pp. 669–673 (2011)
15. A. Ruegamer, F. Foerster, M. Stahl, G. Rohmer, *A flexible and portable multiband GNSS front-end system*, in 25th International Technical Meeting of the Satellite Division of the Institute of Navigation, ION GNSS, pp. 17–21, China (2012)
16. C. Cai, Y. Gao, L. Pan, J. Zhu, Precise point positioning with quad-constellations: GPS, BeiDou, GLONASS and Galileo. Adv. Space Res. **56**(1), 133–143 (2015)
17. G. Hein, T. PANy, S. Wallner, J.-H. Won, Platforms for a future GNSS receiver. Inside GNSS **1**(2), 56–62 (2006)
18. C. Fernández-Prades, J. Arribas, P. Closas, Robust GNSS receivers by array signal processing: theory and implementation. Proc. IEEE **104**(6), 1207–1220 (2016)
19. C. Clemente, J.J. Soraghan, GNSS-based passive bistatic radar for micro-Doppler analysis of helicopter rotor blades. IEEE Trans. Aerosp. Electron. Syst. **50**(1), 491–500 (2014)
20. G. Seco Granados, *Antenna Arrays for Multipath and Interference Mitigation in GNSS Receivers*. Universitat Politècnica de Catalunya (2000)
21. P. Closas, C. Fernandez-Prades, J.A. Fernandez-Rubio, A Bayesian approach to multipath mitigation in GNSS receivers. IEEE J. Sel. Top. Sig. Process. **3**(4), 695–706 (2009)
22. Y. Jiang, S. Zhang, J. Zhang, Q. Hu, X. Sun, Multipath effects on the performance of DLL in GNSS receiver. Chin. J. Electron. **19**(3), 543–547 (2010)
23. C. Cheng, J.-Y. Tourneret, Q. Pan, V. Calmettes, Detecting, estimating and correcting multipath biases affecting GNSS signals using a marginalized likelihood ratio-based method. Sig. Process. **118**, 221–234 (2016)

Chapter 40
Conceptual Design and Analysis of Modern Space Solar Power Satellite and Rectenna Systems

Jonas Don-yelee Dakora, Innocent E. Davidson, and Gulshan Sharma

40.1 Introduction

An increase in the world's population along with the increase in global warming has drawn the attention of the development of a clean and reliable power source to meet the demand of energy. Although numerous alternative energy sources have been pursued, none of the technologies has been successful to meet the world's energy demand [1]. Technological developments have recently made way for space solar power systems as a means to meet energy demands.

Space solar power satellite system is a system whereby large satellites are placed into geostationary orbit to capture and convert a large amount of sunlight into microwave energy which is then transmitted to a rectifying antenna array on Earth [2]. The rectenna then receives and converts the microwave power into electrical energy that is pumped into the terrestrial electric grid for use. The applications of space solar power systems include electric power sources for remote areas and other satellites and a supplement to the terrestrial electric grid and also provide power for certain businesses and domestic use in some countries [3]. To sustain humanity's energy future and to reduce atmospheric CO_2 over time, it is practical to build a feasible power system that will continually harvest solar energy from our sun. The sun has unlimited energy and an SSPS is a system to achieve just that. Since many researchers have focused on the GEO SSPS, as well as the ground rectenna array, this work looked at the microwave beam space environment stretching from a GEO SSPS to Earth or to lower orbiting satellites.

J. D.-y. Dakora (✉) · I. E. Davidson · G. Sharma
Department of Electrical Power Engineering, Durban University of Technology, Durban, South Africa
e-mail: jonasd@dut.ac.za

S. Bauk, S. D. Ilčev (eds.), *The 1st International Conference on Maritime Education and Development*, https://doi.org/10.1007/978-3-030-64088-0_40

The space solar power system is a free energy source, which has to be considered a feasible method to provide electrical generating capacity due to constant sunshine in space [4]. Due to orbit location, the amount of solar energy received by the satellite during the year is very effective and consistent [5]. Some satellites have the same rotational time as the Earth at geosynchronous orbit and therefore are fixed at a particular location at all times which enables the satellite to deliver constant power to a ground-receiving site [6].

40.2 Research Objective

The research objective is to design an economically feasible, solar power satellite system to collect solar energy from space and transmit it to Earth at maximum power efficiency using the microwave wireless power transmission technology. The various space solar power systems would be investigated under new technologies, and the major elements of the ground system would be evaluated to determine an effective viable system design.

40.3 Space Solar Power Satellite System Design Requirements

A Systems Tool Kit (STK) is used to design the SSPS and low Earth orbit satellite system and simulated using software packages such as CST Microwave/Ansys Ansoft HFSS. The CST Microwave Studio is a tool for simulating high-frequency devices, and the Ansys HFSS is a simulation software package used to design and simulate high-frequency antennas, antenna arrays, and microwave systems [7]. Experimental validation will be undertaken using physical models of an actual design system. This will be achieved using field experiments, measured field data, and engagement with the South African National Space Agency (SANSA). The entire system consists of the space system (Spacetenna/SUNSAT) and the ground power system (Rectenna) as shown in Fig. 40.1.

Space Satellite (SUNSAT) System

The SUNSATs are required to produce radio frequency power at a constant rate with minimum interruptions. The main components of the system include the solar cell arrays, the power conversion hardware, and a battery backup system [8]. The solar

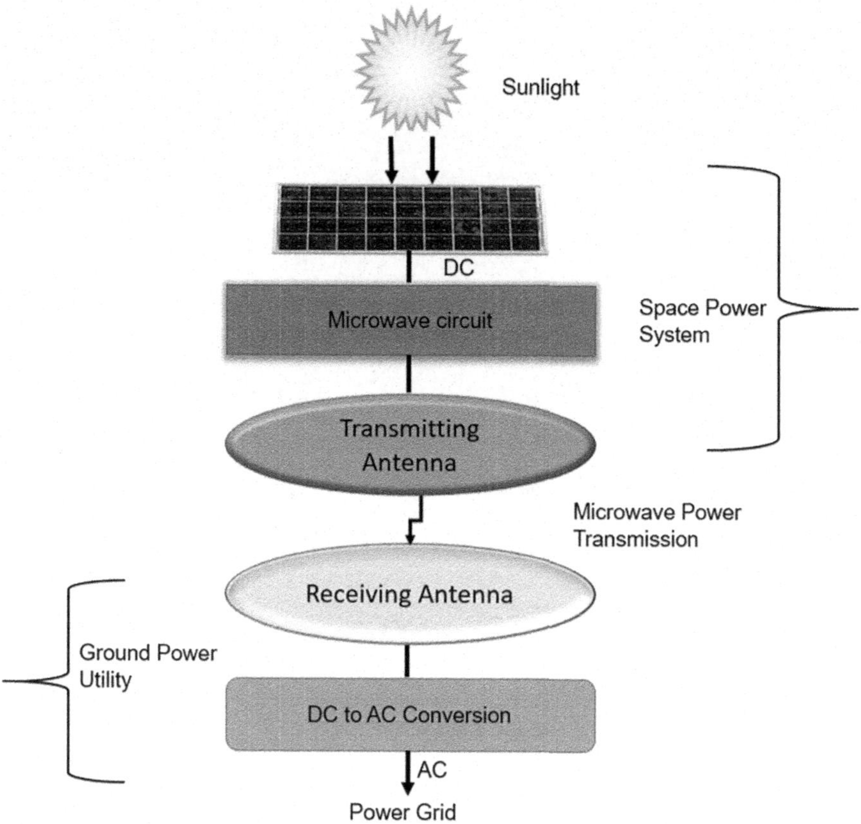

Fig. 40.1 SSPSS block diagram

arrays will be paired with a reflector array to help concentrate solar energy onto the photovoltaic cells. The DC power produced by the SUNSATs will be transmitted to Earth at a frequency of 5.8 GHz. This frequency compared to 2.4 GHz has less attenuation by atmospheric conditions. The higher frequency of 5.8 GHz instead of 2.4 GHz increases the gain of the antennas used in the design and thereby the efficiency of the system. Figure 40.2 shows a solar collector system that would embedded onto the space satellite.

Various types of antennas are used in space solar power systems. Most space solar power systems are designed with the phased array antenna. Figure 40.3 shows the design of an HPGU configuration for experimental 3U CubeSat which can also be implemented in space solar satellite system for experimental studies.

Fig. 40.2 Solar collectors used in SSPS [2]

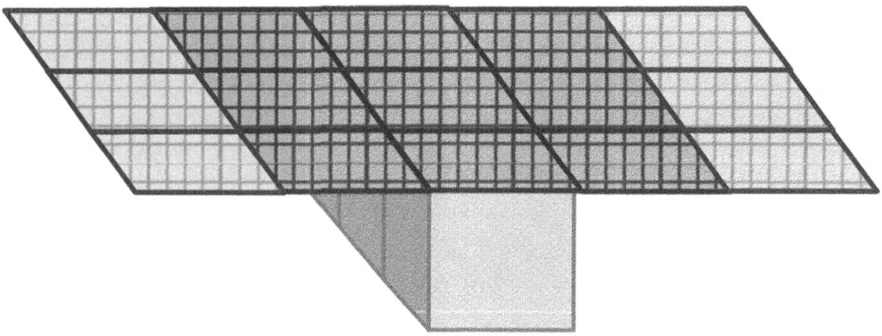

Fig. 40.3 Hybrid solar PV power generation unit configuration for experimental 3U CubeSat

Microwave Power Transmission System

Microwave power transmission is one of the advanced technologies used in SSPS system for efficient power transmission. The SSPS collects solar energy which is then converted into microwave power using magnetron to be transmitted to the rectenna. The output power from the microwave source is controlled by electronic circuits [9]. A waveguide circulator is connected to the microwave power source through a coax-waveguide adaptor to protect the microwave source from reflected power [9]. The directional coupler separates the attenuated signals based on the direction of signal propagation. The SUNSAT then radiates the power uniformly to the rectenna system. The rectenna converts the microwave power back to DC power using Schottky barrier diodes. Figure 40.4 shows a microwave power transmission system block diagram. Figure 40.5 is a simulated microwave signal transmitted from the SUNSAT which is received by the rectenna.

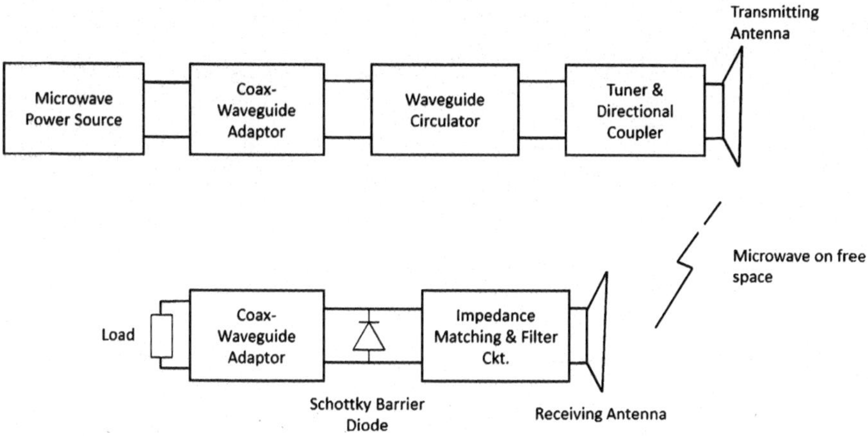

Fig. 40.4 Functional block diagram of microwave power transmission system

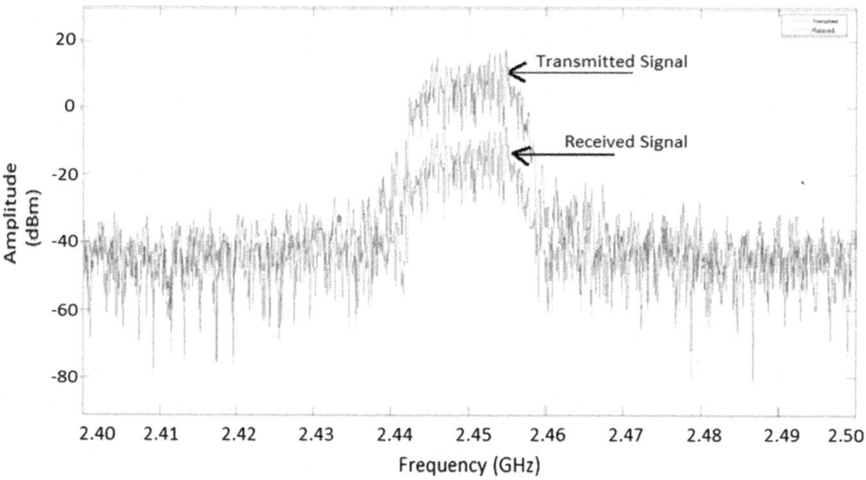

Fig. 40.5 Simulated signal from transmitter/receiver system

The 2.45 GHz or 5.8 GHz frequency falls within the industrial, scientific, and medical radio bands. The 2.45 GHz frequency was the first frequency to be used in the study of SSPS, but current developments in the C-band RF technologies made the 5.8 GHz more convenient frequency for experiments [10]. Microwave power-transmitting devices such as magnetron, klystron, and semiconductor microwave transmitters are usually used in wireless power transmission systems. Magnetron is the most widely used technology for experimentation of wireless power transmission [11]. In this project, the magnetron technology is accessed and employed. Phase and amplitude controlled magnetron system shown in Fig. 40.6 was developed at Kyoto University in Japan to control beam directions in both 2.45 GHz and

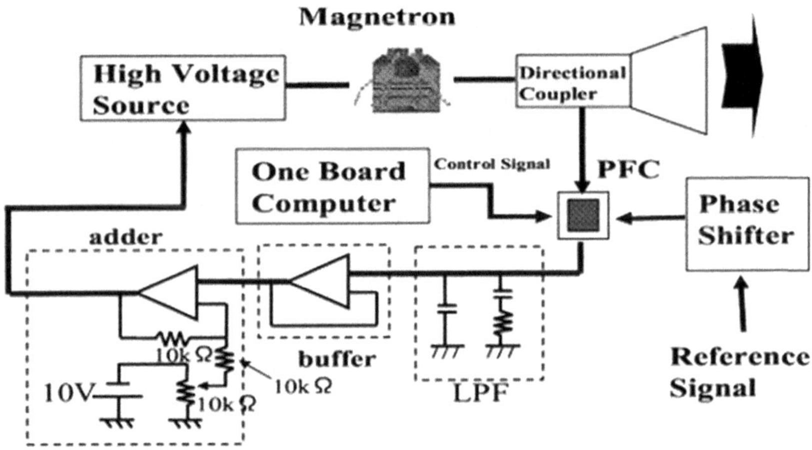

Fig. 40.6 Diagram of magnetron system [12]

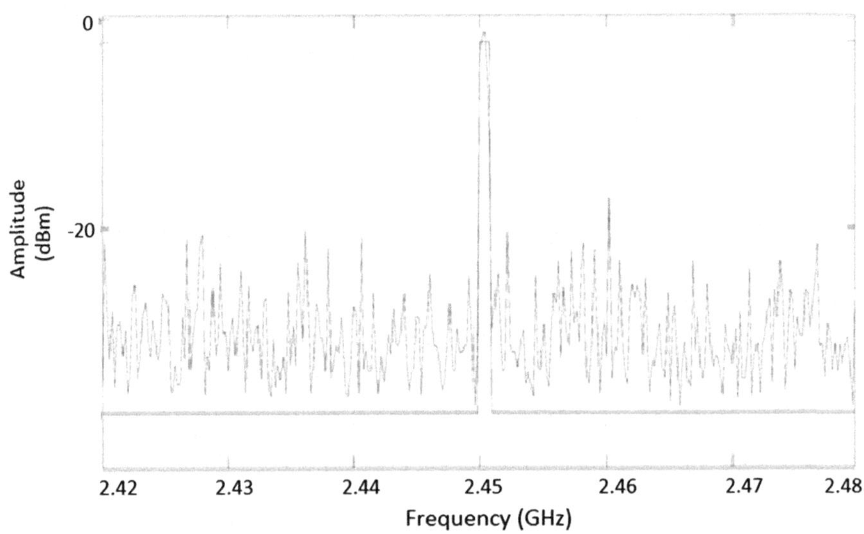

Fig. 40.7 Simulated magnetron spectrum

5.8 GHz frequencies [12]. Figure 40.7 shows a simulated magnetron spectrum of phase-lock loop with a resolution bandwidth of 200 kHz which the spectrum is centered at 2.45 kHz with a reference of 0 dBm.

Ground Rectenna System

For the ground station, a parabolic dish antenna with the ability to track the satellite will be used. This antenna will have the ability to transmit maximum power using the magnetron system [13]. Space-based solar power system design is mainly dependent on the capacity to transmit electrical power from space to Earth efficiently so that a large percentage of the power harvested is recovered on Earth. The rectenna should be able to harvest approximately 85% of the transmitted power and convert the RF energy back into DC power [14]. The type of rectenna to be used will be determined by its efficiency. The rectenna developed by Suh and Chang which is capable of converting RF to DC at 5.8 GHz with 82.7% efficiency was chosen for experimental studies [15]. The design will be based on the use of a printed dipole antenna attached to a series of filters that block the reradiation of higher-order harmonics. After passing through the filters, the signal is then directed through a Schottky diode, capacitor, and load-matched resistor where it is rectified into DC power as shown in Fig. 40.8 [16].

Figure 40.9 shows a schematic diagram of a rectenna system. The rectifying circuit is chosen based on the power density and operational frequency range. The output of the rectifier passes through a low-pass filter to restrict any RF energy from flowing into the power management system [9]. Equation (40.1) gives the average RF power over the operational frequency range of the rectenna, and the DC power can be determined from Equation (40.2) [9]. Table 40.1 gives the performance of different types of printed rectenna:

$$P_{RF}(t) = \frac{1}{f_{high} - f_{low}} \int_{f_{low}}^{f_{high}} \int_{0}^{4\pi} S(\theta,\varphi,f,t)\, A_{eff}(\theta,\varphi,f,)\, d\Omega df \tag{40.1}$$

$$P_{DC} = P_{RF}(f_i,t)\,\eta\left(P_{RF}(f_i,t),\rho,Z_{DC}\right) \tag{40.2}$$

For the safety of microwave radiation, the microwave beam is controlled in a confined area. The power density outside the restricted area is planned to be kept well below the regulatory value of connected load. Figure 40.10 shows a dual-frequency rectenna system developed by Texas A&M that operates both 2.45 GHz and 5.8 GHz frequencies.

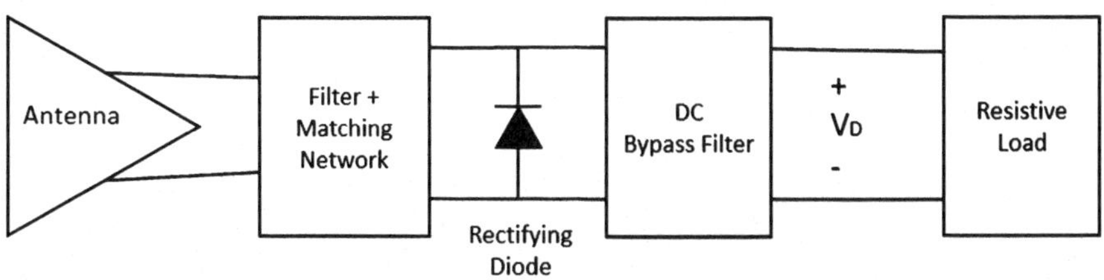

Fig. 40.8 Block diagram of the components of a rectenna

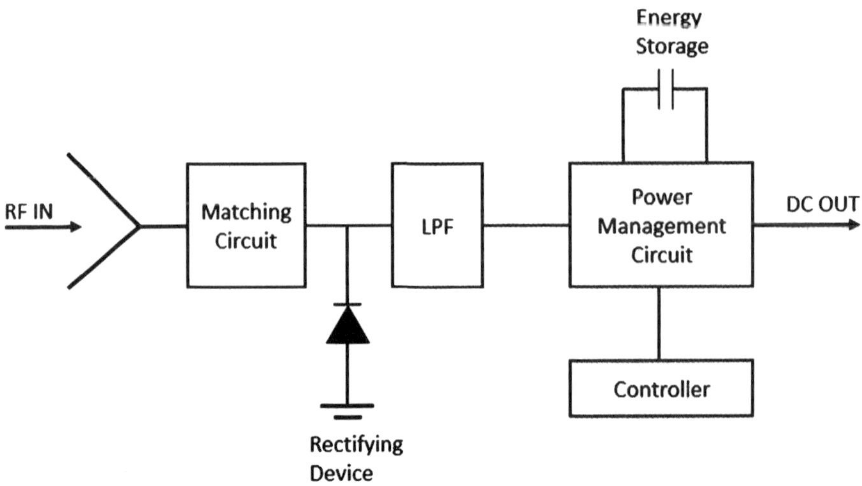

Fig. 40.9 Rectenna schematic diagram [9]

Table 40.1 Different types of printed rectenna performance [17]

Rectenna type	Frequency (GHz)	Efficiency (%)
Printed dipole	2.45	85
Circular patch	2.45	81
Printed dual rhombic	5.6	78
Square patch	8.51	66

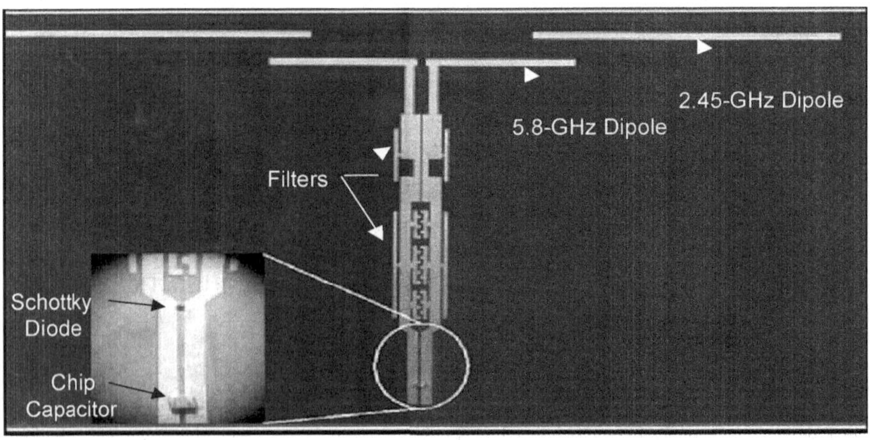

Fig. 40.10 Dual-frequency rectenna for 2.45 GHz and 5.8 GHz [16]

40.4 Conclusion

The world is seeking sustainable source of energy to meet the increasing demand for electricity. This calls for new technologies to be developed. The space-based solar power satellite system is a promising and technically challenging technology to be widely exploited. Space solar power is the most productive and attractive sources of free, constant, and reliable power. SBSP, at the present trend, uses solar mirrors or panels to collect the available sun's energy and deflect it to Earth through wireless power transfer. The collection of limitless and constant solar energy available in space and beaming it downward to Earth to provide electrical power for use is a power source that can replace the crippling conventional energy sources. Although SSPS system has a challenge for very high efficient solar cells for the space solar panels and microwave power transmission technologies, the system can reduce the high demanding power crisis for developing countries. Space-based solar power transmissions are yet to be proven in Africa to alleviate the constant energy crises.

References

1. Consumer Energy Information. "EREC Reference Briefs, https:// www.eere.gov/consumerinfo/rebriefs/123.html, Apr. 15, 2003 [Jun. 20, 2019].
2. H. Matsumoto, Research on solar power satellites and microwave power transmission. IEEE Microwave Magazine **3**(4), 36–45 (2002)
3. S. Potter: "Solar power satellites: an idea whose time has come," https://www.spacefuture.com/archive/solar power satellites an idea whose time has come.shtml, Jan. 12, 2004 [Jun. 20, 2019].
4. A.L. Geoffrey, An evolutionary path to SPS. Proc Space Power **9**(4), 365–371 (1990)
5. B. Oleary, The construction of satellite solar power stations from non-Terrestrial materials: feasibility and economics. Alternative Energy Sources **3**, 1155–1167 (1978)
6. Australian Government. Satellites Geo-stationary Orbits and Solar Eclipses, https://www.sws.bom.gov.au/Educational/5/4/3, Nov. 10, 2011 [Jun. 22, 2019].
7. M. Jones: Ansys High Frequency Structure Simulator (HFSS), https://www.Ansys.com/products/ansys-hfss, Jan. 16, 2017 [Jun. 19, 2019].
8. H.N. Ralph, Ground test program for developing solar power satellites. Space Energy Transport **1**(3), 187–194 (1996)
9. J.L. Volakis, *Antenna Engineering Handbook*, 4th edn. (McGraw Hill, New York, 2007)
10. H. Feingold, A. Friedlander, G. Rauwolf, G. Didinsky, C. Christensen, C. Mullins, E. Gresham, A. Charenia, *Space Solar Power 1998 Concept Definition Study*, Report Number SAIC-99/1016 (Science Applications International Corporation, Schaumburg, 1999)
11. N. Shinohara, J. Fujiwara, H. Matsumoto, Development of active phased array with phase-controlled magnetrons. Proc. ISAP2000 Fukuoka **2**, 713–716 (2000)
12. N. Shinohara, T. Mitani, H. Matsumoto, Development of Phase and Amplitude Controlled Magnetron. Proc. of Sixth International Vacuum Electronics Conference IVEC2005, 61–64 (2005)
13. P. Om, *Gandhi.: Microwave engineering and application* (Pergamon Press, Oxford, 1981), pp. 175–178

14. R. Dickinson: Magnetron directional amplifier space solar power beamer concept design. https://www.trsnew.jpl.nasa.gov/dspace/bitstream/2014/15070/1/00-1077.pdf, May. 4, 2000 [Jun. 22, 2019].
15. Radio Labs. WiFi Antenna High Gain 802.11 Omni Wireless Antenna, https://www.radiolabs.com/products/antennas/2.4gig/high-gain-wifi-antenna.php.50ST1G18, Dec. 2, 2011 [Jun. 22, 2019].
16. Y. Suh, K. Chang, A High-efficiency Dual-frequency Rectenna for 2.45-GHz and 5.8-GHz Wireless Power Transmission. IEEE Transac. Microw. Theor Techniques **50**(7), 1784–1789 (2002)
17. T. W. Yoo: Experimental and theoretical study on 35 GHz RF-to-DC power conversion receiver for millimeter-wave beamed power transmission," Ph.D. dissertation, Dept. of Electrical Engineering, Texas A&M University, (Dec. 1993).

Index

S. Bauk, S. D. Ilčev (eds.), *The 1st International Conference on Maritime Education and Development*, https://doi.org/10.1007/978-3-030-64088-0

Printed by Printforce, the Netherlands